Mathematische Optimierung und Wirtschaftsmathematik | Mathematical Optimization and Economathematics

In der Reihe werden Arbeiten zu aktuellen Themen der mathematischen Optimierung und der Wirtschaftsmathematik publiziert. Hierbei werden sowohl Themen aus Grundlagen, Theorie und Anwendung der Wirtschafts-, Finanz- und Versicherungsmathematik als auch der Optimierung und des Operations Research behandelt. Die Publikationen sollen insbesondere neue Impulse für weitergehende Forschungsfragen liefern, oder auch zu offenen Problemstellungen aus der Anwendung Lösungsansätze anbieten. Die Reihe leistet damit einen Beitrag der Bündelung der Forschung der Optimierung und der Wirtschaftsmathematik und der sich ergebenden Perspektiven.

Lukas Graf

Dynamic Network Flows with Adaptive Route Choice based on Current Information

Springer Spektrum

Lukas Graf
University of Passau
Passau, Germany

Zugleich Dissertation, Universität Augsburg, 2023
Erstgutachter: Prof. Dr. Tobias Harks
Zweitgutachter: Prof. Dr. Martin Skutella
Drittgutachter: Prof. Dr. Roberto Cominetti
Datum der mündlichen Prüfung: 20.09.2023

ISSN 2523-7926 ISSN 2523-7934 (electronic)
Mathematische Optimierung und Wirtschaftsmathematik | Mathematical Optimization and Economathematics
ISBN 978-3-658-44947-6 ISBN 978-3-658-44948-3 (eBook)
https://doi.org/10.1007/978-3-658-44948-3

This Springer Spektrum imprint is published by the registered company Springer Fachmedien Wiesbaden GmbH, part of Springer Nature.
The registered company address is: Abraham-Lincoln-Str. 46, 65189 Wiesbaden, Germany

Many Thanks

Even though this thesis only has a single author, it owes its existence to a much larger group of people, who supported me in many different ways. This, of course, makes it impossible to name everyone who has influenced, shaped and helped me during my studies. Nevertheless, I want to take this opportunity to at least specifically mention and thank a few of those that had the most impact on this thesis:

Starting in chronological order, I am, first and foremost, deeply indebted to my parents who supported me throughout my study making it possible to fully focus on learning and taking all the time I needed. I am thankful to my brother who knows everything about life that I don't.

Next, I want to thank my supervisor Tobias Harks who guided me through my PhD giving me both a lot of freedom to explore on my own and the right amount of nudges forward to not get stuck for too long at any point. I also fondly remember all the fruit- and insightful discussions on specific research questions as well as academic life in general and hope for many more to come.

I am grateful to the German Research Foundation (Deutsche Forschungsgemeinschaft) for their financial support and to our whole community for their welcoming and supportive atmosphere. As just one example on the first day of my first conference, in Potsdam, I was approached by Leon Sering with ideas for a possible cooperation leading to many more interesting discussion and finally my first paper. Other highlights in this regard were further collaborations with Prashant Palkar, Michael Markl and Kostas Kollias as well as the amazing seminars in Schloss Dagstuhl.

I want to thank my colleagues at the chair for Optimization in Augsburg: Manuel Surek and Anja Schedel for helping me get started as a PhD student and always being available whenever I had any questions about it. And Julian Schwarz for the many, many enlightening discussions on the intricate technical details of dynamic flows.

A special thanks goes to my friends and fellow students Kathrin Gimmi, Ingo Blechschmidt and Lukas Stoll both for their friendship and support throughout the years and for their willingness to help proofread this thesis on very last minute notice (quite, literally, in Lukas' case).

Finally, I want to thank Martin Skutella and Roberto Cominetti for writing very in-depth assessments of my thesis and for Mirjam Dür for being part of my defense committee and asking the best question at the end.

Contents

Symbols and Notation

General

$\mathbb{N}_0$	The natural numbers including 0
$\mathbb{N}^*$	The natural numbers excluding 0
$[n]$	The set $\{1, \ldots, n\}$; the empty set for $n = 0$ (cf. Sec. 2.1)
$[n] - k$	The set $\{1 - k, \ldots, n - k\}$
$\mathbb{R}$	The real numbers
$\mathbb{R}_{\geq 0}$	The non-negative real numbers
$\mathbb{R}_{>0}$	The (strictly) positive real numbers
$\tilde{\mathbb{R}}_{\geq 0}$	The extended non-negative numbers including ∞
$\lceil . \rceil$	The ceil function $\lceil x \rceil$ rounds x to the smallest whole number larger or equal to x
$\lfloor . \rfloor$	The floor function $\lfloor x \rfloor$ rounds x to the largest whole number smaller or equal to x
2^A	The power set of A, i.e. the set of subsets of A
$\dot{\cup}$	The disjoint union. Used to emphezize that the two sets combinded in $A \cup B$ are disjoint
$\mathcal{O}(f)$	Big O: The set of functions growing assymptotically at most as fast as f (cf. Def. 2.70)
$\Omega(f)$	Big Omega: The set of functions growing assymptotically at least as fast as f (cf. Def. 2.70)
$v \oplus w$	The direct sum of two vectors (cf. Sec. 2.1)

Topology

$B_r(x)$	The open ball of radius r around x (cf. Def. 2.27)
$\bar{B}_r(x)$	The closed ball of radius r around x (cf. Def. 2.27)
$\|.\|$	The norm function of a normed vector space
$\|.\|_\infty$	The uniform norm on $C(J)$ (cf. Prop. 2.30)
$\|.\|_p$	The p norm on $L^p(J)$ (cf. Prop. 2.31)

Measure Theory

μ	The Lebesgue measure on $\mathbb{R}$ (or some subset of $\mathbb{R}$) (cf. Def. 2.2)
$f =_{\text{a.e.}} g$	The functions f and g are equal for almost all θ in their domain (cf. Def. 2.5)
$f \leq_{\text{a.e.}} g$	The function f is essentially bounded by g, i.e. less or equal to g almost everywhere on their common domain (cf. Def. 2.5)
$\int_J f(\zeta)\mathrm{d}\zeta$	The lebesgue integral of f over J

Functions and function spaces

$\partial_\theta f$	The derivative of f with respect to the argument θ. Index θ can be omitted
$\partial_- f$	The derivative of f from the left side
$\partial_+ f$	The derivative of f from the right side
$f\|_A$	The restriction of function f to subset A of its domain
$\mathbb{1}_A$	The characteristic function of set A (cf. Sec. 2.1)
graph(f)	The graph of a function (cf. Def. 2.55)
$f^{(n)} \overset{w}{\to} f$	Weak convergence: The sequence (f^n) converges to f with respect to the weak topology (cf. Def. 2.33)
$L^p_{\text{loc}}(J)$	The space of locally p-integrable functions on the interval J (cf. Def. 2.16)
$L^p_{\text{loc}}(J\mathbb{R}_{\geq 0})$	The set of locally p-integrable functions from the interval J to $\mathbb{R}_{\geq 0}$ (cf. Def. 2.16)
$AC(J)$	The space of absolutely continuous functions on the interval J (cf. Def. 2.44)
$AC^{\nearrow}(J)$	The set of absolutely continuous non-decreasing functions on the interval J (cf. Def. 2.44)

Graphs

$G = (V, E)$	A directed graph with node set V and edge set E (cf. Def. 2.59)
$E(G)$	Set of edges of graph G; argument G usually ommited
$vw \in E$	An edge from v to w
$V(G)$	Set of nodes of graph G; argument G usually ommited
$\delta_G^+(v)$	The set of all edges $e = vw \in E$ leaving node v; index G usually omitted (cf. Def. 2.59)
$\delta_G^-(v)$	The set of all edges $e = wv \in E$ entering node v; index G usually omitted (cf. Def. 2.59)
$G' \subseteq G$	G' is a subgraph of G (cf. Def. 2.60)
$G[W]$	The induced subgraph on W, i.e. the maximal subgraph of G with node set W (cf. Def. 2.61)
$E[W]$	The edgeset of induced subgraph

Networks

$\mathcal{N}$	A network consisting of a graph with free flow travel times and capacities and commodities with network inflow rates and sink nodes (cf. Def. 3.48)
$\tau_e \in \mathbb{R}_{\geq 0}$	Free flow travel time: The time it takes to traverse edge e without any congestion (cf. Def. 3.48)
$\nu_e \in \mathbb{R}_{>0}$	Capacity: The maximum rate at which particles may traverse edge e
I	Set of commodities; individual commodities are usually denoted by i
$u_{v,i}$	Network inflow rate: A locally integrable function denoting the rate at which particles of commodity i enter the network at node v
$T_i \subseteq V$	Set of sink nodes of commodity i
$S_i \subseteq V$	Set of source nodes of commodity i (cf. Def. 3.48)
$V_i^\dagger \subseteq V$	Set of dead-end nodes of commodity i (cf. Def. 3.48)
$U_{v,i}$	Cumulative network inflow: $U_{v,i}(\theta)$ denotes the total volume of flow of commodity i which has entered the network at node v by time θ (cf. Def. 3.48)
U_i	$U_i(\theta)$ denote the total volume of flow of commodity i that has entered the network by time θ (cf. Prop. 3.53)
$\hat{\theta}_i \in \bar{\mathbb{R}}_{\geq 0}$	Earliest time such that no flow of commodity i enters the network after that time (cf. Def. 3.48)

$\hat{\theta} \in \tilde{\mathbb{R}}_{\geq 0}$	Earliest time such that no flow of any commodity enters the network after that time (cf. Def. 3.48)
$\tau_{\max}$	The largest free flow travel time of any edge in the given network (cf. Cor. 6.9)
$\tau_{p_{\max}}$	The largest free flow travel time along any v, T_i-path in a given network (cf. Subsection 6.1.1)
$\nu_{\min}$	The smallest capacity of any edge in a given network (cf. Subsection 6.1.1)
$U(\mathcal{N})$	The total cumulative network inflow into $\mathcal{N}$ (cf. Thm. 6.15)
$\tau(\mathcal{N})$	Network size, i.e. the sum of all free flow travel times in $\mathcal{N}$ (cf. Thm. 6.15)

Dynamic flows

f_e^+	(Anonymous) inflow rate: A function denoting the inflow rate (of all commodities) into edge e (cf. Def. 3.1)
f_e^-	(Anonymous) outflow rate: A function denoting the outflow rate (of all commodities) from edge e (cf. Def. 3.1)
$f_e = (f_e^+, f_e^-)$	(Anonymous) edge flow: In- and outflow rate describing the (anonymous) flow on a single edge e (cf. Def. 3.1)
F_e^+	(Anonymous) cumulative inflow: A function denoting the cumulative inflow (of all commodities) into edge e (cf. Def. 3.2)
F_e^-	(Anonymous) cumulative outflow: A function denoting the cumulative outflow (of all commodities) from edge e (cf. Def. 3.2)
$F_e^{\Delta}(\theta)$	Edge load: Total volume of flow on edge e at time θ (cf. Def. 3.2)
$\hat{C}_e(\theta)$	Experienced (current) travel time on edge e for interchangeable particles when entering at time θ (cf. Def. 3.7)
$f_{e,i}^+$	(Commodity-specific) inflow rate: A function denoting the inflow rate of commodity i into edge e (cf. Def. 3.25)
$f_{e,i}^-$	(Commodity-specific) outflow rate: A function denoting the outflow rate of commodity i from edge e (cf. Def. 3.25)
$(f_{e,i}^+, f_{e,i}^-)_i$	(Multi-commodity) edge flow: In- and outflow rates for all commodities describing the flow on a single edge e (cf. Def. 3.25)

$F^{+}_{e,i}$	(Commodity-specific) cumulative inflow: A function denoting the cumulative inflow of commodity i into edge e
$F^{-}_{e,i}$	(Commodity-specific) cumulative outflow: A function denoting the cumulative outflow of commodity i from edge e
$F^{\Delta}_{e,i}(\theta)$	(Commodity-specific) edge load, i.e. the total volume of flow of commodity i on edge e at time θ (cf. Subsection 3.1.2)
$\hat{C}_{e,i}(\theta)$	Experienced (current) travel time on edge e for particles of commodity i when entering at time θ (cf. Def. 3.7)
$Q_e(\theta)$	Queue length on edge e at time θ (cf. Def. 3.10)
$C_e(\theta)$	(Expected) current travel time on edge e when entering at time θ (cf. Def. 3.14)
$C_p(\theta)$	(Expected) current travel time expected along some path p when entering at time θ (cf. Def. 3.60)
$T_e(\theta)$	(Expected) exit time from edge e when entering at time θ (cf. Def. 3.14)
$L_{v,i}(\theta)$	(Expected) current distance, i.e. the shortest distance from v to the closest sink of commodity i under the current travel times at time θ (cf. Def. 3.60)
$E_i(\theta)$	The set of active edges for commodity i at time θ (cf. Def. 3.60)
$P_{v,i}(\theta)$	The set of active v, T_i-path for commodity i at time θ (cf. Def. 3.60)
$B_{v,i}(\theta)$	The node balance of commodity i at node v at time θ (cf. 3.50)
$F^{\Delta}_{i}(\theta)$	Commodity-specific network load: The total volume of flow of commodity i in the network at time θ (cf. Def. 3.50)
$F^{\Delta}(\theta)$	Network load: The total volume of flow in the network at time θ (cf. Def. 3.50)
$Z_i(\theta)$	Cumulative network outflow: Flow volume of commodity i that already left the network at time θ (cf. Prop. 3.53)
$\Psi_i(f)$	Makespan: The last arrival time of any particles of commodity i at a sink under f (cf. Def. 3.67)
$\Xi_i(f)$	Total travel time: The sum of travel times of all particles of commodity i under f (cf. Def. 3.67)
$\mathcal{F}(\mathcal{N})$	The set of all dynamic flows in $\mathcal{N}$ (cf. Def. 3.49)
$\mathcal{F}_{pa}(\mathcal{N})$	The set of all partial flows in $\mathcal{N}$ (cf. Def. 4.1)

$\mathcal{F}_{pa}^{\text{IDE}}(\mathcal{N})$	The set of all partial IDE in $\mathcal{N}$ (cf. Def. 4.1)
$\lim_k(f^{(k)}, \xi_k)$	The limit of a sequence $(f^{(1)}, \xi_1) \preceq (f^{(2)}, \xi_2) \preceq \ldots$ of partial flows (cf. Def. 4.8)
Φ_e^{ξ}	Edge loading: A function mapping edge inflow rates to edge outflow rates such that together they form a Vickrey edge flow until ξ (cf. Cor. 3.46)
ψ_e	A function denoting the change of waiting times on edge e depending on the (constant) future inflow rate into that edge (cf. Eq. (4.11))

1 Introduction

This thesis studies dynamic flows comprised of infinitesimally small agents (=flow particles) travelling through a network while continuously adapting their choice of route based on the current state of the network as determined by the flow up to the current time. The main motivation for such models comes from the study of large-scale road traffic wherein individual drivers try to get as fast as possible from their starting point to their destination and adjust their routes on-the-fly whenever they get new information about current congestion events (e.g. via traffic radio or navigational devices). To make this idea clearer, we start with a small, slightly informal example (see Example 3.65 in Chap. 3 for the more detailed, formal version):

Example 1.1. Consider the following network represented as a directed graph with edge costs denoting the number of time units it takes to traverse an edge without congestion. In addition, each edge has a capacity associated with it determining how much flow this edge can handle at once before the traversal time starts to be affected by congestion effects. In our example here, we only have two types of capacities: Small capacities (indicated by thin edges) and large enough capacities such that no congestion will ever take place (indicated by thick edges):

L. Graf, *Dynamic Network Flows with Adaptive Route Choice based on Current Information*, Mathematische Optimierung und Wirtschaftsmathematik | Mathematical Optimization and Economathematics,
https://doi.org/10.1007/978-3-658-44948-3_1

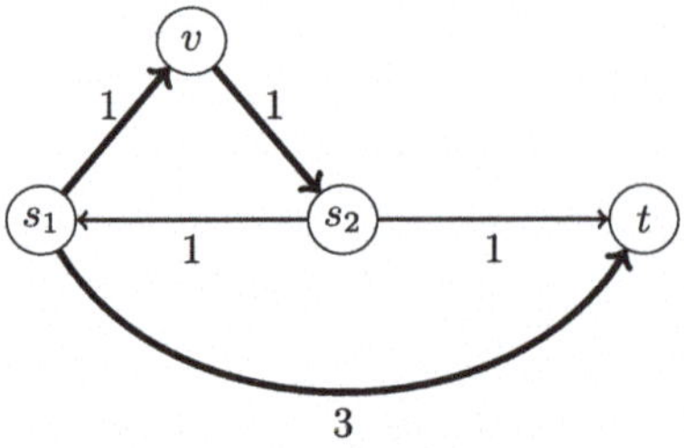

Now, imagine a group of agents (represented by a continuous flow of particles) entering the network at the left-most node s_1 with the goal of reaching the right-most node t. They have two paths to choose from: They may either travel via nodes v and s_2 or along the direct edge from s_1 to t. Since the network is currently empty, both choices seem equally good. So, let us assume that, as these agents start their journey one after the other at node s_1, they all decide to take the upper path.[1]

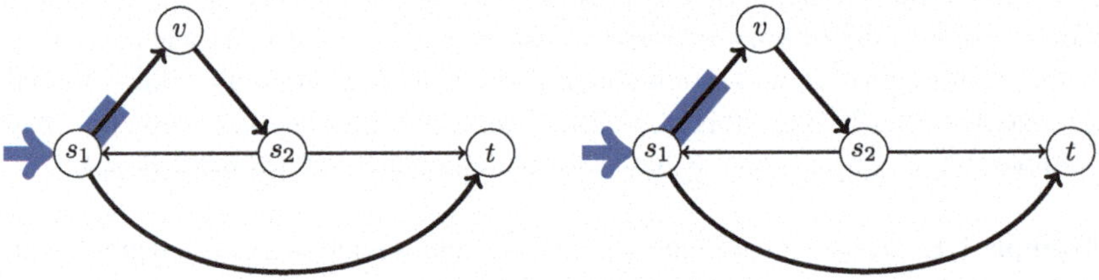

However, just as the last of these agents have entered edge s_1v (and the first ones arrive at node v) a new group of agents arrives in the network: This time at node s_2 but with the same destination t. These agents, again, can choose between two possible paths but, as one of them is strictly faster (the direct edge towards t), they all choose this one. This, though, is more than this edge's capacity allows for and, thus, a traffic jam starts to form on this edge (represented here by a vertical queue at the start of this edge):

[1] Note that time moves continuously here even though our figures only show snapshots of the flow at discrete points in time.

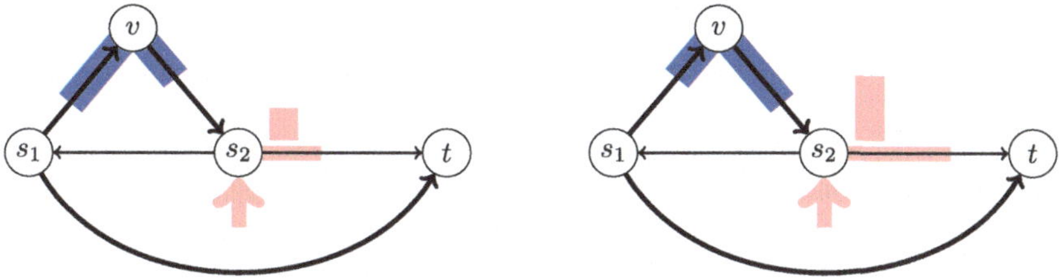

While this is happening, the agents that started from node s_1, currently traversing edge vs_2 and originally planning to take the now congested edge s_2t next, are informed about this new state of the network. Hence, by the time they arrive at node s_2 some of them may reconsider their route choice and opt to avoid the traffic jam by taking a detour via node s_1 while the rest continues on its original route and joins the end of the traffic jam on edge s_2t (we assume here that this split results in the current travel times along both paths staying equal and, so, both choices are valid):

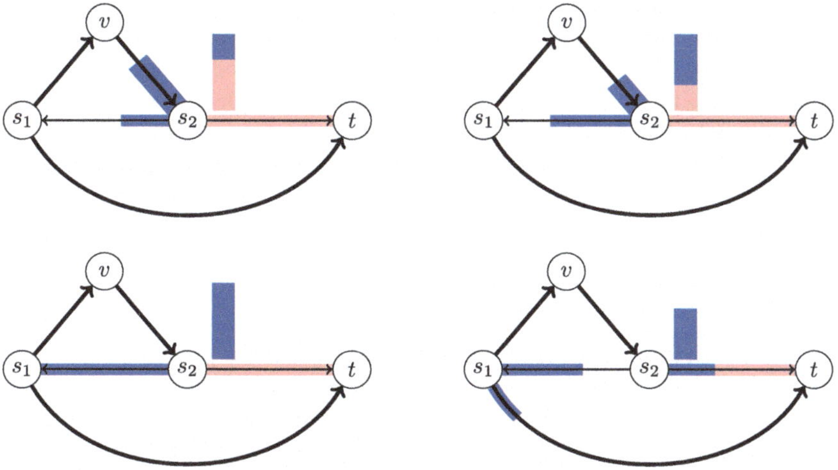

Already in this small, informal example we can observe several interesting properties of the flows arising from our rather simple behavioural assumptions: First of all, these flows are certainly not unique. For example, the agents initially starting at node s_1 could have also chosen the direct edge towards t resulting in a very different flow evolution. Next, and maybe most importantly, the adaptive and, in hindsight, imperfect nature of the individual agents' decisions may lead to situations where flow travels around a cycle. This, in particular, means that solely path-based flow models may be insufficient for capturing the full range of possible flow dynamics. Additionally, this might also lead to quite complex and difficult to analyse flow patterns.

On the other hand, the sequential nature of the flow pattern in the above example suggests that maybe our flows could still exhibit some nice structure which might even allow us to compute such flows in some iterative fashion similarly to how one picture can be deduced from the previous one in our example. However, for larger, more complex networks the complicated interdependence of the choices made at the same time at different nodes in the network makes this far from obvious. Moreover, even when only considering a single node, the distribution of the outflow has to be chosen carefully as it may have an instantaneous impact on the current travel times on the chosen paths. In our example a different flow split of the blue particles at node s_2 may lead to a different congestion on either of the two outgoing edges (as both of them have a small capacity) and, thus, potentially immediately invalidates one of the chosen routes. This indicates that even just showing existence of dynamic flows consistent with our behavioural assumptions might be a non-trivial task.

Finally, all of this, of course, also critically depends on the underlying physically flow model and, in particular, the exact way we chose to model congestion. Since we are studying flows over time, the flow dynamics on a single edge can have ripple effects throughout the whole network both in terms of influencing the behaviour of the individual agents and in how these dynamics effect the physical propagation of the flow particles through the network.

Before we describe in more detail how we will approach these challenges in this thesis, we want to give a brief historic overview over the study of dynamic flows.

1.1 A (Brief) Historic Overview of Dynamic Flows

Right from the very beginning in the middle of the 20th century the study of network flows was motivated by the goal of better understanding transportation and communication networks (see [FF62, p. 1])—a task that has only become more important over the last decades with the rapid growth of the Internet and the increasingly urgent challenge of mitigating the climate crisis to which the transportation sector and, in particular, road traffic is a main contributor (see [Jar+22]).

Static Flows. The field started out with the study of static flows. Here, in the simplest case, we are given a directed graph $G = (V, E)$ with capacities on the edges and two special nodes: A source node s and a sink node t. A flow can then be described completely by a vector $(f_e)_{e \in E} \in \mathbb{R}^E_{\geq 0}$ giving for every edge the amount of flow on this edge. In order to be considered feasible such a flow must be bounded by the edge capacities and satisfy flow conservation at all nodes except for the source and the sink node. A natural question is then how to find a maximal such flow, i.e. one where the amount of flow leaving the source node or, equivalently, entering the sink node is maximal. The famous Max-Flow Min-Cut Theorem which states that the maximum value of such a flow is equal to the minimal s,t-cut was one of the first results in this field ([FF56]). Many extensions of this base model are known and studied to this day (e.g. multiple source/sink nodes, multiple commodities with different source/sink nodes, additional costs on the edges, fixed supplies/demands at sources/sinks, different objectives, ...). Already by 1962 the field had grown enough that Ford and Fulkerson were able to write a whole book on "*Flows in Networks*" [FF62].

Dynamic Flows. While their book predominantly focused on static flows, it also already includes a section on *dynamic* flows ([FF62, Chapter III, Section 9]) which had been introduced four years prior in [FF58] by the same authors. In this model every edge has, in addition to its capacity limiting the amount of flow allowed to enter the edge during any unit time interval, a traversal time which determines how long it takes flow to traverse an edge from one end to the other. Time is partitioned into a finite number of equally sized time periods $1, 2, \dots, T$ during each of which the flow remains constant. Thus, a dynamic flow can be described by a (finite-dimensional) vector $(f_e(\theta))_{\theta\in[T]} \in \mathbb{R}^T$ for every edge e where $f_e(\theta)$ denotes the inflow into edge e during the θ-th time period. The outflow from this edge during the same time period is then described by $f_e(\theta - \tau_e)$ where $\tau_e \in \mathbb{N}_0$ denotes the fixed traversal time of edge e. Similarly to the static case, Ford and Fulkerson then studied the problem of determining a maximal such dynamic flow and showed that this can be done efficiently by reducing the problem to a sequence of static flow problems.

Later on, Fleischer and Tardos showed in [FT98] that this algorithm, as well as several other ones developed in the meantime, can also be applied to the same problems for dynamic flows with *continuous* time. In this setting flow is now described by a function $f_e : [0, T] \to \mathbb{R}_{\geq 0}$ denoting for every time $\theta \in [0, T]$ the rate $f_e(\theta)$ at which flow enters edge e at that time. We refer to [Sku09] for a modern introduction to the study of such dynamic flows. This model is now already quite close to the physical model described in our initial example. However, what is still missing is a proper way of modelling congestion.

The Vickrey Point Queue Model. Since, in the models discussed so far, the traversal times were fixed, the only way of modelling congestion is to have particles wait at nodes. This is often technically allowed in these models but, typically, not used as long as one is only interested in optimal flows. This changes once we view the flow as being comprised of individual agents making their own, selfish decisions (instead of trying to achieve some system optimum). For such a game theoretic setting (though for the case of only a single edge and giving the agents a choice over their departure time instead of over their route) Vickrey discussed in his seminal 1969 paper on congestion theory [Vic69] six different types of congestion and how to model them by suitable edge flow dynamics.

One of those, called "pure bottleneck situation" in his paper, is maybe the most natural extension of the model discussed in the previous paragraph: Here, every edge e is again characterised by two values, a free flow travel time τ_e and a capacity ν_e. As long as particles enter the edge at a rate below this capacity, no congestion takes place and traversing the edge requires exactly the time given by the free flow travel time. If, however, the inflow rate f_e^+ exceeds the capacity, a queue starts to form at the start of the edge and particles have to wait there until they are allowed to enter the edge. Thus, the total time needed to traverse the edge then consists of a waiting time in the queue plus the constant free flow travel time (see Fig. 1.1 for a graphical depiction). This model (and its application to full networks) has been studied and adapted extensively in the past decades (a recent bibliometric review [LHY20] counted over 200 papers published on this topic) and is nowadays also called Vickrey point queue, deterministic queueing or just bottleneck model.

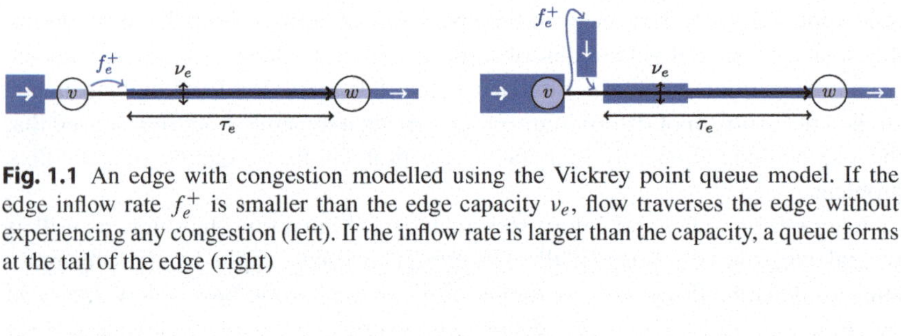

Fig. 1.1 An edge with congestion modelled using the Vickrey point queue model. If the edge inflow rate f_e^+ is smaller than the edge capacity ν_e, flow traverses the edge without experiencing any congestion (left). If the inflow rate is larger than the capacity, a queue forms at the tail of the edge (right)

Dynamic Equilibria. Within such a game theoretic view it now becomes natural to study equilibrium flows, i.e. dynamic flows where at every point in time every particle travels along a route which is compatible with decisions of a selfishly acting agent. Of course, how such selfish decisions look like depends on the information available to those agents. Two main models have been studied here, giving rise to two types of equilibria: Equilibria based on *current information* where, as in our initial example, agents choose their route based on the current state of the network and then adjust their choices on-the-fly to changing travel times, and equilibria based on *full information* where agents make their routing decision already anticipating future changes of travel times and, thus, always choose routes which are best possible in hindsight.

Current Information Equilibria. One of the first to study dynamic flows which could be considered equilibria in a current information setting was Yagar who in [Yag71] proposed the following heuristic approach for simulating road traffic: At first, all flow is just assigned to shortest paths. However, whenever this leads to queues, flow leaving such a queue is then considered as new flow entering the network and, thus, is assigned a new currently shortest path. While no theoretical properties of such flows are shown here and the exact model differs from the ones discussed here in several ways, this can certainly be seen as one of the earliest motivations for studying this type of dynamic equilibrium. Indeed more formal definitions of such equilibria were then studied by Friesz, Luque, Tobin and Wie in [FLTW89, Section 3], Boyce, Ran and Leblanc in [BRL93, BRL95] and Ran and Boyce in [RB96, Chapter 7] who characterised these equilibria as solutions to certain variational inequalities. These works also used more general edge dynamics described by differential equations and introduced the adjective "instantaneous" as a way of referring to travel times/equilibrium states based only on current information. Note, however, that these models are also quite complex and, thus, do not allow for much theoretical analysis (e.g. showing existence of equilibria, studying their quality or providing non-heuristic algorithms for computing them). Moreover, the exact way these models are formalised (often using path-based formulations!) can lead to certain unintuitive effects which call into question how well these formal models actually match their verbal descriptions. For example Boyce, Ran and Leblanc note that "[the] restricted definitions of used links and routes may cause the underlying definition of instantaneous DUO [i.e. current information equilibria] to be vacuous in some cases." ([BRL95, p. 130])

Full Information Equilibria. As agents in the full information setting choose their route with full knowledge over the true future travel times in the network, adaptive route choice is not relevant here and, therefore, a path-based model is much less problematic. In fact the path-based formalisation of full information dynamic equilibria introduced by Friesz, Bernstein, Smith, Tobin and Wie in [Fri+93, Definition 2] has clearly stood the test of time as it is still in use today. Nevertheless, even though this model immediately got a lot of attention in the transportation science community (see [PZ01] and [FH19] for two survey articles on this very active area of research), it took until 2000 that existence of such equilibria for a fairly general class of edge dynamics was established by Zhu and Marcotte in [ZM00]. And even then, this existence result was highly non-constructive as it is based on a characterisation of dynamic equilibria by an infinite dimensional variational inequality and an existence result for solutions of those.

A completely new approach to studying equilibria when using the Vickrey point queue model for the edge dynamics was then opened up by Koch and Skutella with the introduction of so called thin flows in [KS11]. These thin flows are somewhat reminiscent of the early results by Ford and Fulkerson for optimal dynamic flows in that thin flows are static flows which can be used to completely characterise dynamic equilibria (in the full information setting). Koch and Skutella then used this characterisation to study the quality of such equilibria compared to optimal flows. This new structural insight let to an increased interest in this specific model: Cominetti, Correa and Larré [CCL15] used it to show existence and uniqueness (in terms of travel times) for dynamic equilibria, Cominetti, Correa and Olver [CCO22a] to study the long term behaviour and convergence to a steady state, Bhaskar, Fleischer and Anshelevich [BFA15] as well as Correa, Cristi and Oosterwijk [CCO22b] to derive further bounds on their quality and Olver, Sering and Koch [OSK22] to study continuity properties. Moreover, the model was also extended to the multi-source multi sink case by Sering and Skutella [SS18], to a bicriteria setting by Oosterwijk, Schmand and Schröder [OSS22], to include spillback and kinematic waves by Sering and Vargas Koch [SV18] and to allow for time varying capacities and time varying free flow travel times by Pham and Sering [PS20]. We refer to a recent survey article by Schmand [Sch21] for a good overview and to the PhD theses by Koch [Koc12] and Sering [Ser20] for a comprehensive introduction to and study of the base model and several of its extensions, respectively.

Competitive Packet Routing. Finally, we also want to briefly mention the discrete counterpart to the continuous flow model we have been discussing so far: Competitive packet routing. In this model, instead of infinitesimally small flow particles, we have atomic unsplittable packets that have to traverse a given network. This model (and its many variations) is its own large field of research and we refer to the PhD thesis of Vargas Koch [Var20] for a good introduction to it. Interestingly, the study of equilibria in this model seems to be mostly concerned with Nash equilibria, i.e. equilibria of the full information type. One of the few exceptions is Ismaili who in [Ism17, Section 5] studied packet routing games with so called "GPS-agents" which behave in a very similar way to how it is assumed of the infinitesimal agents making up a continuous equilibrium flow in the current information setting discussed before. Ismaili then studies the quality of the routing choices of such agents and, in particular, shows that there exist networks in which they can get trapped in cycles forever ([Ism17, Theorem 8]).

Even though one could argue that such discrete packet routing models are closer to reality when trying to model road traffic (as individual cars are certainly unsplittable and not infinitesimal in size), we will only consider the continuous model here. One reason for that is that in many cases continuous flows tend to be nicer behaved and, so, more open to formal study (e.g. equilibria may not even exist in packing routing models, cf. [Var20, Theorems 19, 28] or [SVZ22, Proposition 1]). Another reason is that, as the number of packets (vehicles) increases, it may become easier to model and, in particular, compute the resulting traffic flow as an aggregate instead of considering all the interactions between individual agents. Moreover, it seems intuitively clear that the two models approximate each other if the packet size only gets small enough (compared to the total flow volume). Nevertheless, formally establishing such a connections has long been an open problem and only very recently achieved for the deterministic queuing model by Sering, Vargas Koch and Ziemke in [SVZ22] and most recently extended by Olver, Sering and Koch [OSK23] to full information equilibria therein (again based on structural insights involving thin flows).

1.2 Thesis Contribution and Organization

Inspired by the recent progress in understanding dynamic equilibria in the full information setting, the goal of this thesis is to gain similar insights for the current information setting. That is, we study dynamic flows in networks with deterministic queueing wherein particles only ever enter edges which lie on currently shortest paths towards their destination. Following the naming conventions of Ran and Boyce we will call those equilibria *instantaneous dynamic equilibria (IDE)*. We have already seen a first example for such an equilibrium flow at the start of this chapter (Example 1.1) where we also discussed some of the interesting questions relating to this equilibrium concept.

We now give a short overview over the following chapters of this thesis as well as its main contributions towards answering those questions:

In **Chap.** 3 we formally introduce first the physical flow model and then our behavioural model. While the physical model is certainly not new (it is the same as the one used e.g. in [KS11, CCL15]), we still describe the model in full detail to ensure that this fundamental aspect of our model is formally sound. We also present several equivalent definitions used for it in the literature and show their equivalence. We then use these equivalent characterisations to deduce five basic properties of the deterministic queueing model. Finally, we establish our definition of instantaneous dynamic equilibria by adapting the label-based definition of full information dynamic equilibria used in [KS11, CCL15] to the current information setting. At the end of this Chapter, in Sect. 3.3, we also provide a summary of the whole model for easier reference.

Following the definition of our model there are three Chapters containing our main results: In **Chap.** 4 we show existence of IDE by reducing this problem to the existence of IDE-extensions and (for certain classes of networks) even further down to the existence of local IDE-extensions (e.g. at a single node). We use this general framework to highlight the commonalities (but also differences) between three different approaches for showing existence of IDE (for different classes of networks): One using an infinite dimension fixed point theorem, one using a finite dimensional fixed point theorem and one using a convex optimization problem. The latter two approaches also make critical use of a variant of thin flows adapted from the full information setting to our setting.

In **Chap.** 5 we then study the computational complexity of finding IDE. On the positive side we show that for single-commodity networks we can compute IDE-thin flows (and, therefore, IDE-extensions) in polynomial time and complete IDE in finite time. On the negative side we give examples for IDE that require an exponential or, in the case of multi-commodity networks, even infinite number of extensions and show NP-hardness of several decision problems involving IDE. The proof of the latter also showcases our first use of gadget-based constructions of IDE-networks which allow us to build and analyse quite complex IDE-instances.

These constructions then also play an important role in the final **Chap.** 6 where we study the quality of IDE. In particular, we show that in single-commodity networks IDE are guaranteed to terminate and give explicit upper and lower bounds on the worst case values of both the time of termination and the total travel time. We then show that in multi-commodity networks termination is not guaranteed anymore by providing a finely tuned instance wherein particles travelling along currently shortest paths get trapped in cycles forever. We conclude this chapter by summarising our results on the quality of IDE in terms of the corresponding Price of Anarchy.

The main results shown in Chaps. 4 to 6 have been published in [GH19, GHS20, GH21, GH23] and [GH20, GH22], respectively. Here, we present most of them in greater generality and with rewritten and extended proofs.

On how this thesis is written: One of our main goals when writing this thesis was to be as mathematically precise as possible, both when it comes to the statements of definitions and theorems and with regards to their proofs. This is mostly born out of the author's intrinsic desire for mathematical rigour but, hopefully, also makes it easier to adapt or even directly use the results presented here to/in related models in the future. The downside of this approach is, of course, that some proofs have become quite long and technical. To mitigate this we tried our best to make the proofs as modularised as possible to allow a reader more interested in the high-level ideas to skip the detailed proofs of more technical lemmas and claims while still getting the important ideas of the main proofs. Moreover, we often give a rough sketch of the intuitive ideas behind a proof before delving into its formal details (often highlighted by boxes like this one).

Additionally, we collected in Chap. 2 all but the most basic definitions from various areas of mathematics that will come up at some point in this thesis together with some key results from these areas that we will make use of. Importantly, this chapter is not meant to be read in its entirety but rather intended as a place to look up definitions/theorems the reader might not be familiar with once they come up in the main part of this thesis.

Preliminaries 2

In this chapter we collect some basic definitions and statements from the areas of measure theory, topology, graph theory, optimization and complexity theory which we will then use in this thesis. Note that this chapter is not intended to be read through in its entirety (except for maybe the first section on general notations) but instead as a reference-chapter when reading the other chapters (together with the lists of definitions and notations in the appendix).

2.1 General Notation

We start by stating some general notations and conventions we will use throughout this thesis.

Sets

We denote by $\mathbb{R}$ the set of real numbers and by $\tilde{\mathbb{R}} := \mathbb{R} \cup \{ \pm\infty \}$ the set of extended real numbers. We use $\mathbb{R}_{>0}$, $\mathbb{R}_{\geq 0}$ and $\tilde{\mathbb{R}}_{\geq 0}$ to refer to the subsets of strictly positive real numbers, non-negative real numbers and non-negative real numbers including ∞, respectively We denote by $\mathbb{N}_0 := \{ 0, 1, 2, \dots \}$ the natural numbers including 0 and by $\mathbb{N}^* := \{ 1, 2, \dots \}$ the natural numbers excluding 0.

For any $x \in \mathbb{R}_{\geq 0}$ we define the numbers $\lceil x \rceil := \min \{ n \in \mathbb{N}_0 \mid n \geq x \}$ and $\lfloor x \rfloor := \max \{ n \in \mathbb{N}_0 \mid n \leq x \}$ and for any natural number $n \in \mathbb{N}_0$ we denote by $[n] := \{ 1, 2, \dots, n \}$ the set of natural numbers from 1 to n. In particular, we use $[0]$ as a synonym for the empty set. To refer to the set of natural numbers from 0 to n we use the notation $[n+1] - 1 = \{ 1, 2, \dots, n+1 \} - 1 := \{ 0, 1, \dots, n \}$.

L. Graf, *Dynamic Network Flows with Adaptive Route Choice based on Current Information*, Mathematische Optimierung und Wirtschaftsmathematik | Mathematical Optimization and Economathematics,
https://doi.org/10.1007/978-3-658-44948-3_2

For any sets A and B we denote by 2^A the power set of A (i.e. the set of subsets of A), by $A \times B$ the product set of A and B (i.e. the set of tuples (a, b) with $a \in A$, $b \in B$) and by A^B the B-fold product of A (i.e. the set of tuples $(a_b)_{b \in B}$). If there is no danger for confusion, we will just write (a_b) instead of $(a_b)_{b \in B}$ to refer to elements of A^B.

For any subset $B' \subseteq B$ we denote by $|_{B'} : A^B \to A^{B'}$ the projection from A^B to $A^{B'}$. If C is a third set and $(a_{b,c})_{b \in B, c \in C}$ an element of $A^{B \times C}$, we use $(a_{\hat{b},\cdot}) := (a_{b,c})\big|_{\{\hat{b}\} \times C} = (a_{\hat{b},c})_{c \in C} \in A^C$ as a shorthand for the projection of $(a_{b,c})$ to $\{\hat{b}\} \times C$. We write $B \mathbin{\dot{\cup}} C$ for the union of B and C if the two sets are disjoint (and we want to emphasise this. If, in such a situation, we have two tuples $(a_b) \in A^B$ and $(a_c) \in A^C$, we use $(a_b) \oplus (a_c)$ to refer to the tuple $(a'_x)_{x \in B \dot{\cup} C} \in A^{B \cup C}$ defined by $a'_x := a_b$ if $x = b \in B$ and $a'_x := a_c$ if $x = c \in C$.

For sums we make use of the shorthand $\sum_{a \in (a_b)} a := \sum_{b \in B} a_b$ and the convention $\sum_{b \in \emptyset} a_b = 0$.

Functions

For any function $f : A \to B$ and any subset $A' \subseteq A$ we denote the restriction of f to A' by $f|_{A'} : A' \to B$. We write the characteristic function of a subset $A' \subseteq A$ by

$$\mathbb{1}_{A'} : A \to \mathbb{R}, a \mapsto \begin{cases} 1, & \text{if } a \in A' \\ 0, & \text{else.} \end{cases}$$

If $f : J \to \mathbb{R}$ is a function on some interval $J \subseteq \mathbb{R}$, we use ∂f to denote the derivative of f (we sometimes also write $\partial_\theta f$ to emphasise the variable θ with respect to which the derivative is taken). Moreover we use $\partial_- f$ and $\partial_+ f$ two denote the left and right derivatives of f, respectively. Note, that we will never use f' to denote a derivative and instead reserve $'$ for use in variable names.

Definition 2.1. Let $f : J \to \mathbb{R}$ be any function on some interval $J \subseteq \mathbb{R}$. We then say that f is

- *right-constant* if for any $\theta \in J$ there exists some $\varepsilon > 0$ such that $f|_{[\theta,\theta+\varepsilon)}$ is constant,
- *non-decreasing* if it satisfies $\theta \leq \zeta \implies f(\theta) \leq f(\zeta)$ for all $\theta, \zeta \in J$ and
- *non-increasing* if it satisfies $\theta \leq \zeta \implies f(\theta) \geq f(\zeta)$ for all $\theta, \zeta \in J$.

If f is non-decreasing we say that it is *strictly increasing (from the left) at* $\theta \in J$ if it satisfies

$$\zeta < \theta \implies f(\zeta) < f(\theta) \text{ for all } \zeta \in J.$$

2.2 Measure Theory

As flows in our model will be described by (Lebesgue)-measurable functions on $\mathbb{R}_{\geq 0}$, we will need some basic definitions and results from measure theory. We follow here mostly the introduction in [RF10].

The Lebesgue Measure

Definition 2.2. Let $J \subseteq \mathbb{R}$ be any non-empty interval. Then the *outer (Lebesgue) measure* μ^* on J is defined by

$$\mu^* : 2^J \to \tilde{\mathbb{R}}_{\geq 0}, A \mapsto \inf \left\{ \sum_{k=1}^{\infty}(b_k - a_k) \;\middle|\; a_k, b_k \in \mathbb{R} : A \subseteq \bigcup_{k=1}^{\infty}(a_k, b_k) \right\}.$$

For any set $A \subseteq J$ we say that

- A is a *(Lebesgue) null set* if it has measure zero, i.e. $\mu^*(A) = 0$
- A is *(Lebesgue) measurable* if it satisfies $\mu^*(A) = \mu^*(A \cap B) + \mu^*(A \setminus B)$ for any set $B \subseteq J$.

The restriction of μ^* to only the Lebesgue measurable subsets of J is called the *Lebesgue measure* and we will denote it by μ. Since μ is the only measure we use in this thesis, we will usually omit the word Lebesgue and just talk about *the* measure on J.

Observation 2.3. We observe some basic properties of null sets:

- Any countable set $A \subseteq J$ is a null set.
- Any null set $N \subseteq J$ is measurable.
- The countable union of null sets is again a null set.

Definition 2.4. Let $J \subseteq \mathbb{R}$ be any interval. A function $f : J \to \tilde{\mathbb{R}}$ is called *(Lebesgue) measurable* if for any $C \in \mathbb{R}$ the set $\{\theta \in J \mid f(\theta) \leq C\} \subseteq J$ is measurable.

Definition 2.5. Let $f, g : J \to \tilde{\mathbb{R}}$ be two functions on some interval $J \subseteq \mathbb{R}$.

- We say that f and g are *equal almost everywhere* if there exists null set $N \subseteq J$ such that $f|_{J \setminus N} = g|_{J \setminus N}$. We denote this by $f =_{\text{a.e.}} g$.
- We say that f is *bounded by g almost everywhere* if there exists some null set $N \subseteq J$ such that $f|_{J \setminus N} \leq g|_{J \setminus N}$. We denote this by $f \leq_{\text{a.e.}} g$. If g is a constant function we also say that f is *essentially bounded*.

Observation 2.6
- $=_{\text{a.e.}}$ is a equivalence relations on the set of functions from J to $\tilde{\mathbb{R}}$.
- $=_{\text{a.e.}}$ is compatible with addition, multiplication and scalar multiplication.

Proposition 2.7 ([RF10, Chapter 3, Proposition 5]). *Let $f, g : J \to \tilde{\mathbb{R}}$ be two functions with $f =_{\text{a.e.}} g$. If f is measurable, then so is g.*

Proposition 2.8 ([RF10, Chapter 3, Proposition 9]). *Let $f_n : J \to \tilde{\mathbb{R}}$ be a sequence of measurable functions converging pointwise almost everywhere to a function $f : J \to \tilde{\mathbb{R}}$. Then f is measurable.*

The Morse-Sard Theorem

The Morse-Sard Theorem states that for sufficiently smooth maps f the set of critical values $f(\{\theta \mid \partial f(\theta) = 0\})$ has Lebesgue measure zero (cf. e.g. [Hir76, Chapter 3, Theorem 1.3]). In the one-dimensional case this even holds without the smoothness-condition. It is often stated for (absolutely) continuous functions where it can be derived from Vitali's Covering Theorem (cf. [Mar22, Theorem 3.2.8]) but the following fundamental lemma by Varberg ([Var65]) actually implies that it holds for any function $f : \mathbb{R} \to \mathbb{R}$:

Proposition 2.9 ([Var65, Fundamental Lemma]). *Let $f : J \to \mathbb{R}$ be any function from some compact interval $J \subseteq \mathbb{R}$ to $\mathbb{R}$. Then for any number $K \geq 0$ and any set $S \subseteq J$ such that the derivative of f exists for all $\theta \in S$ and satisfies $|\partial f(\theta)| \leq K$, we have*

$$\mu^*(f(S)) \leq K \cdot \mu^*(S).$$

Proposition 2.10. *Let $f : J \to \mathbb{R}$ be any function.*

a) The set of images of points where the derivative of f exists and is zero is a null set, i.e.

$$\mu(f(\{\theta \in J \mid \partial f(\theta) = 0\})) = 0.$$

b) If f is non-decreasing, then we have $|f^{-1}(\theta)| \leq 1$ for almost all $\theta \in \mathbb{R}$.

Proof. a): If J is bounded, then this follows immediately from Proposition 2.9 by choosing $S := \{\theta \in J \mid \partial f(\theta) = 0\}$ and $K = 0$. Otherwise we can cover J by a countable sequence of compact intervals and apply Proposition 2.9 for each of those. As any countable union of null sets is still a null set, this shows the claim.

b): This can be deduced from a) (see e.g. the proof of [Mar22, Corollary 3.2.9]) but we can also show this directly: For any $n \in \mathbb{N}^*$ denote by

$$S_n := \{x \in \mathbb{R} \mid \exists a, b \in J : |b - a| \geq \tfrac{1}{n} \text{ and } f(a) = f(b) = x\}$$

the set of values of f such that an interval of length at least $\frac{1}{n}$ gets mapped to x by f. Note, that any such set S_n contains at most countably many points as f is non-decreasing and, thus, any two intervals $[a, b]$ and $[a', b']$ attesting for two different points $x, x' \in S_n$ must be disjoint. Thus, the same is true for the set

$$S := \bigcup_{n \in \mathbb{N}^*} S_n.$$

Since this set contains all values $x \in \mathbb{R}$ with at least two preimages, this shows the claim. □

The Lebesgue Integral

Let $J \subseteq \mathbb{R}$ be an interval, $f : J \to \tilde{\mathbb{R}}$ a measurable function and $J' = [a, b] \subseteq J$ some bounded subinterval. Then we denote by $\int_{J'} f(\zeta)\mathrm{d}\zeta = \int_a^b f(\zeta)\mathrm{d}\zeta \in \tilde{\mathbb{R}}$ the Lebesgue integral of f on J' (if it exists). We recall the following properties of the Lebesgue integral (see [RF10, Chapter 4]):

Proposition 2.11. *Let $f, g : J \to \mathbb{R}$ be two measurable functions.*

- *We have $f =_{\text{a.e.}} g$ if and only if $\int_J |f(\zeta) - g(\zeta)|\,\mathrm{d}\zeta = 0$.*
- *Taking integrals is compatible with $=_{\text{a.e.}}$, i.e. for any $a, b \in J$ such that $\int_a^b f(\zeta)\mathrm{d}\zeta$ exists we have*

$$f =_{\text{a.e.}} g \implies \int_a^b f(\zeta)\mathrm{d}\zeta = \int_a^b g(\zeta)\mathrm{d}\zeta.$$

- *The Lebesgue integral is monotone, i.e. for any $a, b \in J$ such that $\int_a^b f(\zeta)\mathrm{d}\zeta$ and $\int_a^b g(\zeta)\mathrm{d}\zeta$ exist we have*

$$f \leq_{a.e.} g \implies \int_a^b f(\zeta)\mathrm{d}\zeta \leq \int_a^b g(\zeta)\mathrm{d}\zeta.$$

Definition 2.12. Let $f : J \to \tilde{\mathbb{R}}$ be a measurable function and $p \in [1, \infty)$. We say that f is *p-integrable* if $\int_J |f(\zeta)|^p \mathrm{d}\zeta < \infty$. We call it *locally p-integrable* if it is p-integrable on every compact subset of J. For $p = 1$ we usually just say (locally) integrable instead of (locally) 1-integrable.

We denote the sets of all (locally) p-integrable functions on J by $\mathcal{L}^p_{\mathrm{loc}}(J)$ and $\mathcal{L}^p(J)$, respectively.

Note that, for compact intervals J, the notions of locally p-integrable and p-integrable are equivalent.

We now collect several useful statements for locally integrable functions:

Proposition 2.13. *Let $f, g : J \to \mathbb{R}$ be two measurable functions with $|f| \leq_{a.e.} |g|$. Then*

$$g \in \mathcal{L}^p_{\mathrm{loc}}(J) \implies f \in \mathcal{L}^p_{\mathrm{loc}}(J).$$

This follows directly from the definition of local p-integrability and the monotonicity of the Lebesgue integral.

Proposition 2.14. *Let $f : J \to \mathbb{R}$ be a measurable, locally essentially bounded function. Then f is locally p-integrable for any $p \in [1, \infty)$.*

This follows from the previous proposition by observing that locally constant functions are locally p-integrable.

The following proposition is a very useful tool for showing that some equality for a locally integrable function holds almost everywhere on some given interval:

Proposition 2.15 ([CCL15, Lemma 8]). *Let $f : \mathbb{R} \to \mathbb{R}_{\geq 0}$ be any non-negative locally integrable function and $\{ [a_k, b_k) \}_{k \in K}$ be any family of half open intervals (possibly uncountably many). Then the following two properties are equivalent*

a) f vanishes almost everywhere on $[a_k, b_k)$ for every $k \in K$ (separately!),
b) f vanishes almost everywhere on $\bigcup_{k \in K} [a_k, b_k)$.

L^p-Spaces and Right Constant Functions

Definition 2.16. Let $J \subseteq \mathbb{R}$ be any interval and $p \geq 1$. Then we define the space

$$L^p_{\text{loc}}(J) := \mathcal{L}^p_{\text{loc}}(J)/{=_{\text{a.e.}}} \text{ and } L^p(J) := \mathcal{L}^p(J)/{=_{\text{a.e.}}}.$$

Moreover, we denote by $L^p_{\text{loc}}(J, \mathbb{R}_{\geq 0}) := \{ [f] \in L^p_{\text{loc}}(J) \mid 0 \leq_{\text{a.e.}} f \}$ the set of (equivalence classes of) non-negative locally p-integrable functions.

Observation 2.17. Both $L^p_{\text{loc}}(J)$ and $L^p(J)$ are (infinite dimensional) real vector spaces (with pointwise addition and scalar multiplication).

Note that $L^p_{\text{loc}}(J)$ contains equivalence classes of functions, i.e. elements of the form $[f : J \to \mathbb{R}]$. To keep our notation simple we will usually omit the square brackets and even call elements of $L^p_{\text{loc}}(J)$ 'functions' instead of 'equivalence classes of functions'. Consequently, we will usually also refer to $L^p_{\text{loc}}(J)$ as the 'set of locally p-integrable functions'. However, it is important to keep in mind that we usually cannot evaluate such 'functions' at any individual point (as the value of "$[f](\theta)$" would depend on the specific representative chosen for the evaluation). Instead we may only talk about integrals of f or (in)equalities of two such functions which hold *almost everywhere*.

We will, however, make one exception to this rule – whenever an equivalence class $[f] \in L^p_{\text{loc}}(J)$ contains some representative g which is constant on some proper interval $[a, b)$ then we will choose g as the canonical function representing $[f]$ on this interval. In particular, evaluating $[f]$ on this interval will be defined as evaluating g.

Definition 2.18. Let $[f] \in L^p_{\text{loc}}(J)$ and $[a, b) \subseteq J$ be a proper interval. We say that $[f]$ is *constant on* $[a, b)$ if there exists some $c \in \mathbb{R}$ such that we have $f(\theta) = c$ for almost all $\theta \in [a, b)$

Proposition 2.19. *Let $[f] \in L^p_{\text{loc}}(J)$ and $[a, b) \subseteq J$ such that $[f]$ is constant on $[a, b)$ with constant $c \in \mathbb{R}$. Then there exists some representative $g \in [f]$ with $g(\theta) = c$ for all $\theta \in [a, b)$.* □

In the same way, we can define right-constant $[f] \in L^p_{\text{loc}}(J)$ for which we then also have a canonical representative (which we can use for evaluation at individual points):

Definition 2.20. Let $J \subseteq \mathbb{R}$ be a right-open interval. Then $[f] \in L^p_{\text{loc}}(J)$ is called *right-constant* if there exists a covering of J with proper intervals $[a, b)$ such that $[f]$ is constant on each of those intervals.

Proposition 2.21. *If* $[f] \in L^p_{\text{loc}}(J)$ *is right-constant, then it has a unique right-constant representative* $g \in [f]$. □

2.3 Topology

We recall several standard concepts from topology – for a more detailed introduction we refer to standard textbooks like e.g. [Jän84] or [AB06, Chapter 2,5,6].

Definition 2.22. Let X be a topological space. We then say that

- X is *Hausdorff* if for any two points $x \neq y \in X$ there exist two disjoint open neighbourhoods $U, V \subseteq X$ of x and y, respectively.
- X is *compact* if every family of open sets $(U_i)_{i \in I}$ covering X has a finite subset covering X.
- X is *sequentially compact* if every sequence in X has a convergent subsequence.
- X is a *topological vector space* if X is also a real vector space and both addition and scalar multiplication are continuous mappings.
- X is *locally convex* if it is a topological vector space and every neighbourhood of $0 \in X$ contains a convex neighbourhood of 0.

Proposition 2.23 ([Jän84, Chapter I, Section §8, Note]). *Limit points in Hausdorff spaces are unique.*

Proposition 2.24 ([Jän84, Chapter I, Section §8, Lemma]). *Compact subsets of Hausdorff spaces are closed.*

Proposition 2.25. *Let* X *be a sequentially compact space,* $(x_n) \in X^{\mathbb{N}^*}$ *a sequence and* $x \in X$ *some point. If every convergent subsequence of* (x_n) *converges to* x*, then* (x_n) *itself converges to* x.

Proof. We show this via contraposition: So, let $(x_n) \in X^{\mathbb{N}^*}$ be a sequence not converging to x. Then, there exists some neighbourhood U of x such that there are infinitely many $x_n \notin U$. In particular, there exists a subsequence $(x_{n_k})_{k \in \mathbb{N}^*}$

with $x_{n_k} \in X \setminus U$ for all $k \in \mathbb{N}^*$. Since X is sequentially compact, this sequence must have a convergent subsequence $(x_{n_{k_j}})_{j \in \mathbb{N}^*}$. And, as we have $x_{n_{k_j}} \notin U$ for all elements of this sequence, it has to converge to a point different to x – which proves the proposition. □

Normed Vector Spaces

The topological spaces we use in this thesis will mostly be special types of topological vector spaces:

Definition 2.26. A *(real) normed vector space* is a (real) vector space V together with a *norm* map $\|.\| : V \to \mathbb{R}_{\geq 0}$ satisfying

- Positive definiteness: $\|x\| = 0 \iff x = 0$ for all $x \in V$,
- Absolute homogeneity: $\|\alpha x\| = |\alpha| \cdot \|x\|$ for all $\alpha \in \mathbb{R}, x \in V$ and
- Triangle inequality: $\|x + y\| \leq \|x\| + \|y\|$ for all $x, y \in V$.

We will consider two different topologies on such spaces: The first (most natural) one is the topology directly induced by the norm which we will call the strong topology on $(V, \|.\|)$:

Definition 2.27. Let $(V, \|.\|)$ be a normed space, $x \in V$ and $\alpha > 0$. Then we denote by $B_\alpha(x) := \{ y \in V \mid \|x - y\| < \alpha \}$ the open ball of radius α around x and by $\bar{B}_\alpha(x) := \{ y \in V \mid \|x - y\| \leq \alpha \}$ the closed ball of radius α around x.

We call the topology on $(V, \|.\|)$ induced by its open balls (as a base) its *(strong) topology*.

Definition 2.28. We call a normed space $(V, \|.\|)$ a *Banach space* if it is complete with respect to the strong topology, i.e. every Cauchy sequence in $(V, \|.\|)$ converges.

In this thesis we will encounter three such Banach spaces:

- the real numbers $\mathbb{R}$ together with the absolute value $|.| : \mathbb{R} \to \mathbb{R}_{\geq 0}$,
- the set $C(J) := \{ f : J \to \mathbb{R} \mid f \text{ continuous} \}$ of continuous functions on a compact interval together with the uniform norm

$$\|.\|_\infty : C(J) \to \mathbb{R}_{\geq 0}, \ f \mapsto \sup_{\theta \in J} |f(\theta)|$$

- and the set of p-integrable functions $L^p(J)$ together with the p-norm

$$\|.\|_p : L^p(J) \to \mathbb{R},\, f \mapsto \left(\int_J |f(\zeta)|^p \mathrm{d}\zeta\right)^{1/p}.$$

Proposition 2.29 ([RF10, Chapter I, Theorem 17]). $(\mathbb{R}, |.|)$ *is a Banach space.*

Proposition 2.30 ([Con19, Chapter III, Example 1.6]). *Let $J \subseteq \mathbb{R}$ be any compact interval. Then $(C(J), \|.\|_\infty)$ is a Banach space.*

Proposition 2.31 ([RF10, Chapter 7, Riesz–Fischer Theorem]). *Let $J \subseteq \mathbb{R}$ be any compact interval and $p \geq 1$. Then $(L^p(J), \|.\|_p)$ is a Banach space.*

The other topology on a normed space (which we will call the weak topology) is the topology induced by its dual space.

Definition 2.32. For any normed vector space $(V, \|.\|)$ we define its *dual space* $(V', \|.\|')$ where V' is the set of continuous linear functions $\varphi : V \to \mathbb{R}$ with the natural vector space structure and the norm $\|.\|'$ is defined by

$$\|\varphi\|' := \sup_{\|x\| \leq 1} \{\varphi(x)\} \text{ for any } \varphi \in V'.$$

Definition 2.33. Let $(V, \|.\|)$ a normed space and V' its dual space. Then we call the topology induced by the sets (as a subbase)

$$U(\varphi, x, \varepsilon) := \{\, y \in V \mid |\varphi(x) - \varphi(y)| < \varepsilon \,\} \text{ for } \varphi \in V', x \in V, \varepsilon > 0$$

its *weak topology*.

Note that, since we now have two topologies on a normed space, we have to always ensure to make clear to which topology we are referring to whenever we talk about subsets of those spaces or continuous maps between them. Whenever there is any danger of confusion, we will therefore explicitly say "weakly open", "weakly compact", ... when referring to the respective property with respect to the weak topology. For functions we will use e.g. "weak-strong" continuous to refer to a function $f : X \to Y$ which is continuous with respect to the weak topology on X and the strong topology on Y. For a sequence $(x_n)_n$ converging weakly to some x we write $x_n \xrightarrow[n \to \infty]{w} x$.

Proposition 2.34 ([AB06, Section 6.1], [RF10, Chapter 14, Proposition 21]). *Let X be a normed space. Then it is a locally convex, Hausdorff topological vector space both with respect to the strong and the weak topology on X.*

Proposition 2.35 ([AB06, 6.34 Eberlein–Šmulian Theorem]). *Let X be a normed space. Then a subset $K \subseteq X$ is weakly compact if and only if it is sequentially weakly compact.*

Proposition 2.36 ([RF10, Chapter 8, Proposition 6]). *Let $p, q \in (1, \infty)$ with $\frac{1}{p} + \frac{1}{q} = 1$. Then, a sequence $(f_n) \in L^p(J)^{\mathbb{N}_0}$ converges weakly to some $f \in L^p(J)$ if and only if we have*

$$\lim_{n \in \mathbb{N}_0} \int_J f_n(\zeta) \cdot g(\zeta) \mathrm{d}\zeta = \int_J f(\zeta) \cdot g(\zeta) \mathrm{d}\zeta \text{ for all } g \in L^q(J).$$

Proposition 2.37. *Let $p > 1$, $J \subseteq \mathbb{R}$ some interval and $(f_n)_n \in L^p(J)^{\mathbb{N}_0}$ a sequence of functions converging weakly to some function $f \in L^p(J)$. Furthermore, let $g \in L^p(J)$ be another p-integrable function and $M \subseteq \mathbb{R}$ some measurable subset. If $f_n|_M \leq_{a.e.} g|_M$ holds for almost all n, then we also have $f|_M \leq_{a.e.} g|_M$.*

Proof This is an immediate consequence of Proposition 2.36: Assume, for contradiction, that $f|_M \not\leq_{\text{a.e.}} g|_M$. Then there exists some measurable subset $S \subseteq M$ with finite measure such that we have $f(\zeta) > g(\zeta)$ for all $\zeta \in S$. Since $\mathbb{1}_S$ is certainly q-integrable this leads to the following contradiction

$$\begin{aligned}\int_J g(\zeta) \cdot \mathbb{1}_S(\zeta) \mathrm{d}\zeta < \int_J f(\zeta) \cdot \mathbb{1}_S(\zeta) \mathrm{d}\zeta &\overset{\text{Prop.2.36}}{=} \lim_n \int_J f_n(\zeta) \cdot \mathbb{1}_S(\zeta) \mathrm{d}\zeta \\ &\leq \int_J g(\zeta) \cdot \mathbb{1}_S(\zeta) \mathrm{d}\zeta.\end{aligned}$$

□

Proposition 2.38 ([GHP22, Lemma A.1]). *Let $A, B \subseteq \mathbb{R}$ be two subsets of real numbers, $g_n : A \to B$ a sequence of functions converging uniformly to some function $g : A \to B$ and $f_n : B \to \mathbb{R}$ another sequence of functions converging uniformly to some continuous function $f : B \to \mathbb{R}$. Then the sequence $f_n \circ g_n : A \to \mathbb{R}$ converges pointwise to the function $f \circ g : A \to \mathbb{R}$.*

For finite dimensional vector spaces the canonical mapping from itself to its double dual is always an isomorphism. For infinite dimensional Banach spaces this need not be true anymore. If it is true, such a space is called reflexive.

Definition 2.39. Let $(V, \|.\|)$ be a Banach space. $(V, \|.\|)$ is called *reflexive* if the canonical mapping from V to its double dual V''

$$V \to V'', x \mapsto (V' \to \mathbb{R}, (f : V \to \mathbb{R}) \mapsto f(x))$$

is surjective.

Proposition 2.40 ([RF10, Chapter 14, Proposition 20]). *For $p > 1$ the space $L^p(J)$ is a reflexive Banach space and its dual can be identified with $L^q(J)$ where $q > 1$ is chosen such that $\frac{1}{p} + \frac{1}{q} = 1$.*

Proposition 2.41 ([Hun13, Corollary 7.32]). *Let X be a reflexive Banach space. Then, any convex, bounded and (strongly) closed subset $K \subseteq X$ is weakly compact.*

Product Spaces

We will often consider products of topological/normed spaces. These then become topological/normed spaces in a natural way as well:

Definition 2.42. Let X, Y be two topological spaces. Then the *product topology* on $X \times Y$ is the topology generated (as a base) by the sets $U \times V$ with $U \subseteq X$, $V \subseteq Y$ open.

Let $(X, \|.\|_X)$, $(Y, \|.\|_Y)$ be two normed spaces. Then the *product norm* on $X \times Y$ is defined by

$$\|.\|_{X \times Y} : X \times Y \to \mathbb{R}_{\geq 0}, (x, y) \mapsto \|x\|_X + \|y\|_Y .$$

As we will only need finite products, all the relevant properties of those spaces easily carry over to their product as well:

Proposition 2.43. *Let X, Y be two topological spaces.*

- *If X and Y are Hausdorff, so is $X \times Y$.*
- *If X and Y are (sequentially) compact, so is $X \times Y$.*
- *If X and Y are locally convex, so is $X \times Y$.*

Now let $(X, \|.\|_X)$, $(Y, \|.\|_Y)$ be two normed spaces. Then $(X \times Y, \|.\|_{X\times Y})$ is a normed space. Moreover, the following properties hold:

- *The strong/weak topology on $(X \times Y, \|.\|_{X\times Y})$ is the product topology of the strong/weak topologies on $(X, \|.\|_X)$ and $(Y, \|.\|_Y)$.*
- *If X and Y are Banach spaces, so is $X \times Y$.*
- *If X and Y are reflexive Banach spaces, so is $X \times Y$.* □

Absolutely Continuous Functions

Definition 2.44. Let $J \subseteq \mathbb{R}$ be some interval. A function $F : J \to \mathbb{R}$ is *absolutely continuous* if for any $\varepsilon > 0$ there exists some $\delta > 0$ such that

$$\sum_{k=1}^{n} |b_k - a_k| < \delta \implies \sum_{k=1}^{n} |F(b_k) - F(a_k)| < \varepsilon$$

for every finite sequence of pairwise disjoint subintervals $(a_k, b_k) \subseteq J$.

We denote the set of all such functions by $AC(J)$ and by $AC^{\nearrow}(J)$ the set of all non-decreasing absolutely continuous functions.

Observation 2.45. Absolute continuity is related as follows to the more standard types of continuity:

Lipschitz-continuous $\implies$ absolutely continuous $\implies$ uniformly continuous.

Observation 2.46. The set $AC(J)$ with pointwise addition and scalar multiplication forms a real vector space.

Proposition 2.47 ([BS20, Corollary 4.3.5], [Var65, Theorem 2]). *Let $F : [a, b] \to \mathbb{R}$ be an absolutely continuous function. Then F also has the following properties:*

a) F is differentiable almost everywhere.
b) F maps null sets to null sets, i.e. for every set $N \subseteq [a, b]$ we have

$$\mu(N) = 0 \implies \mu(F(N)) = 0.$$

In [CCL15] Cominetti, showed a helpful characterisation of non-negativity for absolutely continuous functions which we state here in slightly altered form:

Proposition 2.48. *Let $F : [a, b) \to \mathbb{R}$ be any absolutely continuous function with $a \in \mathbb{R}$, $b \in \tilde{\mathbb{R}}$ and $F(a) \geq 0$. Then the following conditions are equivalent:*

a) $\forall \theta \in [a, b) : F(\theta) \geq 0$,
b) *For almost all* $\theta \in [a, b)$ *we have* $F(\theta) < 0 \implies \partial F(\theta) \geq 0$.
c) *For almost all* $\theta \in [a, b)$ *we have* $F(\theta) \leq 0 \implies \partial F(\theta) = 0$.

Proof This is essentially the same statement as [CCL15, Lemma 9]. The only differences are our assumption $F(a) \geq 0$ instead of $F(a) = 0$ in [CCL15] and statement b) which is slightly stronger in [CCL15] than here ($F(\theta) \leq 0$ instead of $F(\theta) < 0$). However, it is easy to see that the same proof also applies to our version of this lemma. □

The Fundamental Theorem of Calculus

Proposition 2.49 (Fundamental Theorem of Calculus). *Let J be a non-trivial interval and $c \in J$ some fixed point in this interval. Then we have:*

$$
\begin{array}{ccc}
L^1_{\text{loc}}(J) \times \mathbb{R} & \underset{\tilde{\partial}}{\overset{\tilde{\int}}{\rightleftarrows}} \ 1:1 & AC(J) \\
\cup\!\!| & & \cup\!\!| \\
L^1_{\text{loc}}(J, \mathbb{R}_{\geq 0}) \times \mathbb{R} & \underset{\tilde{\partial}}{\overset{\tilde{\int}}{\rightleftarrows}} \ 1:1 & AC^{\nearrow}(J)
\end{array}
$$

where the two mappings are defined by

$$\tilde{\int} : L^1_{\text{loc}}(J) \times \mathbb{R} \to AC(J), (f, C) \mapsto \left(J \to \mathbb{R}, \theta \mapsto \int_c^\theta f(\zeta) \mathrm{d}\zeta + C \right)$$

and

$$\tilde{\partial} : AC(J) \to L^1_{\text{loc}}(J), F \mapsto (\partial F, F(c)).$$

Proof This follows directly from [BS20, Theorem 4.4.2] and [BS20, Corollary 4.4.4] together with Proposition 2.48. □

Because of the above stated fundamental theorem of calculus, we will usually use capital letters to refer to absolutely continuous functions. We then use corresponding lower case letters to refer to their derivative.

Proposition 2.50. *Let $J = [a, b]$ be a compact interval and $p > 1$. Then the mapping*

$$\int : L^p(J) \to C(J), f \mapsto (\theta \mapsto \int_a^\theta f(\zeta)\mathrm{d}\zeta)$$

is sequentially weak-strong continuous.

Proof. The integration mapping is a compact operator from $L^p(J) \to C(J)$ (cf. [BS20, Example 6.9.4(iv)]). According to [Con19, Chapter VI, Proposition 3.3] any such operator is completely continuous, i.e. maps weakly convergent sequences to strongly convergent sequences. □

We can now also state three useful rules for evaluating integrals: The chain rule, integration by parts and the change of variable formula:

Proposition 2.51 ([SV69, Corollary 4]). *Let $G : [a, b] \to [c, d]$ be a monotone function and $F : [c, d] \to \mathbb{R}$ an absolutely continuous function. Then, $F \circ G$ is differentiable almost everywhere and we have*

$$\partial(F \circ G)(\theta) = \partial F(G(\theta)) \cdot \partial G(\theta)$$

for almost all $\theta \in [a, b]$.

Proposition 2.52 ([BS20, Corollary 4.4.5]). *Let $F, G : [a, b] \to \mathbb{R}$ be two absolutely continuous functions. Then we have*

$$\int_a^b \partial F(\zeta) \cdot G(\zeta)\mathrm{d}\zeta = [F(\zeta)G(\zeta)]_a^b - \int_a^b F(\zeta) \cdot \partial G(\zeta)\mathrm{d}\zeta,$$

where we use the notation $[F(\zeta)G(\zeta)]_a^b := F(b)G(b) - F(a)G(a)$.

Proposition 2.53 ([SV69, Corollaries 6 and 7]). *Let $G : [a, b] \to [c, d]$ be absolutely continuous and $f \in L^1([c, d])$. If G is monotone or f is bounded, then we have*

$$\int_{G(a)}^{G(b)} f(\zeta)\mathrm{d}\zeta = \int_a^b f(G(\zeta)) \cdot \partial G(\zeta)\mathrm{d}\zeta.$$

Additionally, we can use the fundamental theorem of calculus to give several sufficient conditions for non-decreasing, absolutely continuous functions for being strictly increasing in terms of its derivative:

Proposition 2.54. *Let $F : J \to \mathbb{R}$ be an absolutely continuous, non-decreasing function with derivative $f : J \to \mathbb{R}$ almost everywhere. Then we have*

a) F is not strictly increasing at $\theta \in J$ if and only if there exists some $\varepsilon > 0$ such that $[\theta - \varepsilon, \theta] \subseteq J$ and we have $f(\zeta) = 0$ for almost all $\zeta \in [\theta - \varepsilon, \theta]$.
b) If F is differentiable at θ and $\partial F(\theta) > 0$ then F strictly increases at θ.
c) For almost all θ with $f(\theta) > 0$ we have that F strictly increases at θ.
d) Given two times $\xi < \zeta$ with $F(\xi) < F(\zeta)$ then there must be a subset $J \subseteq [\xi, \zeta]$ of positive measure such that F strictly increases at all times $\theta \in J$.

Proof. a): If there exists some interval $[\theta - \varepsilon, \theta]$ on which f is zero almost everywhere then we have $F(\theta) = F(\theta - \varepsilon) + \int_{\theta-\varepsilon}^{\theta} f(\zeta)d\zeta = F(\theta - \varepsilon) + 0$. Thus, F is not strictly increasing at θ.
If, on the other hand, F is not strictly increasing at θ, then there must be some $\vartheta < \theta$ with $F(\vartheta) = F(\theta)$. Since F is non-decreasing, this implies that it is constant on $[\vartheta, \theta]$ and we have $f(\zeta) = 0$ for almost all $\zeta \in [\vartheta, \theta]$.
b): This follows directly from a).
c): This follows from b) since we have $f(\theta) = \partial F(\theta)$ for almost all θ.
d): Since $\int_{\vartheta}^{\theta} f(\zeta)\mathrm{d}\zeta = F(\theta) - F(\vartheta) > 0$, there must be a subset $S \subseteq [\vartheta, \theta]$ of positive measure with $f(\zeta) > 0$ for all $\zeta \in S$. By c) F is strictly increasing for almost all these $\zeta \in S$. □

2.4 Two General Existence-Results

The two following results will be key ingredients for our existence results for IDE.

A Fixed Point Theorem for Infinite Dimensional Spaces

Fixed point theorems like Brouwer's and Kakutani's Fixed Point Theorem are a standard tool for showing existence of equilibria in game theory. We will make use here of a version of Kakutani's Fixed Point Theorem which also applies to infinite dimensional topological vector spaces.

Definition 2.55. Let X be a topological space and $\Gamma : X \to 2^X$ a correspondence (i.e. a function from a set to its power set). We denote the *graph of* Γ by

$$\mathrm{graph}(\Gamma) := \{ (x, y) \in X \times X \mid y \in \Gamma(x) \}.$$

We say that Γ has a *closed graph* if $\mathrm{graph}(\Gamma) \subseteq X \times X$ is a closed subset.

Theorem 2.56 (Kakutani–Fan–Glicksberg, [AB06, Corollary 17.55]). *Let X be a locally convex Haussdorff space, $K \subseteq X$ a non-empty, convex and compact subset and $\Gamma : K \to 2^K$ a correspondence with closed graph and non-empty convex values.*

Then the set of fixed points $\{ x \in K \mid x \in \Gamma(x) \}$ of Γ is compact and non-empty.

Zorn's Lemma

Another important tool for our existence proofs for IDE will be Zorn's Lemma. We follow here the presentation in [nLa23a, nLa23b]:

Definition 2.57. A binary relation $\preceq$ on a set X is called

- *reflexive* if it satisfies $x \preceq x$ for all $x \in X$,
- *transitive* if $x \preceq y \wedge y \preceq z \implies x \preceq z$ for all $x, y, z \in X$,
- *antisymmetric* if $x \preceq y \wedge y \preceq x \implies x = y$ for all $x, y \in X$,
- *total* if we have $x \preceq y \vee y \preceq x$ for all $x, y \in X$,
- a *preorder* if it is reflexive and transitive and
- a *total order* if it is reflexive, transitive, antisymmetric and total.

Given a preorder $\preceq$ on X we call $(X, \preceq)$ a *preordered set* and

- $x \in X$ *maximal* if it satisfies $y \preceq x$ for all $y \in X$ with $x \preceq y$,
- $Y \subseteq X$ a *chain* if the restriction of $\preceq$ to Y is a total order and
- $x \in X$ an *upper bound* to $Y \subseteq X$ if it satisfies $y \preceq x$ for all $y \in Y$.

Lemma 2.58 (Zorn's Lemma). *Let $(X, \preceq)$ be a preordered set where every chain has an upper bound. Then X has a maximal element (with respect to $\preceq$).*

This version of Zorn's Lemma is from [nLa23b] where they also provide a proof, i.e. show that it follows from the Axiom of Choice ([nLa23b, Theorem 2.3]).

2.5 Graph Theory and Optimization

We recall some basic definitions and results from graph theory and optimization. For a more detailed introduction we refer to standard text books on these topics like e.g. [Jun12, Jun14] where we also take most of our notations from.

Basic Definitions from Graph Theory

Definition 2.59. A *(directed) graph* $G = (V(G), E(G))$ consists

- A finite set $V(G)$ of *nodes* and
- a finite set $E(G) \subseteq V(G) \times V(G) \setminus \{ (v, v) \mid v \in V \}$ of *edges*.

For any node $v \in V$ we then denote by $\delta_G^+(v) := \{ e = (v, w) \in E(G) \}$ the set of all edges leaving v and by $\delta_G^-(v) := \{ e = (w, v) \in E(G) \}$ the set of all edges entering v. We say that e is an edge from v to w if $e \in \delta_G^+(v) \cap \delta_G^-(w)$. We also extend this notation to subsets of nodes by setting $\delta_G^+(W) := \{ e = (w, v) \in E(G) \mid w \in W, v \notin W \}$ and $\delta_G^-(W) := \{e = (v, w) \in E(G) | v \notin W, w \in W\}$ for any subset $W \subseteq V$.

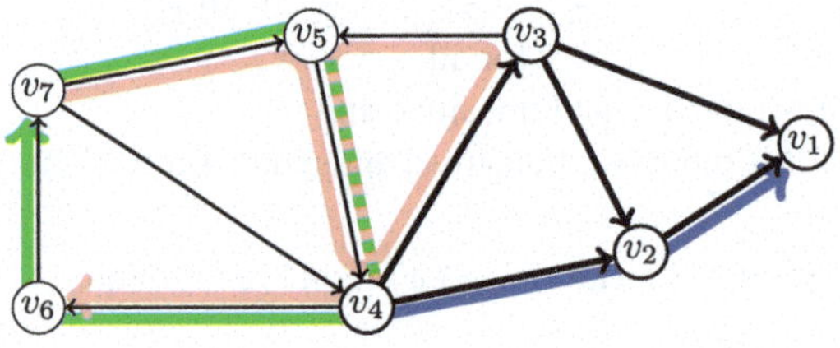

Fig. 2.1 An example graph showcasing the various definition from graph theory: The thick nodes and edges form an induced subgraph induced by the node set $W = \{ v_1, v_2, v_3, v_4 \}$. This subgraph $G[W]$ is an acyclic graph and $v_4 \prec v_3 \prec v_2 \prec v_1$ defines a topological order on its nodes. The blue arrow indicates the v_4, v_1-path v_4v_2, v_2v_1, the red arrow indicates the v_7, v_6-walk $v_7v_5, v_5v_4, v_4v_3, v_3v_5, v_5v_4, v_4v_6$ and the green arrow indicates the cycle $v_7v_5, v_5v_4, v_4v_6, v_6v_7$

Whenever the underlying graph G is clear from the context, we may omit the argument/index G from all the notation introduced in this definition. Moreover, we usually write vw instead of (v, w) to refer to an edge from v to w.

Note that our definition only allows simple graphs, i.e. does not allow parallel edges or loops. This is, however, only a notational convenience (in particular it allows us to uniquely identify edges by their head and tail nodes). All the results from this thesis also hold for multigraphs (and can be shown using the same proofs) (Fig. 2.1).

Definition 2.60. A graph G' is a *subgraph* of another graph G if we have $V(G') \subseteq V(G)$ and $E(G') \subseteq E(G')$. We then denote this by $G' \subseteq G$.

Definition 2.61. For any node subset $W \subseteq V(G)$ the *induced subgraph on* W is the graph $G[W]$ defined by

- the node set $V(G[W]) := W$ and
- the edge set $E(G[W]) := E(G)[W] := \{ vw \in E(G) \mid v, w \in W \}$

In other words, the induced subgraph on W is the maximal subgraph of G using W as its set of nodes.

Definition 2.62. Let $G = (V, E)$ be a graph, $n \in \mathbb{N}_0$ and a finite sequence of edges

$$p := (e_1, e_2, \dots, e_n) = (v_1 w_1, v_2 w_2, \dots, v_n w_n) \in E^n.$$

Then

- p is a *walk* if it consists of consecutive edges, i.e. we have $w_k = v_{k+1}$ for all $k \in [\max \{ n - 1, 0 \}]$,
- p is a *path* if it is a walk and it visits no node twice, i.e. we have $v_k \neq v_\ell$ for all $k \neq \ell \in [n]$ and $v_1 \neq w_n$ (if $n \geq 1$),
- p is a *cycle* if it is a walk of length at least 2, ends at its start node and visits no node in between twice, i.e. we have $n \geq 2$, $v_k \neq v_\ell$ for all $k \neq \ell \in [n]$ and $w_n = v_1$.

We call n the *length* of p. If $p = (v_1w_1, v_2w_2, \ldots, v_nw_n)$ is a walk/path, we call v_1 the start node of p, w_n its end node and p itself a v_1, w_n-walk or v_1, w_n-path, respectively. We consider the empty path () a v, v-path for any node $v \in V$.

For any two node subsets $U, W \subseteq V$ a U, W-walk/U, W-path is any v, w-walk/ v, w-path with $v \in U$, $w \in W$. We will also use v, W-path as shorthand for $\{v\}, W$-path.

Similar to our simplified notation for edges we will usually omit the parentheses when referring paths/walks, i.e. we write $e_1, e_2, \ldots, e_n$ instead of $(e_1, e_2, \ldots, e_n)$. This also allows us to write the concatenation of two walks p and q as just p, q.

Definition 2.63. A graph is called *acyclic* if it contains no cycles.

Definition 2.64. A binary relation $\prec$ on V is a *topological order* on a graph $G = (V, E)$ if $\preceq$ is a total order and $\prec$ satisfies

$$vw \in E \implies v \prec w$$

for all edges $vw \in E$, i.e. edges always go "upwards" (with respect to the topological order).

Proposition 2.65 ([Jun12, Theorem 2.6.3]). *A graph admits a topological order if and only if it is acyclic.*

Node Labels in Graphs with Edge Costs

In this thesis we will usually consider graphs with additional data. In particular, we will usually have some cost $\gamma_e \in \mathbb{R}$ associated with each edge $e \in E$ and some set $T \subseteq V$ of terminal nodes. We are then interested in the cheapest paths from any node in the network to any terminal node.

Definition 2.66. Let $G = (V, E)$ be a graph with edge costs $\gamma = (\gamma_e)_{e \in E} \in \mathbb{R}^E$ and a set of terminal nodes $T \subseteq V$. We then define *node labels* $\lambda_v \in \tilde{\mathbb{R}}$ (with respect to γ) by setting

$$\lambda_v := \inf \{ \gamma_p \mid p \text{ a } v, T\text{-path} \},$$

where $\gamma_p := \sum_{e \in p} \gamma_e$ denotes the cost of a path p. We say that a v, T-path p is *efficient* (with respect to γ) if $\gamma_p = \lambda_v$ and that an edge $e = vw \in E$ is *efficient* (with respect to γ) if it satisfies $\lambda_v \geq \gamma_e + \lambda_w$. Finally, we denote by

$$V^\dagger := \{ v \in V \mid \nexists\, v, T\text{-path} \}$$

the set of *dead-end nodes*, i.e. nodes from which there exists no path towards any terminal node.

It is well-known (cf. e.g. [Jun12, Theorem 3.5.3]) that for graphs without cycles of non-positive cost the node labels can be equivalently defined by Bellman's equations:

$$\lambda_v = \begin{cases} 0, & \text{if } v \in T \\ \inf \{ \gamma_e + \lambda_w \mid e = vw \in \delta^+(v) \}, & \text{else} \end{cases}. \tag{2.1}$$

For graphs with zero cost cycles these equation do not necessarily have a unique solution anymore. However, as long as there are no cycles of negative cost, the node labels are still the unique maximal solution to (2.1) (cf. [Mar22, Section 2.2]). We collect these facts together with several other helpful properties of such node labels in the following proposition:

Proposition 2.67. *Let $G = (V, E)$ be a graph with edge costs $\gamma \in \mathbb{R}^E$, terminal nodes $T \subseteq V$ and corresponding node labels $\lambda \in \tilde{\mathbb{R}}^V$. Then we have:*

a) $\lambda_v = \infty \iff v \in V^\dagger$.
b) $\lambda_v < \infty$ *if and only if there exists at least one efficient v, T-path.*
c) $\forall t \in T : \lambda_t \leq 0$.
d) The first edge of an efficient path is always efficient.
e) Any non-terminal node with at least one outgoing edge also has an efficient outgoing edge.

If we have $\gamma_c \geq 0$ for all cycles c, then the node labels also satisfy:

f) $\forall e = vw \in E : \lambda_v \leq \gamma_e + \lambda_w$.
g) An edge $e = vw \in E$ is efficient if and only if $\lambda_v = \gamma_e + \lambda_w$.
h) $\gamma_c = 0$ *for every cycle c consisting only of efficient edges.*
i) Every edge on an efficient path is efficient itself.

If, additionally, we have $\gamma_p \geq 0$ for any T, T-path p, then the node labels further satisfy:

j) $\forall t \in T : \lambda_t = 0$.
k) $\gamma_p = 0$ *for every efficient T, T-path p.*
l) A v, T-path is efficient if and only if all its edges are efficient.

m) The vector $(\lambda_v)_v \in \tilde{\mathbb{R}}^V$ *is the unique maximal solution to* (2.1).

If, additionally, we have $\gamma_c > 0$ *for any cycle c, then the node labels also satisfy:*

n) There exists no cycle consisting only of efficient edges.
o) The vector $(\lambda_v)_v \in \tilde{\mathbb{R}}^V$ *is the unique solution to* (2.1).

Proof. a),b): Since γ_e is always finite, we have $\lambda_v = \infty$ if and only if we take the infimum over the empty set, i.e. if there are no paths from v to any terminal node. If, on the other hand, this set is non-empty, the infimum is attained since the set of v, T-paths is always finite.

c): For any terminal node $t \in T$ the trivial path is a t, T-path with cost 0.

d): Let $p = e, p'$ be an efficient path and $e = vw$ its first edge. Then, p' is a w, T-path and, therefore,

$$\lambda_v = \gamma_p = \gamma_e + \gamma_{p'} \geq \gamma_e + \lambda_w.$$

e): We consider two cases: If there exists at least one v, T-path, then there is also a (non-trivial) efficient path and, thus, an efficient edge leaving v by d). If there is no v, T-path, then take any edge $e = vw \in \delta^+(v)$. Then, there can also be no w, T-path and, thus, we have

$$\lambda_v \overset{\text{a)}}{=} \infty = \gamma_e + \infty \overset{\text{a)}}{=} \gamma_e + \lambda_w.$$

f): If $\lambda_w = \infty$, then the inequality trivially holds. Otherwise, let p be any efficient w, T-path (which exists by b)). Then e, p is a v, T-walk and, thus, contains a v, T-path p' differing from e, p in at most a cycle. Since such a cycle has non-negative cost, this implies

$$\lambda_v \leq \gamma_{p'} \leq \gamma_{e,p} = \gamma_e + \gamma_p = \gamma_e + \lambda_w.$$

g): This follows directly from f).

h): Let c be any cycle consisting only of efficient edges. Then we have

$$0 \leq \gamma_c = \sum_{e \in c} \gamma_e \overset{\text{g)}}{=} \sum_{e = vw \in c} (\lambda_v - \lambda_w) = 0.$$

i): Let p be an efficient v, T-path and $e = v'w' \in p$ some edge on this path, i.e. $p = p_1, e, p_2$ for some subpaths p_1 and p_2. Then a) and b) imply that there exists some efficient v', T-path p'. As p_1, p' is a v, T-walk, we can turn it into a v, T-path $\tilde{p}$ by removing cycles. Since all cycles have non-negative cost, this implies

$$\gamma_{p_1,p'} \geq \gamma_{\tilde{p}} \geq \gamma_p = \gamma_{p_1,e,p_2}$$

and, therefore, $\gamma_{p'} \geq \gamma_{e,p_2}$. This now gives us

$$\lambda_{v'} = \gamma_{p'} \geq \gamma_{e,p_2} = \gamma_e + \gamma_{p_2} \geq \gamma_e + \lambda_{w'}$$

which proves that $e = v'w'$ is efficient.

j): This follows directly from c) together with our additional assumption that all T, T-paths have non-negative cost.

k): Let p be an efficient t, t'-path with $t, t' \in T$. Then we have

$$\gamma_p = \lambda_t \overset{\text{j)}}{=} 0.$$

l): Let p be a v, t-path with $t \in T$ consisting only of efficient edges. Then we have

$$\gamma_p = \sum_{e \in p} \gamma_e \overset{\text{g)}}{=} \sum_{e=v'w' \in p} (\lambda_{v'} - \lambda_{w'}) = \lambda_v - \lambda_t \overset{\text{j)}}{=} \lambda_v.$$

Thus, p is efficient. The other direction already holds due to i).

m): We first show that $(\lambda_v)_v$ is indeed a solution to (2.1): Take any node $v \in V$. If $v \in T$, then we have $\lambda_v = 0$ by j). If $v \notin T$ and $\delta^+(v) = \emptyset$, then we have $\lambda_v = \infty$ by a) while the right side of (2.1) is also clearly infinite. Finally, in the remaining case e) guarantees the existence of an efficient edge $e = vw \in \delta^+(v)$ and we have $\lambda_v = \gamma_e + \lambda_w$ by g). At the same time we have $\lambda_v \leq \gamma_e + \lambda_{w'}$ for any edge $e = vw' \in \delta^+(v)$ by f). Together, this shows

$$\lambda_v = \min\{\gamma_e + \lambda_w \mid e = vw \in \delta^+(v)\}.$$

Now, to show that $(\lambda_v)_v$ is maximal among all solutions to (2.1), let $(\lambda'_v)_v$ be any solution to (2.1). Then, for any node v without any v, T-path we have $\lambda'_v \leq \infty \overset{\text{a)}}{=} \lambda_v$. For any other node v let p be an efficient v, T-path (which exists by b)) and denote this path by $p = v_1w_1, v_2w_2, \ldots, v_kw_k$. with $v_1 = v$ and $w_k \in T$. Since $(\lambda'_v)_v$ is a solution to (2.1), we then have

$$\lambda'_v \leq \lambda'_{v_2} + \gamma_{v_1 w_1} \leq \lambda'_{v_3} + \gamma_{v_2 w_2} + \gamma_{v_1 w_1} \leq \cdots \leq \lambda'_{v_k} + \sum_{e \in p} \gamma_e \overset{\text{j)}}{=} 0 + \gamma_p = \lambda_v.$$

n): This follows directly from our additional assumption together with h).

o): We already know from m) that $(\lambda_v)_v$ is a solution to (2.1). So, it remains to show that it is also the only solution. We assume for contradiction that there exists another, different solution (λ'_v). Since, according to m), $(\lambda_v)_v$ is the maximal solution, this means there is some node $v_0 \in V$ with $\lambda'_{v_0} < \lambda_{v_0}$. We now consider three different cases for this node v_0 and show that they all lead to a contradiction:

1. **Case: $v_0 \in T$:** Then (2.1) together with j) imply $\lambda'_{v_0} = 0 = \lambda_{v_0}$, which is a contradiction to $\lambda'_{v_0} < \lambda_{v_0}$.
2. **Case: $v_0 \notin T$ and $\delta^+(v_0) = \emptyset$:** Here, (2.1) gives us $\lambda'_{v_0} = \infty$ which is an immediate contradiction to $\lambda'_{v_0} < \lambda_{v_0}$.
3. **Case: $v_0 \notin T$ and $\delta^+(v_0) \neq \emptyset$:** In this case (2.1) guarantees the existence of some edge $v_0 v_1$ with $\lambda'_{v_0} = \gamma_{v_0 v_1} + \lambda'_{v_1}$. For the label at node v_1 we now get

$$\begin{aligned} \lambda'_{v_1} &= \lambda'_{v_0} - \gamma_{v_0 v_1} < \lambda_{v_0} - \gamma_{v_0 v_1} \\ &\overset{\text{m)}}{=} \inf \{ \gamma_e + \lambda_w \mid e = v_0 w \in \delta^+(v_0) \} - \gamma_{v_0 v_1} \\ &\leq \gamma_{v_0 v_1} + \lambda_{v_1} - \gamma_{v_0 v_1} = \lambda_{v_1} \end{aligned}$$

Thus, we can now make the same case distinction for node v_1 as we just did for node v_0 resulting either in a contradiction (in the first two cases) or in the existence of another node v_2 with $\lambda'_{v_1} = \gamma_{v_1 v_2} + \lambda'_{v_2}$ and $\lambda'_{v_2} < \lambda_{v_2}$. Continuing on this way we get a sequence of nodes $v_0, v_1, \ldots, v_k$ ending either with a contradiction for some node v_k in one of the first two cases or a cycle (i.e. $v_k = v_0$). In the latter case we have found a cycle c for which we have

$$\gamma_c = \sum_{e \in c} \gamma_e = \sum_{j=0}^{k-1} \gamma_{v_j v_{j+1}} = \sum_{j=0}^{k-1} \left(\lambda'_j - \lambda'_{j+1} \right) = \lambda'_0 - \lambda'_k = 0,$$

a contradiction to the assumption that all cycles have strictly positive cost. □

Tools from Optimization

A generic optimization problem is given by some set S of feasible solutions and an objective function $f : S \to \mathbb{R}$. The goal is then to find (local) minima/maxima of f over S, i.e. a feasible point $x \in S$ with $f(x) = {}^{\min}/_{\max}\{ f(x') \mid x' \in S\}$ (or $f(x) = {}^{\min}/_{\max}\{ f(x') \mid x' \in U \}$ for some neighbourhood $U \subseteq S$ of x).

If S is compact and f continuous, then such a solution is guaranteed to exist:

Proposition 2.68 ([AB06, Corollary 2.35]). *Let $f : S \to \mathbb{R}$ be a continuous function on a non-empty compact set. Then f attains its minimum and its maximum.*

The KKT-conditions discovered by Karush, Kuhn and Tucker provide necessary conditions for (locally) optimal points of certain classes of optimization problems. We will use them for problems of the form

$$\begin{aligned} \min\ & f(x) \\ \text{s.t. } & Ax \leq a \\ & Bx = b \\ & x \in \mathbb{R}^n \end{aligned} \tag{MIN}$$

where $f : \mathbb{R}^n \to \mathbb{R}$ is a function, $A \in \mathbb{R}^{k \times n}, B \in \mathbb{R}^{m \times n}$ two matrices and $a \in \mathbb{R}^k, b \in \mathbb{R}^m$ two vectors. The KKT-conditions for such a problem are then as follows:

Proposition 2.69 ([Jun14, Satz 6.1.5]). *Let $x \in S$ be a (local) minimum of (MIN) such that f is differentiable at x. Then there exist vectors $\lambda \in \mathbb{R}^k$ and $\mu \in \mathbb{R}^m$ such that*

$$\nabla f(x) + \lambda^\top A + \mu^\top B = 0, \quad \lambda^\top (Ax - a) = 0 \quad \text{and} \quad \lambda \geq 0.$$

Here, $\nabla f(x)$ denotes the gradient of f at x, i.e. the row-vector $(\partial_{x_i} f(x))_{i \in [n]}$.

2.6 Complexity Theory

For analysing the quality of both IDE themselves and algorithms for computing them, we will make use of two helpful tools from complexity theory: The Landau notation to describe asymptotic bounds and the concept of NP-hardness to be able to express formally that certain problems are difficult. For a more formal introduction to these topics we refer to [GJ79, Pap95, CLRS22].

Definition 2.70. Let $f : \mathbb{R}^n_{\geq 0} \to \mathbb{R}_{\geq 0}$ be any non-negative function. Then we denote by

$$\mathcal{O}(f) := \left\{ g : \mathbb{R}^n \to \mathbb{R} \;\middle|\; \begin{array}{l} \exists N, C \in \mathbb{N}_0 : \forall x \in \mathbb{R}^n_{\geq 0} : \\ x_1, \ldots, x_n \geq N \implies g(x) \leq C \cdot f(x) \end{array} \right\}$$

the set of functions which are asymptotically upper bounded by f and by

$$\Omega(f) := \left\{ g : \mathbb{R}^n \to \mathbb{R} \;\middle|\; \begin{array}{l} \exists N, c \in \mathbb{N}_0 : \forall x \in \mathbb{R}^n_{\geq 0} : \\ x_1, \ldots, x_n \geq N \implies c \cdot g(x) \geq f(x) \end{array} \right\}$$

the set of functions which are asymptotically lower bounded by f.

Note, that for $n = 1$ this reduces to the standard definitions for the Landau notation (cf. e.g. [CLRS22, Section 3.2]). For the case of multiple variables, however, this is just one of several different and incompatible definitions of $\mathcal{O}$ and Ω which are in use – see [How08].

A decision problem consists of a class of instances (problems) and a question which for each of the instances has either "yes" or "no" as the answer. An example would be

ACYCLIC GRAPH:
- **Input:** A directed graph G
- **Question:** Is G acyclic?

The class P contains all decision problems for which there exists a deterministic polynomial time algorithm, i.e. an algorithm which for any instance I correctly identifies whether the answer is "yes" or "no" and has a runtime in $\mathcal{O}(p(\langle I \rangle))$ for some polynomial p (where $\langle I \rangle$ stands for the encoding size of the instance I).

The class NP contains all decision problem for which there exists a non-deterministic polynomial time algorithm, i.e. an algorithm as in the above case of the class P but which additionally is allowed to use an oracle which for any "yes"-instance provides a polynomial size proof for the answer being "yes".

Clearly, the class P is contained in the class NP. However, whether it is a proper inclusion or whether the two classes are actually the same, is still an open problem: The famous "P = NP?" problem which is one of the six unsolved millennium problems (cf. [Bom+06]).

Definition 2.71. Let Π_1 and Π_2 be two decision problems. We say that Π_1 is harder than Π_2 (written $\Pi_1 \geq \Pi_2$) if there exists a transformation T from instances of Π_1 to instances of Π_2 such that for any instance I of Π_1 we have

- I is a "yes"-instance for Π_1 if and only if $T(I)$ is a "yes"-instance for Π_2 and
- there is an algorithm which constructs $T(I)$ in polynomial time (in the encoding size of I).

A decision problem Π is *NP-hard* if we have $\Pi \geq \Pi'$ for all decision problems Π' in NP. It is *NP-complete* if it is NP-hard and itself contained in NP.

Proposition 2.72 ([CLRS22, Lemma 34.8]). *Let Π be any decision problem. If we have $\Pi \geq \Pi'$ for some NP-complete problem Π', then Π is NP-hard.*

A famous NP-complete decision problem is satisfiability ([GJ79, Theorem 2.1]): Here we consider a boolean formula in conjunctive normal form, i.e. a formula which is a conjunction of clauses which in turn are disjunctions of literals (i.e. variables x or their negations $\neg x$). An example for such a formula is then

$$(x_1 \vee x_3 \vee \neg x_4) \wedge (\neg x_1) \wedge (\neg x_3 \vee \neg x_2) \wedge (x_2 \vee x_3). \tag{2.2}$$

Such a formula is said to be satisfiable if there exists an interpretation of the variables such that the formula becomes true, i.e. a function $\beta : \{x_1, \ldots, x_n\} \rightarrow \{\text{TRUE}, \text{FALSE}\}$ such that every clause contains at least one literal which is true. For example, the formula in (2.2) is satisfiable as the interpretation $\beta(x_1) = \beta(x_3) = \beta(x_4) = \text{FALSE}$, $\beta(x_2) = \text{TRUE}$ makes the whole formula true.

A more restricted version of this problem is 3SAT where we only allow formulas where every clause contains exactly three literals:

3SAT:
- **Input:** A boolean formula ϕ in conjunctive normal form where every clause has exactly three literals
- **Question:** Is ϕ satisfiable?

Proposition 2.73 ([GJ79, Theorem 3.1]). 3SAT *is NP-complete.*

Model 3

Using the formalisms introduced in [KS11, CCL15] we describe in this chapter the model we will then study in this thesis. This description consists of two main parts: First, we have a physical model describing how flow propagates throughout the network (which, in turn, is mostly governed by the flow dynamics on individual edges). Second, we have a behavioural model which describes how the individual particles making up the flow choose their route through the network.

3.1 Physical Model

In this section we formally define the physical flow model. This model is based on the deterministic queuing model introduced by Vickrey but extended to multi-commodity flows and whole networks. On our way of describing this model we will also show five fundamental properties of the edge flow dynamics induced by this model:

Existence and Uniqueness: For every inflow rate there exists a unique outflow rate such that together they form an edge flow satisfying the constraints of our model (Corollaries 3.21 and 3.43).

Continuity: The resulting mapping from inflow to outflow rates as well as to current travel times is continuous with respect to appropriately chosen topologies on the respective function spaces (Corollaries 3.46 and 3.45). This will be important for our general existence result in Chap. 4.

L. Graf, *Dynamic Network Flows with Adaptive Route Choice based on Current Information*, Mathematische Optimierung und Wirtschaftsmathematik | Mathematical Optimization and Economathematics,
https://doi.org/10.1007/978-3-658-44948-3_3

Structure preservation Right-constant inflow rates lead to a right constant outflow rates (Proposition 3.22 and Corollary 3.44). This property will be particularly important when we want to construct flows, i.e. in Chap. 5, but also for the more specialized existence results in Chap. 4.

Monotonicity: A larger (cumulative) inflow can only lead to a larger or equal (cumulative) outflow (Corollary 3.23). This will be helpful when we have to show that a whole range of flows satisfies some given bounds in one of our gadget constructions in Chap. 6.

No idling: Flow cannot stay on an edge for too long, i.e. whenever there is flow on an edge, flow of that volume will leave the edge "soon" after (Corollary 3.24). This will be the basic ingredient for showing upper bounds on the termination time of flows in Chap. 6.

We refer to [Koc12, Chapters 3 and 7] for a much more extensive study of both the physical model used here as well as more general models for flow dynamics (for the single-commodity case).

3.1.1 Anonymous Edge Flows

Let e be an edge characterised by a *free flow travel time* $\tau_e \in \mathbb{R}_{\geq 0}$ and a *capacity* $\nu_e \in \mathbb{R}_{>0}$. The free flow travel time denotes the time it takes a particle to traverse edge e if there is no congestion. The capacity is the maximal rate at which particles may travel along this edge (without causing congestion). We start by considering flows on such an edge where all particles are interchangeable, i.e. we cannot distinguish between different particles and, in particular, all particles belong to the same commodity.

Definition 3.1. An *anonymous edge flow* $f_e = (f_e^+, f_e^-)$ on an edge e consists of two functions $f_e^+, f_e^- \in L^1_{\text{loc}}(\mathbb{R}_{\geq 0}, \mathbb{R}_{\geq 0})$ where $f_e^+(\theta)$ denotes the *(edge) inflow rate* into edge e at time θ and $f_e^-(\theta)$ denotes the *(edge) outflow rate* from edge e at time θ.

For any such edge flow defined by in- and outflow functions we can then also define corresponding *cumulative* flow functions denoting for every time θ the *volume* of flow which has entered/left edge e up to that time θ.

Definition 3.2. For any anonymous edge flow the associated *cumulative anonymous edge flow* $F_e = (F_e^+, F_e^-)$ is defined by:

- the *cumulative inflow* $F_e^+ : \mathbb{R}_{\geq 0} \to \mathbb{R}_{\geq 0}, \theta \mapsto F_e^+(\theta) := \int_0^\theta f_e^+(\zeta)d\zeta$ and
- the *cumulative outflow* $F_e^- : \mathbb{R}_{\geq 0} \to \mathbb{R}_{\geq 0}, \theta \mapsto F_e^-(\theta) := \int_0^\theta f_e^-(\zeta)d\zeta$.

Moreover, we define the *edge load* at any time $\theta \in \mathbb{R}_{\geq 0}$ by $F_e^\Delta(\theta) := F_e^+(\theta) - F_e^-(\theta)$.

An important property of these cumulative flow functions is that they are non-decreasing and absolutely continuous. Moreover, their derivatives are again the underlying flow rates (almost everywhere). This follows directly from Proposition 2.49:

Proposition 3.3. *The cumulative in- and outflow are non-decreasing, absolutely continuous functions satisfying*

$$\partial F_e^+(\theta) = f_e^+(\theta) \text{ and } \partial F_e^-(\theta) = f_e^-(\theta)$$

for almost all $\theta \in \mathbb{R}_{\geq 0}$. □

Remark 3.4. According to Proposition 2.49 there is actually a one-to-one correspondence between edge flow rates and cumulative edge flow functions. Thus, one could equivalently use the cumulative versions as the definition for an edge flow. In fact, this is the approach taken e.g. in [Koc12, HFY13, OSK23]. Note that in [HFY13, OSK23] absolute continuity is replaced by left- or right-continuity making their model more general as it then allows for flow of positive measure to enter an edge at a single time. However, since we will not allow waiting at nodes, this extension will not be relevant for our model and we can just freely switch between the two different view points.

Now, given an edge flow f_e we would like to determine the current travel time on edge e (as a function over time). A natural way of accomplishing that would be to look at a flow particle entering the edge at time θ and then measuring the time until it leaves the edge again. However, since we cannot distinguish individual particles, this is not possible. What we can do instead is to take the cumulative inflow at time θ (i.e. $F_e^+(\theta)$) and then measure the time until a flow of equal volume has left the edge. Assuming that all particles are treated equally and, in particular, enter and leave the edge in accordance with the first-in first-out (FIFO) principle this then gives us the actual travel time experienced by all particles that entered the edge at a specific time.

In order to make this into a formal definition, we also need to be able to say at which points in time flow actually enters an edge (under a given edge flow f_e). Since the flow rates are, formally, equivalence classes of functions and, therefore, only defined up to changes on any set of measure zero, it would not be well defined to say that flow enters edge e at time θ if $f_e^+(\theta) > 0$. Instead, we will use the cumulative inflow function here (we can then use Proposition 2.54 to translate between this definition and properties of the inflow rate):

Definition 3.5. We say that *flow enters edge e at time* $\theta > 0$ if F_e^+ is strictly increasing at θ, i.e. if for all $\theta' < \theta$ we have $F_e^+(\theta') < F_e^+(\theta)$.

Remark 3.6. Note that this definition is very similarly to that of the monotonicity set of F_e^+ used in [Koc12, Definition 2.24] for essentially the same purpose. Namely, we have

$$\begin{aligned}\text{Monotonicity Set of } F_e^+ = &\left\{ \theta \in \mathbb{R}_{\geq 0} \;\middle|\; \text{flow enters } f \text{ at } \theta \right\} \\ &\cup \left\{ \max \{ \theta \geq 0 \mid F_e^+(\theta) = F_e^+(0) \} \right\}.\end{aligned}$$

Definition 3.7. For any time θ at which flow enters edge e we define the *experienced current travel time* on edge e by

$$\hat{C}_e(\theta) := \inf \{ \zeta \geq -\theta \mid F_e^-(\theta + \zeta) \geq F_e^+(\theta) \}.$$

Up to now, there is no connection between edge-flows on an edge e and the edge's free flow time and capacity. Thus, as our next step we want to define *feasible* or "physically possible" flows: At a minimum, an edge flow should respect the edge's capacity and free flow travel time, i.e. it may not leave an edge at a higher rate than ν_e and not earlier than τ_e time units after it entered the edge. These two conditions can be formalised as follows:

Definition 3.8. An anonymous edge flow f_e satisfies *weak flow conservation until* $\xi \in \tilde{\mathbb{R}}_{\geq 0}$ if

$$F_e^-(\theta + \tau_e) \leq F_e^+(\theta) \text{ for all } \theta < \xi \tag{3.1}$$

and *strong flow conservation until* ξ if

$$F_e^-(\theta + \tau_e) = F_e^+(\theta) \text{ for all } \theta < \xi.$$

Definition 3.9. An anonymous edge flow f_e *respects the capacity of edge e until* $\xi \in \tilde{\mathbb{R}}_{\geq 0}$ if

$$f_e^-(\theta + \tau_e) \leq \nu_e \text{ for almost all } \theta < \xi. \tag{3.2}$$

Some Background: While, in the end, we usually want these (and all further) properties of dynamic flows to hold for all times (or almost all times in case of constraints on the flow rates) we will make these definitions more general in the sense that we always define properties "until some time $\xi \in \tilde{\mathbb{R}}_{\geq 0}$". This will be especially helpful in Chap. 4, where we will construct dynamic flows with certain properties by starting with an arbitrary flow (which trivially satisfies the desired property until $\xi = 0$) and then iteratively adjust it in such a way that it satisfies the property until some later and later time. A property which is satisfied until $\xi = \infty$ holds for all time. In this case we will usually drop the "until ξ" and just say that the given flow satisfies this property.

Also note that, as we can already see in the two definitions above, a property up to some time ξ will typically impose some restriction on the edge inflow during $[0, \xi)$ but during $[\tau_e, \xi + \tau_e)$ on the edge outflow. In particular, constraint (3.2) does not restrict the outflow rate during $[0, \tau_e)$. However, for any $\xi > 0$ constraint (3.1) already ensures that the outflow rate is zero during this time interval.

By combining the previous two definitions we get two main ways of modelling congestion: Respecting capacity and strong flow conservation or respecting capacity and weak flow conservation. In the first case we effectively limit the inflow into an edge and, thus, congestion happens before the edge, e.g. on the previous node. This is the model used by Ford and Fulkerson when they first introduced dynamic flows in [FF58]. In the second case we allow flow to enter the edge at any rate and limit the rate at which it can leave the edge instead. Hence, in this model congestion happens on the edge. This is the approach proposed by Vickrey in [Vic69] and which we will follow here as well.

More precisely, we will handle congestion by a point queue at the beginning of the edge, i.e. flow particles currently on an edge can only be in one of two states: Waiting in the queue at the beginning of the edge or travelling on the edge itself without experiencing any further congestion delays.

Definition 3.10. For any anonymous edge flow f_e we define the *queue length function* of an edge e by

$$Q_e : \mathbb{R}_{\geq 0} \to \mathbb{R}, \theta \mapsto Q_e(\theta) := F_e^+(\theta) - F_e^-(\theta + \tau_e). \tag{3.3}$$

We say that the queue of edge e *starts empty* if $Q_e(0) = 0$.

Note, that even though the notation makes no reference to the underlying edge flow f_e, the queue length does of course, depend on this flow. In the rare occasions where it is not obvious from the context which flow the queue length (or other flow dependent objects we will define later on) are derived from, we will make this explicit by a superscript in the notation. E.g. if there are two edge flows f_e and g_e, we may use Q_e^f and Q_e^g to refer to the queue length functions of f_e and g_e, respectively.

Remark 3.11. One could, of course, also place the queue at the end of the edge (by defining $Q_e(\theta) := F_e^+(\theta - \tau_e) - F_e^-(\theta)$). For the physical flow model this would not make any difference except for a time shift in the notation. The same is true for the behavioural model in a full information setting (and, in fact, both conventions are in use there). In the current information setting, however, this choice makes a subtle but important difference. Namely, it determines when other agents in the network get informed about congestion: Already when particles enter an edge or only once they reach the end of that edge. Placing the queue at the beginning of the edge, thus, gives the agents in the network earlier access to congestion information about that edge.

Note that, if we are allowed to use edges of zero free flow travel time, then we can always model queues at the end of edges using two consecutive edges with queues at the beginning and vice versa.

Proposition 3.12. *Let f_e be an anonymous edge flow and $\xi \in \tilde{\mathbb{R}}_{\geq 0}$ some positive time. Then the following properties hold:*

a) The queue length function is absolutely continuous.
b) The queue on edge e starts empty if and only if $f_e^-|_{[0,\tau_e]} =_{a.e.} 0$.
c) The queue is non-negative during $[0, \xi)$ if and only if f_e satisfies weak flow conservation until ξ.
d) If f_e satisfies weak flow conservation until some time $\xi > 0$, then the queue on edge e starts empty.

Proof. All four properties follow directly from the definitions (and Proposition 3.3 in case of the first one). □

Using the queue length function we can now formally state Vickrey's deterministic queuing model, i.e. a queue only forms if the edge inflow exceeds the edge's capacity and whenever there is a queue, then particles leave this queue (and start traversing the edge itself) at a rate of ν_e:

Definition 3.13. A queue *operates at capacity until* $\xi \in \tilde{\mathbb{R}}_{\geq 0}$ if it satisfies

$$f_e^-(\theta + \tau_e) = \begin{cases} \nu_e, & \text{if } Q_e(\theta) > 0 \\ \min\{ f_e^+(\theta), \nu_e \}, & \text{else} \end{cases} \quad \text{for almost all } \theta < \xi \tag{3.4}$$

and, if $\xi > 0$, starts empty[1].

Intuitively, it should be clear that a queue operating at capacity leads to a travel time of $\frac{Q_e(\theta)}{\nu_e} + \tau_e$ (waiting time in the queue plus traversal time for the edge itself) for particles entering the edge at time θ. This leads to the following definitions of *expected* current travel and exit times:

Definition 3.14. For any edge flow f_e we define the *(expected) current travel time* by

$$C_e : \mathbb{R}_{\geq 0} \to \mathbb{R}, \theta \mapsto \frac{Q_e(\theta)}{\nu_e} + \tau_e.$$

and the *(expected) current exit time* by

$$T_e : \mathbb{R}_{\geq 0} \to \mathbb{R}, \theta \mapsto \theta + C_e(\theta) = \theta + \frac{Q_e(\theta)}{\nu_e} + \tau_e.$$

Observation 3.15. The current travel time and current exit time function are both absolutely continuous as as linear combinations of absolutely continuous functions.

[1] Excluding the case $\xi = 0$ here is only a notational convenience as it ensures that any edge flow has a queue operating at capacity until time $\xi = 0$.

We collect some standard properties of the current exit time function, which can also be found in similar form in many other works on Vickrey flows (e.g. [CCL15, Ser20, Mar22]).

Proposition 3.16. *Let f_e be an anonymous edge flow which respects capacity until $\xi \in \mathbb{R}_{\geq 0}$. Then the following properties hold:*

a) The current exit time function is non-decreasing on $[0, \xi]$.
b) We have $\partial T_e(\theta) = 0 \implies f_e^+(\theta) = 0$ for almost all $\theta \in [0, \xi)$.
c) If, additionally, f_e satisfies weak flow conservation until ξ, then it also satisfies the following flow conservation condition:

$$F_e^-(T_e(\theta)) \leq F_e^+(\theta) \tag{3.5}$$

for every $\theta < \xi$ with $T_e(\theta) < \xi + \tau_e$.

Intuitive explanation: Property a) essentially states that the (expected) current travel time of an edge flow respecting capacity satisfies the following weak first-in first-out (FIFO)defindexl condition: Entering the edge later never leads to an earlier (expected) exit time. On the other hand, it is easy to see (cf. Example 3.17) that such a flow need not obey strict FIFO, i.e. entering an edge later can lead to the same exit time. This is in contrast to other flow models like linear edge delays where an even stronger FIFO condition is satisfied, namely that there exists some constant $\gamma_e > 0$ such that one always has $T_e(\theta_2) - T_e(\theta_1) \geq \gamma_e(\theta_2 - \theta_1)$ for all $\theta_2 \geq \theta_1$ (cf. [ZM00, Theorem 5.1]).

Together with b) we do, however, get strict FIFO for our model at least for all times where flow enters the edge.

Finally, property c) states that the flow is not faster than the expected travel time (whenever there is proper inflow). Note here that, even though (3.5) looks like a stronger condition than weak flow conservation, it does not imply weak flow conservation (even under the assumption that the flow respects capacity) – see Example 3.18.

Proof. All three properties follow from the definitions by mostly straight forward computations:

a) Take any two times $\theta \leq \theta' < \xi$. Then we have

$$T_e(\theta) = \theta + \frac{Q_e(\theta)}{\nu_e} + \tau_e \leq \theta + \frac{\int_\theta^{\theta'} f_e^-(\zeta + \tau_e)\mathrm{d}\zeta + Q_e(\theta')}{\nu_e} + \tau_e$$
$$\overset{(3.2)}{\leq} \theta + \frac{\nu_e(\theta' - \theta) + Q_e(\theta')}{\nu_e} + \tau_e = \theta + (\theta' - \theta) + \frac{Q_e(\theta')}{\nu_e} + \tau_e = T_e(\theta').$$

b) Since F_e^+ and F_e^- are both absolutely continuous, it suffices to show the claim for only those $\theta < \xi$ where we have $\partial F_e^+(\theta) = f_e^+(\theta)$ and $\partial F_e^-(\theta + \tau_e) = f_e^-(\theta + \tau)$. Now, for almost all such times we have

$$\partial T_e(\theta) = 1 + \tfrac{\partial Q_e(\theta)}{\nu_e} = 1 + \tfrac{f_e^+(\theta) - f_e^-(\theta + \tau_e)}{\nu_e} \overset{(3.2)}{\geq} 1 + \tfrac{f_e^+(\theta) - \nu_e}{\nu_e} = \tfrac{1}{\nu_e} \cdot f_e^+(\theta).$$

Hence, for almost all such times θ we have $\partial T_e(\theta) = 0 \implies f_e^+(\theta) = 0$.

c) Let $\theta < \xi$ be any time with $T_e(\theta) < \xi + \tau_e$. Then we have:

$$F_e^-(T_e(\theta)) = F_e^-(\theta + \tau_e) + \int_\theta^{\theta + \frac{Q_e(\theta)}{\nu_e}} f_e^-(\zeta + \tau_e)\mathrm{d}\zeta$$
$$\overset{(3.1),(3.2)}{\leq} F_e^-(\theta + \tau_e) + \int_\theta^{\theta + \frac{Q_e(\theta)}{\nu_e}} \nu_e \mathrm{d}\zeta = F_e^-(\theta + \tau_e) + Q_e(\theta) = F_e^+(\theta).$$

□

Example 3.17. Consider an edge with $\tau_e = \nu_e = 1$ and an edge flow defined by $f_e^+ := 2 \cdot \mathbb{1}_{[0,1]}$ and $f_e^- := \mathbb{1}_{[1,3]}$. This flow clearly satisfies weak flow conservation and respects capacity. During $[1, 2]$ the (expected) current exit time is constant, e.g. entering edge e at any point in this interval leads to the same exit time. Note, however, that under the given flow no flow actually enters during this interval (as guaranteed by Proposition 3.16b)).

Example 3.18. Consider an edge with $\tau_e = \nu_e = 1$ and an edge flow defined by $f_e^+ := \mathbb{1}_{[0,1]}$ and $f_e^- := \mathbb{1}_{[1,\infty)}$ (cf. Fig. 3.1). This flow clearly violates weak flow conservation after time $\theta = 1$. However, constraint (3.5) is satisfied as we have

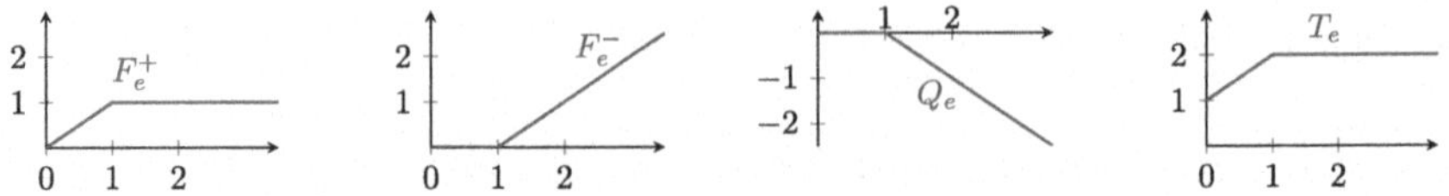

Fig. 3.1 An edge flow on an edge with $\tau_e = \nu_e = 1$ which satisfies (3.5) but not weak flow conservation

$$T_e(\theta) = \theta + \frac{Q_e(\theta)}{\nu_e} + \tau_e = \theta + F_e^+(\theta) - F_e^-(\theta + 1) + 1$$
$$= \begin{cases} \theta + \theta - \theta + 1 = \theta + 1, & \text{for } \theta \leq 1 \\ \theta + 1 - \theta + 1 = 2, & \text{for } \theta > 1 \end{cases}$$

and, therefore,

$$F_e^-(T_e(\theta)) = \begin{cases} F_e^-(\theta + 1) = F_e^+(\theta), & \text{for } \theta \leq 1 \\ F_e^-(2) = 1 = F_e^+(\theta), & \text{for } \theta > 1. \end{cases}$$

We now want to show that for edge flows with a queue operating at capacity expected and experienced current travel time do, in fact, coincide (whenever they are both defined). On our way to proving this, we will also show the equivalence of several alternative definitions of the deterministic queueing model used in literature (e.g. [HFY13, CCL15, Mar22]):

Proposition 3.19. *Let $f_e = (f_e^+, f_e^-)$ be an anonymous edge flow and $\xi \in \tilde{\mathbb{R}}_{\geq 0}$ some time. Then the following properties are equivalent:*

a) The queue of edge e operates at capacity until ξ.
b) The queue of edge e satisfies

$$\partial Q_e(\theta) = \begin{cases} f_e^+(\theta) - \nu_e, & \text{if } Q_e(\theta) > 0 \\ \max\{\, f_e^+(\theta) - \nu_e, 0\,\}, & \text{else} \end{cases} \quad \text{for almost all } \theta < \xi \tag{3.6}$$

and, if $\xi > 0$, starts empty.
c) The flow satisfies weak flow conservation and respects capacity on edge e until ξ and the cumulative outflow satisfies for every $\theta < \xi$ the following equation

$$F_e^-(\theta + \tau_e) = F_e^+(\bar{\theta}) + (\theta - \bar{\theta})\nu_e, \tag{3.7}$$

where $\bar{\theta} := \max\{\theta' \leq \theta \mid Q_e(\theta') = 0\}$ *is the last time before* θ *with empty queue.*

d) *The cumulative edge outflow is completely determined by the cumulative edge inflow in the following way for every* $\theta < \xi$*:*

$$F_e^-(\theta + \tau_e) = \min_{\theta' \leq \theta} \left(F_e^+(\theta') + (\theta - \theta')\nu_e\right) \tag{3.8}$$

e) *The flow satisfies weak flow conservation, respects the capacity until* ξ *and satisfies*

$$F_e^-(T_e(\theta)) = F_e^+(\theta) \text{ for all } \theta < \xi \text{ with } T_e(\theta) < \xi + \tau_e. \tag{3.9}$$

f) *The flow satisfies weak flow conservation and respects the capacity on edge e until* ξ *and for every* $\theta < \xi$ *with* $T_e(\theta) < \xi + \tau_e$ *where flow enters the edge we have* $C_e(\theta) = \hat{C}_e(\theta)$.

Intuitive explanation: Property f) shows that experienced and expected current travel time coincide for flows with queues operating at capacity. Moreover, it states that the following three bounds completely determine the dynamics of an edge flow: Weak flow conservation gives a lower bound of τ_e on the travel time, (3.9) gives an upper bound of $\tau_e + \frac{Q_e(\theta)}{\nu_e}$ and respecting the capacity enforces the creation of a queue if the inflow rate exceeds ν_e. This characterisation is essentially the same as in [CCL15, Proposition 1].

Property e) is a reformulation of f) and – in the form stated in Proposition 3.40b) – often used for multi-commodity flows (e.g. [CCL15, eq. (25)]).

The characterisation in d) is particularly helpful for showing that for queues operating at capacity the outflow is completely determined by the inflow. This is exactly how Cominetti, Correa and Larré use this formulation (stated in the form $Q_e(\theta) = \max_{\theta' \leq \theta} \int_{\theta'}^{\theta} f_e^+(\zeta) - \nu_e \mathrm{d}\zeta)$ in [CCL15, Section 2.2] and how we will use it in Corollary 3.21. In [HFY13] this is used directly as definition for the flow dynamics in the Vickrey model (cf. [HFY13, constraint (4.14)]).

Markl showed in [Mar22, section 3.4] that the equivalence of characterisation d) to the constraint that the queue operates at capacity even holds in the more general case of time-varying capacities. He also introduced characterisation c).

Statement b) is just a restatement of the constraints of queues operating at capacity as a differential equation using the derivative of the defining equation for the queue length functions, i.e. $\partial Q_e(\theta) = f_e^+(\theta) - f_e^-(\theta + \tau_e)$.

Proof. Proofs for most of the implications contained in this proposition can be found in [CCL15] and [Ser20, Lemma 3.1]. We will prove them here by showing a) $\iff$ b) $\implies$ c) $\implies$ d) $\implies$ e) $\implies$ f) $\implies$ a).

a) $\iff$ b): Here we follow the proof of [Ser20, Lemma 3.1 (vii)]: Since both F_e^+ and F_e^- are absolutely continuous they are differentiable with $\partial F_e^+(\theta) = f_e^+(\theta)$ and $\partial F_e^-(\theta + \tau_e) = f_e^-(\theta + \tau_e)$ for almost all $\theta < \xi$. For all those θ we then have by the definition of the queue length function

$$\partial Q_e(\theta) \overset{(3.3)}{=} \partial F_e^+(\theta) - \partial F_e^-(\theta + \tau_e) = f_e^+(\theta) - f_e^-(\theta + \tau_e).$$

This now immediately implies the equivalence of the constraints (3.4) and (3.6).

b) $\implies$ c): If $\xi = 0$, there is nothing to show. Otherwise we have $Q_e(0) = 0$ since the queue starts empty and $\partial Q_e(\theta) = \max\{ f_e^+(\theta) - \nu_e, 0\} \geq 0$ for all times $\theta < \xi$ with $Q_e(\theta) \leq 0$ from (3.6). Thus, we can apply Proposition 2.48 to deduce that the queue is non-negative during $[0, \xi)$ which, by Proposition 3.12, implies that weak flow conservation holds until ξ. Moreover, (3.6) also implies

$$f_e^-(\theta + \tau_e) = \partial F_e^-(\theta + \tau_e) = \partial F_e^+(\theta) - \partial Q_e(\theta) = f_e^+(\theta) - \partial Q_e(\theta) \overset{(3.6)}{\leq} \nu_e$$

for almost all $\theta < \xi$, i.e. that the flow respects capacity until ξ. Finally, in order to show (3.7) we first note that the time $\bar{\theta} := \max\{\theta' \leq \theta \mid Q_e(\theta') = 0\}$ is well defined for any fixed $\theta < \xi$ since the queue length function is continuous and starts at 0. As we have $Q_e(\zeta) > 0$ for all $\zeta \in (\bar{\theta}, \theta)$ we then get

$$
\begin{aligned}
F_e^-(\theta+\tau_e) &= F_e^+(\theta) - Q_e(\theta) = F_e^+(\theta) - Q_e(\bar\theta) - \int_{\bar\theta}^{\theta} \partial Q_e(\zeta)\mathrm{d}\zeta \\
&\overset{(3.6)}{=} F_e^+(\theta) - 0 - \int_{\bar\theta}^{\theta} \left(f_e^+(\zeta) - \nu_e\right)\mathrm{d}\zeta \\
&= F_e^+(\theta) - \int_{\bar\theta}^{\theta} f_e^+(\zeta)\mathrm{d}\zeta + (\theta-\bar\theta)\nu_e = F_e^+(\bar\theta) + (\theta-\bar\theta)\nu_e.
\end{aligned}
$$

c) $\implies$ d): First, note that the minimum is well defined since F_e^+ is continuous. From (3.7) we directly get

$$
F_e^-(\theta+\tau_e) = F_e^+(\bar\theta) + (\bar\theta-\theta)\nu_e \geq \min_{\theta'\leq\theta}\left(F_e^+(\theta') + (\theta-\theta')\nu_e\right).
$$

At the same time, respecting capacity and weak flow conservation imply that for every $\theta' \leq \theta$ we have

$$
\begin{aligned}
F_e^-(\theta+\tau_e) &= F_e^-(\theta'+\tau_e) + \int_{\theta'}^{\theta} f_e^-(\zeta+\tau_e)\mathrm{d}\zeta \overset{(3.2)}{\leq} F_e^-(\theta'+\tau_e) + \int_{\theta'}^{\theta} \nu_e \mathrm{d}\zeta \\
&= F_e^+(\theta') - Q_e(\theta') + (\theta-\theta')\nu_e \overset{(3.1)}{\leq} F_e^+(\theta') - 0 + (\theta-\theta')\nu_e.
\end{aligned}
$$

Together, this shows that (3.8) holds.

d) $\implies$ e): Choosing $\theta' = \theta$ in (3.8) gives us $F_e^-(\theta+\tau_e) \leq F_e^+(\theta)$, i.e. weak flow conservation. As F_e^- is absolutely continuous, it is differentiable for almost all θ and satisfies $\partial F_e^-(\theta+\tau_e) = f_e^-(\theta+\tau_e)$. We will show that f_e respects the capacity for all those θ: So, fix such a time $\theta < \xi$ and define $\hat\theta := \arg\min_{\theta'\leq\theta}\left(F_e^+(\theta') + (\theta-\theta')\nu_e\right)$. Then, using (3.8), we have $F_e^-(\theta+\tau_e) = F_e^+(\hat\theta) + (\theta-\hat\theta)\nu_e$ and

$$
F_e^-(\theta+\varepsilon+\tau_e) = \min_{\theta'\leq\theta+\varepsilon}\left(F_e^+(\theta') + (\theta+\varepsilon-\theta')\nu_e\right) \leq F_e^+(\hat\theta) + (\theta+\varepsilon-\hat\theta)\nu_e
$$

for any $\varepsilon < \xi - \theta$. Thus, we get

$$
\begin{aligned}
f_e^-(\theta + \tau_e) = \partial F_e^-(\theta + \tau_e) &= \lim_{\varepsilon \searrow 0} \frac{1}{\varepsilon} \left(F_e^-(\theta + \varepsilon + \tau_e) - F_e^-(\theta + \tau_e)\right) \\
&\leq \lim_{\varepsilon \searrow 0} \frac{1}{\varepsilon} \left(F_e^+(\hat{\theta}) + (\theta + \varepsilon - \hat{\theta})\nu_e - (F_e^+(\hat{\theta}) + (\theta - \hat{\theta})\nu_e)\right) \\
&= \lim_{\varepsilon \searrow 0} \frac{1}{\varepsilon} \varepsilon \nu_e = \nu_e,
\end{aligned}
$$

i.e. the flow respects the capacity for all $\theta < \xi$ with $\partial F_e^-(\theta + \tau_e) = f_e^-(\theta + \tau_e)$ which is almost all.
Finally, in order to show (3.9), take any $\theta < \xi$ with $T_e(\theta) < \xi + \tau_e$ or, equivalently, $\theta + \frac{Q_e(\theta)}{\nu_e} < \xi$ and observe that for any $\theta' \in [\theta, \theta + \frac{Q_e(\theta)}{\nu_e})$ we have

$$
\begin{aligned}
F_e^+(\theta') + (\theta - \theta')\nu_e \geq F_e^+(\theta) + (\theta - \theta')\nu_e &= F_e^-(\theta + \tau_e) + Q_e(\theta) + (\theta - \theta')\nu_e \\
&\geq F_e^-(\theta + \tau_e) + Q_e(\theta) - \frac{Q_e(\theta)}{\nu_e}\nu_e = F_e^-(\theta + \tau_e).
\end{aligned}
$$

Together with (3.8) this gives us

$$
F_e^-(\theta + \tau_e) = \min_{\theta' \leq \theta + \frac{Q_e(\theta)}{\nu_e}} \left(F_e^+(\theta') + (\theta - \theta')\nu_e\right). \tag{3.10}
$$

Using (3.8) again, now for $\theta + \frac{Q_e(\theta)}{\nu_e}$, we obtain (3.9) as follows:

$$
\begin{aligned}
F_e^-(T_e(\theta)) = F_e^-\left(\theta + \frac{Q_e(\theta)}{\nu_e} + \tau_e\right) &\overset{(3.8)}{=} \min_{\theta' \leq \theta + \frac{Q_e(\theta)}{\nu_e}} \left(F_e^+(\theta') + \left(\theta + \frac{Q_e(\theta)}{\nu_e} - \theta'\right)\nu_e\right) \\
&= \min_{\theta' \leq \theta + \frac{Q_e(\theta)}{\nu_e}} \left(F_e^+(\theta') + (\theta - \theta')\nu_e\right) + Q_e(\theta) \\
&\overset{(3.10)}{=} F_e^-(\theta + \tau_e) + Q_e(\theta) = F_e^+(\theta).
\end{aligned}
$$

e) $\Longrightarrow$ f): Let $0 < \theta < \xi$ be a time with $T_e(\theta) < \xi + \tau_e$ where flow enters the edge, i.e. where F_e^+ is strictly increasing. Since we have $F_e^-(\theta + C_e(\theta)) = F_e^-(T_e(\theta)) = F_e^+(\theta)$, we immediately get $\hat{C}_e(\theta) \leq C_e(\theta)$. Now assume for contradiction that $\hat{C}_e(\theta) < C_e(\theta)$. Due to the continuity of C_e this implies that there exists some $\theta' < \theta$ with $\hat{C}_e(\theta) + \theta \leq C_e(\theta') + \theta'$. Using the fact that both F_e^- and T_e are non-decreasing gives us

$$
F_e^+(\theta') \overset{(3.9)}{=} F_e^-(\theta' + C_e(\theta')) \geq F_e^-(\theta + \hat{C}_e(\theta)) \geq F_e^+(\theta).
$$

This implies that F_e^+ is constant on $[\theta', \theta]$ (since F_e^+ is non-decreasing as well) which is a contradiction to F_e^+ being strictly increasing at θ.

f) $\implies$ a): By Proposition 3.12 weak flow conservation already implies that the queue starts empty. So, it remains for us to show that constraint (3.4) holds. We will do this by following the approach taken in the second half of the proof of [CCL15, Proposition 1]: We start by defining the set $Q^+ := \{\, \theta < \xi \mid Q_e(\theta) > 0 \,\}$ of all times with non-empty queue. We now have to show that

(i) for almost all $\theta \in Q^+$ we have $f_e^-(\theta + \tau_e) = \nu_e$ and
(ii) for almost all $\theta \in [0, \xi) \setminus Q^+$ we have $f_e^-(\theta + \tau_e) = \min\{\, f_e^+(\theta), \nu_e \,\}$.

For (ii) we can apply Proposition 2.48 to the queue length function on $[0, \xi)$ (since we know that this function is absolutely continuous, the queue starts empty and is non-negative due to weak flow conservation). Thus, for almost all $\theta \in [0, \xi) \setminus Q^+$ we have

$$0 \overset{\text{Prop. 2.48c)}}{=} \partial Q_e(\theta) = \partial F_e^+(\theta) - \partial F_e^-(\theta + \tau_e) = f_e^+(\theta) - f_e^-(\theta + \tau_e).$$

Together with the fact that the flow respects the edge capacity this implies $f_e^-(\theta + \tau_e) = f_e^+(\theta) = \min\{\, f_e^+(\theta), \nu_e \,\}$ for almost all such θ.

For (i) we denote by $W(\theta) := [\theta, \theta + C_e(\theta) - \tau_e)$ the current expected waiting period of particles entering the queue at any time $\theta < \xi$. Furthermore, we define the set

$$I^+ := \left\{\, \theta < \xi \;\middle|\; F_e^+ \text{ is strictly increasing at } \theta \text{ and } T_e(\theta) < \xi + \tau_e \,\right\}$$

as the set of all points where $C_e(\theta) = \hat{C}_e(\theta)$ holds according to f).

We will now first show that (i) holds for almost all times in $W(\theta)$ for $\theta \in I^+$ and then that those $W(\theta)$ cover all of Q^+. By Proposition 2.15 this suffices to prove that (i) holds for almost all times in Q^+.

Claim 3.1. *For any $\theta \in I^+$ we have $f_e^-(\zeta + \tau_e) - \nu_e = 0$ for almost all $\zeta \in W(\theta)$.*

Proof. For $\theta \in I^+$ we have $W(\theta) \subseteq [0, \xi)$ and, therefore, $f_e^-(\zeta + \tau_e) - \nu_e \leq 0$ holds almost everywhere on $W(\theta)$ since the flow respects the capacity until ξ. Thus, the claim is equivalent to showing that $\int_{W(\theta)} f_e^-(\zeta + \tau_e) - \nu_e \mathrm{d}\zeta \geq 0$.

Using the definition of $\hat{C}_e(\theta)$ as well as the assumption that $\hat{C}_e(\theta) \overset{f)}{=} C_e(\theta)$ we can show this as follows:

$$\begin{aligned}
\int_{W(\theta)} \big(f_e^-(\zeta + \tau_e) - \nu_e\big)\, d\zeta &= \int_\theta^{\theta + C_e(\theta) - \tau_e} \big(f_e^-(\zeta + \tau_e) - \nu_e\big)\, d\zeta \\
&= \int_\theta^{\theta + C_e(\theta) - \tau_e} f_e^-(\zeta + \tau_e) d\zeta - (C_e(\theta) - \tau_e)\nu_e \\
&= F_e^-(\theta + \hat{C}_e(\theta)) - F_e^-(\theta + \tau_e) - (C_e(\theta) - \tau_e)\nu_e \\
&\geq F_e^+(\theta) - F_e^-(\theta + \tau_e) - \frac{Q_e(\theta)}{\nu_e} \cdot \nu_e = Q_e(\theta) - Q_e(\theta) = 0.
\end{aligned}$$

■

Claim 3.2. *We have* $Q^+ \subseteq \bigcup_{\theta \in I^+} W(\theta)$.

Proof. Take any $\theta^+ \in Q^+$. Since we have $T_e(0) = \tau_e < \tau_e + \xi$ as well as $T_e(\theta^+) = \theta^+ + \frac{Q_e(\theta^+)}{\nu_e} + \tau_e > \theta^+ + \tau_e$ and T_e is continuous, there exists some $\theta' \leq \theta^+$ with

$$\theta^+ + \tau_e < T_e(\theta') < \xi + \tau_e.$$

For this time θ' we clearly have $F_e^+(\theta') \geq Q_e(\theta') > 0$ and, thus, Proposition 2.54d) guarantees the existence of some time $\theta \leq \theta'$ at which flow entered the edge. Therefore,

$$\bar{\theta} := \sup \{ \theta \leq \theta' \mid \text{ flow enters } e \text{ at } \theta \} > -\infty$$

and $F_e^+(\bar{\theta}) = F_e^+(\theta')$, again by Proposition 2.54d). Moreover, there exists some $\hat{\theta} \leq \bar{\theta}$ such that flow enters edge e at $\hat{\theta}$ and we have

$$F_e^+(\bar{\theta}) - F_e^+(\hat{\theta}) < Q_e(\theta') - (\theta^+ - \theta')\nu_e \tag{3.11}$$

(the latter being possible since F_e^+ is continuous and we chose θ' such that $\theta^+ < \theta' + \frac{Q_e(\theta')}{\nu_e}$). See Fig. 3.2 for an overview of how all the different times defined here relate to each other. Note that, in particular, we have $T_e(\hat{\theta}) \leq T_e(\theta') < \xi + \tau_e$ since the flow respects the capacity until ξ and, therefore, T_e is non-decreasing on $[0, \xi)$ by Proposition 3.16a). This implies $\hat{\theta} \in I^+$ and, thus, we have $C_e(\hat{\theta}) = \hat{C}_e(\hat{\theta})$ by f).

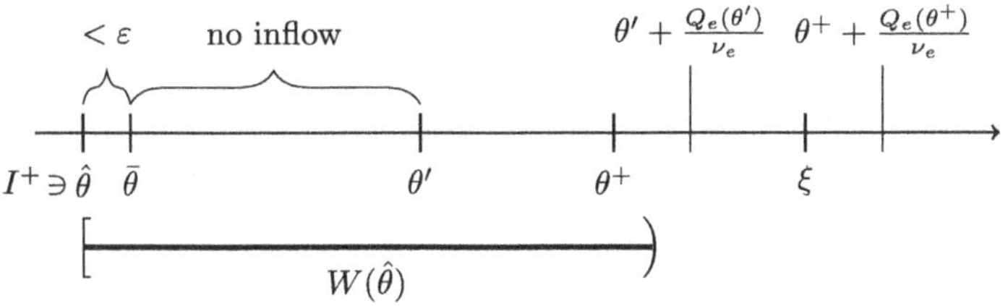

Fig. 3.2 An overview over the different times used in the proof of Claim 3.2 as well as their relationships to each other. Note that in general ξ could also be to the right of $\theta^+ + \frac{Q_e(\theta^+)}{\nu_e}$ but the case shown here is the "more difficult one". The general idea of the proof is as follows: We want to find some $\theta \in I^+$ such that $\theta^+ \in W(\theta)$. As I^+ only contains times θ with $\theta + \frac{Q_e(\theta)}{\nu_e} < \xi$ (and this might not be the case at θ^+) we first go to the left until we reach a time θ' where this is the case (but we still have $\theta^+ < \theta' + \frac{Q_e(\theta')}{\nu_e}$ to ensure that $\theta^+ \in W(\theta')$ is still possible). As, additionally, I^+ only contains times at which there is inflow into e we the go further to the left until we reach such a point $\hat{\theta}$ which is very close to the upper bound of such points $\bar{\theta}$. We then conclude the proof by showing that it is not possible that all flow which had entered the edge by time $\hat{\theta}$ has left it by time $\theta^+ + \tau_e$

This now implies

$$\begin{aligned}
F_e^+(\hat{\theta}) &= F_e^+(\bar{\theta}) - (F_e^+(\bar{\theta}) - F_e^+(\hat{\theta})) = F_e^+(\theta') - (F_e^+(\bar{\theta}) - F_e^+(\hat{\theta})) \\
&= Q_e(\theta') + F_e^-(\theta' + \tau_e) - (F_e^+(\bar{\theta}) - F_e^+(\hat{\theta})) \\
&\overset{(3.11)}{>} F_e^-(\theta' + \tau_e) + (\theta^+ - \theta')\nu_e \\
&\overset{(*)}{\geq} F_e^-(\theta' + \tau_e) + \int_{\theta'}^{\theta^+} f_e^-(\zeta + \tau_e)\mathrm{d}\zeta = F_e^-(\theta^+ + \tau_e),
\end{aligned}$$

i.e. $F_e^-(\hat{\theta} + \theta^+ + \tau_e - \hat{\theta}) < F_e^+(\hat{\theta})$. Here, for $(*)$ we used the fact that the flow respects the capacity until ξ and we have $\theta^+ < \xi$. Thus, we have

$$C_e(\hat{\theta}) = \hat{C}_e(\hat{\theta}) > \theta^+ + \tau_e - \hat{\theta},$$

which implies $\theta^+ < \hat{\theta} + C_e(\hat{\theta}) - \tau_e$ and, finally $\theta^+ \in W(\hat{\theta})$. ■

Using these two claims together with Proposition 2.15 now directly implies that $f_e^-(\theta + \tau_e) - \nu_e$ vanishes almost everywhere on Q^+. Thus, the queue operates at capacity until ξ. □

We can now use these different characterisations to deduce several key properties of the edge dynamics induced by the Vickrey point queue model:

Uniqueness follows directly from Proposition 3.19d) as every outflow has to satisfy (3.8). For existence it remains to show that a function defined by (3.8) is guaranteed to be a cumulative outflow, i.e. it is absolutely continuous and non-decreasing. This will be a direct consequence of the following lemma:

Lemma 3.20. *Let* $g : \mathbb{R}_{\geq 0} \to \mathbb{R}$ *be any continuous non-decreasing function and define*

$$h : \mathbb{R}_{\geq 0} \to \mathbb{R}, \theta \mapsto \min_{\theta' \leq \theta} \left(g(\theta') + (\theta - \theta')\nu\right)$$

for some constant $\nu > 0$. *Then* h *is non-decreasing and Lipschitz-continuous.*

Proof. First, note that h is well-defined as g is continuous and $[0, \theta]$ is compact for any $\theta \in \mathbb{R}_{\geq 0}$.

We now show that h is non-decreasing: For any $a \leq b$ we have

$$\begin{aligned}
h(b) &= \min_{\theta' \leq b} \left(g(\theta') + (b - \theta')\nu\right) \\
&= \min \left\{ \min_{\theta' \leq a} \left(g(\theta') + (b - \theta')\nu\right), \min_{\theta' \in [a,b]} \left(g(\theta') + (b - \theta')\nu\right) \right\} \\
&\geq \min \left\{ \min_{\theta' \leq a} \left(g(\theta') + (a - \theta')\nu\right), \min_{\theta' \in [a,b]} g(\theta') \right\} \\
&\overset{(*)}{\geq} \min \left\{ \min_{\theta' \leq a} \left(g(\theta') + (a - \theta')\nu\right), g(a) + (a - a)\nu \right\} \\
&= \min_{\theta' \leq a} \left(g(\theta') + (a - \theta')\nu\right) = h(a),
\end{aligned}$$

where $(*)$ holds since g is non-decreasing.

Now, using this monotonicity of h we can show Lipschitz-continuity as follows: Again, take $a \leq b$ and choose $\theta_a \in [0, a]$ such that $h(a) = g(\theta_a) + (a - \theta_a)\nu$. Then we have

$$\begin{aligned}
|h(b) - h(a)| = h(b) - h(a) &= \min_{\theta' \leq b} \left(g(\theta') + (b - \theta')\nu\right) - \left(g(\theta_a) + (a - \theta_a)\nu\right) \\
&\leq g(\theta_a) + (b - \theta_a)\nu - g(\theta_a) - (a - \theta_a)\nu = (b - a)\nu = \nu\,|b - a|.
\end{aligned}$$

□

Corollary 3.21. *For any locally integrable function $g_e^+ : \mathbb{R}_{\geq 0} \to \mathbb{R}_{\geq 0}$ there exists a locally integrable function $g_e^- : \mathbb{R}_{\geq 0} \to \mathbb{R}_{\geq 0}$ such that (g_e^+, g_e^-) is an edge flow that operates at capacity.*

Furthermore, if (f_e^+, f_e^-) and (g_e^+, g_e^-) are two edge flows operating at capacity until $\xi \in \tilde{\mathbb{R}}_{\geq 0}$ and $\xi' \leq \xi$ some time with $T_e(\xi') \geq \xi + \tau_e$ then

$$f_e^+|_{[0,\xi')} =_{a.e.} g_e^+|_{[0,\xi')} \implies f_e^-|_{[0,\xi+\tau_e)} =_{a.e.} g_e^-|_{[0,\xi+\tau_e)}.$$

Proof. For the first part, we define $G_e^+ : \mathbb{R}_{\geq 0} \to \mathbb{R}_{\geq 0}, \theta \mapsto \int_0^\theta g_e^+(\zeta)\mathrm{d}\zeta$ which is a well defined absolutely continuous non-decreasing function. We then define another function $G_e^- : \mathbb{R}_{\geq 0} \to \mathbb{R}_{\geq 0}$ by setting

$$G_e^-(\theta) := \begin{cases} 0, & \text{if } \theta \leq \tau_e \\ \min_{\theta' \leq \theta - \tau_e} \left(G_e^+(\theta') + (\theta - \theta')\nu_e\right), & \text{else} \end{cases}.$$

By Lemma 3.20 and the fact that $G_e^+(0) = 0$ this function is absolutely continuous and non-decreasing. Thus, it has a locally integrable derivative $g_e^-(\theta) := \partial G_e^-(\theta)$ almost everywhere. Hence, (g_e^+, g_e^-) is an anonymous edge flow that clearly satisfies Proposition 3.19d) and, therefore, operates at capacity.

For the second part we first note that we have $F_e^+(\theta) = G_e^+(\theta)$ for all $\theta \in [0, \xi')$ and, thus, Proposition 3.19d) implies $F_e^-(\theta) = G_e^-(\theta)$ for all $\theta \in [0, \xi' + \tau_e)$. Now, for all times $\theta \in [\xi', \xi)$ we have

$$\begin{aligned} Q_e^f(\theta) &\geq Q_e^f(\xi') - \int_{\xi'}^{\theta} f_e^-(\zeta + \tau_e)\mathrm{d}\zeta \overset{(3.2)}{\geq} Q_e^f(\xi') - (\theta - \xi')\nu_e \\ &> Q_e^f(\xi') - (\xi - \xi')\nu_e = \nu_e \cdot \left(\xi' + \frac{Q_e^f(\xi')}{\nu_e} + \tau_e - \xi - \tau_e\right) \\ &= \nu_e \cdot \left(T_e^f(\xi') - (\xi + \tau_e)\right) \geq 0 \end{aligned}$$

and, with the same proof, $Q_e^g(\theta) > 0$. Thus, we have

$$\bar{\theta} := \max\{\theta' \leq \theta \mid Q_e^f(\theta') = 0\} = \max\{\theta' \leq \theta \mid Q_e^g(\theta') = 0\} \leq \xi'$$

and, therefore, Proposition 3.19c) implies

$$F_e^-(\theta+\tau_e) \overset{\text{Prop. 3.19c)}}{=} F_e^+(\bar{\theta})+(\theta-\bar{\theta})\nu_e = G_e^+(\bar{\theta})+(\theta-\bar{\theta})\nu_e \overset{\text{Prop. 3.19c)}}{=} G_e^-(\theta+\tau_e)$$

for all such $\theta \in [\xi', \xi)$. □

For the case of right-constant inflow rates Proposition 3.19c) even allows us to explicitly describe the corresponding outflow rate for an edge flow operating at capacity (see Fig. 3.3 for a visual representation of this proposition):

Proposition 3.22. *Let (f_e^+, f_e^-) be an edge flow operating at capacity until ξ and $0 \leq a < b \leq \xi$ times with $T_e(b) \leq \xi + \tau_e$. If f_e^+ is constant on $[a, b)$ then f_e^- is right-constant on $[T_e(a), T_e(b))$ with at most one jump. More precisely:*

- *If $Q_e(a) = 0$ or $f_e^+(a) \geq \nu_e$ then*

$$\begin{aligned} Q_e(\theta) &= Q_e(a) + (\theta - a)\max\{\, f_e^+(a) - \nu_e, 0\,\} && \text{for all } \theta \in [a, b) \quad \text{and} \\ f_e^-(\theta) &= \min\{\, f_e^+(a), \nu_e\,\} && \text{for all } \theta \in [T_e(a), T_e(b)). \end{aligned}$$

- *If $Q_e(a) > 0$ and $f_e^+(a) < \nu_e$ then*

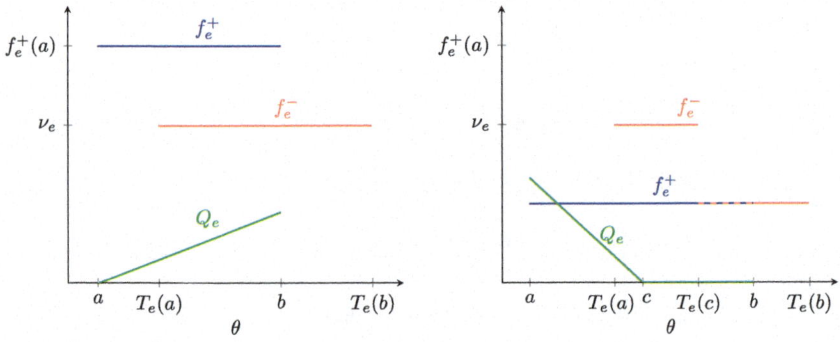

Fig. 3.3 Visualization for the outflow rate of an edge with queue operating at capacity for a constant inflow rate according to the two cases in Proposition 3.22

$$Q_e(\theta) = \begin{cases} Q_e(a) + (\theta - a)(f_e^+(a) - \nu_e) & \textit{for all } \theta \in [a, c) \\ 0 & \textit{for all } \theta \in [c, b) \end{cases} \quad \textit{and}$$

$$f_e^-(\theta) = \begin{cases} \nu_e & \textit{for all } \theta \in [T_e(a), T_e(c)) \\ f_e^+(a) & \textit{for all } \theta \in [T_e(c), T_e(b)) \end{cases}$$

where $c := \min\left\{ a + \frac{Q_e(a)}{\nu_e - f_e^+(a)}, b \right\}$ *is the break point of the queue length function.*

Proof. We show this by a case distinction on whether the queue is empty or not at time a and whether the inflow rate is smaller or larger than the capacity. In each case we first show during which parts of the relevant time interval the queue can or cannot be empty which allows us to derive the outflow rate from Proposition 3.19c). This then, in turn, gives us the exact form of the queue length function.

1. Case: $Q_e(a)=0$, $f_e^+(a) \leq \nu_e$: Here, for any $\theta \in [a, b)$ we have

$$\bar{\theta} := \max\{ \theta' \leq \theta \mid Q_e(\theta') = 0 \} \geq a$$

and, thus,

$$\begin{aligned} Q_e(\theta) &= F_e^+(\theta) - F_e^-(\theta + \tau_e) \overset{(3.7)}{=} F_e^+(\theta) - \left(F_e^+(\bar{\theta}) + (\theta - \bar{\theta})\nu_e\right) \\ &= \int_{\bar{\theta}}^{\theta} f_e^+(\zeta)\mathrm{d}\zeta - (\theta - \bar{\theta})\nu_e \leq \int_{\bar{\theta}}^{\theta} \nu_e \mathrm{d}\zeta - (\theta - \bar{\theta})\nu_e = 0. \end{aligned}$$

This already shows $Q_e(\theta) = 0 = Q_e(a) + (\theta - a) \cdot \max\{ f_e^+(a) - \nu_e, 0 \}$ for all $\theta \in [a, b)$ as well as $T_e(a) = a + \tau_e$ and $T_e(b) = b + \tau_e$.
Using Proposition 3.19c) once more, this gives us

$$F_e^-(\theta + \tau_e) \overset{(3.7)}{=} F_e^+(\theta) = F_e^+(a) + \int_a^{\theta} f_e^+(\zeta)\mathrm{d}\zeta = F_e^+(a) + (\theta - a) f_e^+(a)$$

for all $\theta \in [a, b)$ and, therefore, $f_e^-(\theta + \tau_e) = \partial F_e^-(\theta + \tau_e) = f_e^+(a) = \min\{ f_e^+(a), \nu - e \}$ for almost all $\theta \in [a, b)$.

2. Case: $Q_e(a)$=0, $f_e^+(a) > \nu_e$: For any $\theta \in (a, b)$ we have

$$Q_e(\theta) = Q_e(a) + \int_a^\theta f_e^+(\zeta)\mathrm{d}\zeta - \int_a^\theta f_e^-(\zeta + \tau_e)\mathrm{d}\zeta$$
$$\overset{(3.2)}{\geq} 0 + (\theta - a) f_e^+(a) - (\theta - a)\nu_e > 0.$$

Furthermore, for any $\theta \in [b, T_e(b) - \tau_e)$ we also have

$$Q_e(\theta) \geq Q_e(b) - \int_b^\theta f_e^-(\zeta + \tau_e)\mathrm{d}\zeta \overset{(3.2)}{\geq} Q_e(b) - (\theta - b)\nu_e$$
$$> Q_e(b) - (T_e(b) - \tau_e - b)\nu_e = Q_e(b) - \frac{Q_e(b)}{\nu_e}\nu_e = 0.$$

Thus, we have $\bar{\theta} = a$ for all $\theta \in [a, T_e(b) - \tau_e)$. Hence, we can apply Proposition 3.19c) to obtain

$$F_e^-(\theta + \tau_e) \overset{(3.7)}{=} F_e^+(a) + (\theta - a)\nu_e$$

for all $\theta \in [a, T_e(b) - \tau_e)$ and, therefore, $f_e^-(\theta + \tau_e) = \partial F_e^-(\theta + \tau_e) = \nu_e = \min\{\, f_e^+(a), \nu - e \,\}$ for almost all $\theta \in [a, T_e(b) - \tau_e) \supseteq [T_e(a) - \tau_e, T_e(b) - \tau_e)$.

3. Case: $Q_e(a) > 0$, $f_e^+(a) \geq \nu_e$: For any $\theta \in (a, b)$ we have

$$Q_e(\theta) = Q_e(a) + \int_a^\theta f_e^+(\zeta)\mathrm{d}\zeta - \int_a^\theta f_e^-(\zeta + \tau_e)\mathrm{d}\zeta$$
$$\overset{(3.2)}{\geq} Q_e(a) + (\theta - a) f_e^+(a) - (\theta - a)\nu_e \geq Q_e(a) > 0.$$

Furthermore, for any $\theta \in [b, T_e(b) - \tau_e)$ we also have $Q_e(\theta) > 0$ by the same proof as in the previous case. Thus, we have $\bar{\theta} = \bar{a} \leq a$ for all $\theta \in [a, T_e(b) - \tau_e)$. We can now, once more, apply Proposition 3.19c) to get

$$F_e^-(\theta + \tau_e) \overset{(3.7)}{=} F_e^+(\bar{a}) + (\theta - \bar{a})\nu_e$$

for all $\theta \in [a, T_e(b) - \tau_e)$ and, therefore, $f_e^-(\theta + \tau_e) = \partial F_e^-(\theta + \tau_e) = \nu_e = \min\{\, f_e^+(a), \nu - e \,\}$ for almost all $\theta \in [a, T_e(b) - \tau_e) \supseteq [T_e(a) - \tau_e, T_e(b) - \tau_e)$.

4. Case: $Q_e(a) > 0$, $f_e^+(a) < \nu_e$: For any $\theta \in [a, c)$ we have

$$\begin{aligned} Q_e(\theta) &= Q_e(a) + \int_a^\theta f_e^+(\zeta)\mathrm{d}\zeta - \int_a^\theta f_e^-(\zeta + \tau_e)\mathrm{d}\zeta \\ &\overset{(3.2)}{\geq} Q_e(a) + (\theta - a)f_e^+(a) - (\theta - a)\nu_e = Q_e(a) > 0. \end{aligned}$$

As in previous cases we also get $Q_e(\theta) > 0$ for all $\theta \in [c, T_e(c) - \tau_e)$. Thus, we have $\bar{\theta} = \bar{a} \leq a$ for all $\theta \in [a, T_e(c) - \tau_e)$. Applying Proposition 3.19c) then gives us

$$F_e^-(\theta + \tau_e) \overset{(3.7)}{=} F_e^+(\bar{a}) + (\theta - \bar{a})\nu_e$$

for all $\theta \in [a, T_e(c) - \tau_e)$ and, therefore, $f_e^-(\theta + \tau_e) = \partial F_e^-(\theta + \tau_e) = \nu_e$ for almost all $\theta \in [a, T_e(c) - \tau_e)$. If $c = b$ we are done. Otherwise, we have

$$\begin{aligned} Q_e(c) &= Q_e(a) + \int_a^c f_e^+(\zeta)\mathrm{d}\zeta - \int_a^c f_e^-(\zeta + \tau_e)\mathrm{d}\zeta \\ &= Q_e(a) + (c - a)f_e^+(a) - (c - a)\nu_e = Q_e(a) + (c - a)(f_e^+(a) - \nu_e) \\ &= Q_e(a) + \frac{Q_e(a)}{\nu_e - f_e^+(a)}(f_e^+(a) - \nu_e) = 0. \end{aligned}$$

Thus, for the remaining interval $[c, b)$ we can just apply case 1 in order to obtain $f_e^-(\theta + \tau_e) = f_e^+(c) = f_e^+(a)$ for almost all $\theta \in [c, T_e(b) - \tau_e) = [T_e(c) - \tau_e, T_e(b) - \tau_e)$.

Finally, we see that in all cases the queue length function has exactly the form given in the proposition on $[a, b)$. □

A consequence of Proposition 3.19d) is that flow dynamics of an edge flow with a queue operating at capacity are monotone in the following sense:

Corollary 3.23. *Let f_e, g_e be two edge flows with queues operating at capacity until some time $\xi \in \mathbb{R}_{\geq 0}$ and satisfying $F_e^+(\theta) \leq G_e^+(\theta)$ for all $\theta < \xi$. Then they also satisfy $F_e^-(\theta + \tau_e) \leq G_e^-(\theta + \tau_e)$ for all $\theta < \xi$.*

Proof. This essentially follows from Proposition 3.19d) by a direct computation: For any $\theta < \xi$ choose $\tilde{\theta} \leq \theta$ such that $G_e^-(\theta + \tau_e) = G_e^+(\tilde{\theta}) + (\theta - \tilde{\theta})\nu_e$. Then we have

$$F_e^-(\theta + \tau_e) \overset{\text{Prop. 3.19d)}}{=} \min_{\theta' \leq \theta} \left(F_e^+(\theta') + (\theta - \theta')\nu_e\right) \leq F_e^+(\tilde{\theta}) + (\theta - \tilde{\theta})\nu_e$$
$$\leq G_e^+(\tilde{\theta}) + (\theta - \tilde{\theta})\nu_e = G_e^-(\theta + \tau_e)$$

□

Finally, Proposition 3.19e) gives us an upper bound on how long it can take for flow on an edge with a queue operating at capacity to leave that edge. Namely, all flow that has entered the edge by some time θ will have left the edge by time $T_e(\theta)$. However, we can also use Proposition 3.19 to derive a finer upper bound which also provides us a lower bound on how much of this flow has left the edge at any time between $\theta + \tau_e$ and $T_e(\theta)$.

Corollary 3.24. *For any edge flow f_e operating at capacity until some time ξ, any time $\theta < \xi$ and all values $x \in [0, F_e^\Delta(\theta)]$ with $\theta + \frac{x}{\nu_e} < \xi$ we have*

$$F_e^-\left(\theta + \frac{x}{\nu_e} + \tau_e\right) - F_e^-(\theta) \geq x.$$

Intuitive explanation: This corollary can be read as stating that the following is the worst possible case for upper bounding the time it can take before flow currently on an edge actually leaves the edge: At time θ all flow currently on edge e is actually in its queue. Since the queue operates at capacity, this flow will then immediately start to leave the queue at a rate of ν_e and start to leave the edge τ_e time units later at the same rate.

And while it is not possible to actually achieve this worst case (since that would require all flow to enter the edge at once at time θ), it still shows that the bound given here is tight unless we have any bounds on the rate at which flow starts at the tail node of edge e.

Proof. We distinguish two cases:

1. **Case: $x \geq Q_e(\theta)$:** Here we can first use the monotonicity of F_e^- and then directly apply Proposition 3.19e) to obtain

$$F_e^-(\theta + \tfrac{x}{\nu_e} + \tau_e) \geq F_e^-(\theta + \tfrac{Q_e(\theta)}{\nu_e} + \tau_e) = F_e^-(T_e(\theta)) \overset{(3.9)}{=} F_e^+(\theta).$$

This then implies

$$F_e^-(\theta + \tfrac{x}{\nu_e} + \tau_e) - F_e^-(\theta) \geq F_e^+(\theta) - F_e^-(\theta) = F_e^\Delta(\theta) \geq x.$$

2. Case: $x < Q_e(\theta)$: Since the queue operates at capacity, Proposition 3.19b) implies that ∂Q_e is lower bound by $-\nu_e$ which gives us

$$Q_e(\vartheta) \geq Q_e(\theta) - \nu_e(\vartheta - \theta) > Q_e(\theta) - x > 0$$

for all $\vartheta \in [\theta, \theta + \frac{x}{\nu_e})$. This implies $f_e^-(\vartheta + \tau_e) = \nu_e$ for almost all such ϑ and, thus,

$$F_e^-(\theta + \tfrac{x}{\nu_e} + \tau_e) = F_e^-(\theta + \tau_e) + \int_\theta^{\theta + \frac{x}{\nu_e}} f_e^-(\zeta + \tau_e)\mathrm{d}\zeta = F_e^-(\theta + \tau_e) + \frac{x}{\nu_e} \cdot \nu_e.$$

From this, using the monotonicity of F_e^-, we immediately get

$$F_e^-(\theta + \tfrac{x}{\nu_e} + \tau_e) - F_e^-(\theta) \geq F_e^-(\theta + \tau_e) + \frac{x}{\nu_e} \cdot \nu_e - F_e^-(\theta) \geq x.$$

□

Note, that we will not show any continuity property of edge flows here since the proof would just be exactly the same as the one for showing continuity in the multi-commodity setting (see Corollaries 3.45 and 3.46).

3.1.2 Multi-Commodity Edge Flows

We now introduce an edge flow wherein not all particles are interchangeable with each other. Instead we are given a finite set of commodities I and every particle belongs to exactly one of these commodities. Particles of the same commodity are still indistinguishable from each other but particles from different commodities can be distinguished. Formally, such a multi-commodity edge flow is then given by one edge inflow rate and one edge outflow rate for each commodity:

Definition 3.25. Let I be a finite set of commodities and e an edge with associated travel time $\tau_e \geq 0$ and capacity $\nu_e > 0$. Then a *(multi-commodity) edge flow* on this edge consists of

- a vector $(f^+_{e,i})_{i\in I} \in L^1_{\text{loc}}(\mathbb{R}_{\geq 0}, \mathbb{R}_{\geq 0})^I$ where $f^+_{e,i}(\theta)$ is the *(edge) inflow rate* of commodity $i \in I$ into edge e at time θ and
- a vector $(f^-_{e,i})_{i\in I} \in L^1_{\text{loc}}(\mathbb{R}_{\geq 0}, \mathbb{R}_{\geq 0})^I$ where $f^-_{e,i}(\theta)$ is the *(edge) outflow rate* of commodity $i \in I$ into edge e at time θ.

In the same way as for anonymous flows we can again define cumulative versions of those flow rates by setting

$$F^+_{e,i}(\theta) := \int_0^\theta f^+_{e,i}(\zeta)\mathrm{d}\zeta \text{ and } F^-_{e,i}(\theta) := \int_0^\theta f^-_{e,i}(\zeta)\mathrm{d}\zeta$$

for all $i \in I$ and $\theta \in \mathbb{R}_{\geq 0}$. Additionally, we also define the *commodity specific edge load* by $F^\Delta_{e,i}(\theta) := F^+_{e,i}(\theta) - F^-_{e,i}(\theta)$.

Proposition 3.26. *The commodity-specific cumulative in- and outflow functions are non-decreasing, absolutely continuous and satisfy*

$$\partial F^+_{e,i}(\theta) = f^+_{e,i}(\theta) \text{ and } \partial F^-_{e,i}(\theta) = f^-_{e,i}(\theta)$$

for almost all $\theta \in \mathbb{R}_{\geq 0}$.

Proof. As in the case of anonymous edge flows, this follows directly from Proposition 2.49. □

Also note that any multi-commodity flow has an associated anonymous flow obtained by summing the in- and outflow rates for all commodities.

Definition 3.27. For any multi-commodity edge flow $(f^+_{e,i}, f^+_{e,i})$ we define its *associated anonymous edge flow* (f^+_e, f^-_e) by setting

$$f^+_e := \sum_{i\in I} f^+_{e,i} \text{ and } f^-_e := \sum_{i\in I} f^-_{e,i}.$$

Observation 3.28. For the corresponding cumulative versions of both multi-commodity and anonymous edge flow we clearly have the same relation, i.e.

$$F_e^+ = \sum_{i \in I} F_{e,i}^+, \quad F_e^- = \sum_{i \in I} F_{e,i}^- \quad \text{and} \quad F_e^\Delta = \sum_{i \in I} F_{e,i}^\Delta.$$

This way we can transfer all definitions for anonymous edge flows (i.e. objects like queue length or exit time but also properties like working at capacity or weak flow conservation) to multi-commodity edge flows. Hence, whenever we refer to the queue length of a multi-commodity edge flow, what we mean is the queue length of its associated anonymous edge flow.

From now on we will use the short form "edge flow" to refer to a multi-commodity edge flow. Whenever the set of commodities is a singleton (i.e. $I = \{*\}$) we will call such an edge-flow a *single-commodity* edge flow (and will then not distinguish between $(f_{e,*}^+, f_{e,*}^-)$ and (f_e^+, f_e^-)).

We can now define flow conservation as well as expected travel time for multi-commodity flows in essentially the same way as for anonymous flows.

Definition 3.29. We say that an edge flow satisfies *weak flow conservation for commodity i until* $\xi \in \tilde{\mathbb{R}}_{\geq 0}$ if

$$F_{e,i}^-(\theta + \tau_e) \leq F_{e,i}^+(\theta) \text{ for all } \theta < \xi. \tag{3.12}$$

and *strong flow conservation until* ξ if

$$F_{e,i}^-(\theta + \tau_e) = F_{e,i}^+(\theta) \text{ for all } \theta < \xi.$$

Observation 3.30. If an edge flow satisfies weak/strong flow conservation for every commodity $i \in I$ until ξ then its associated anonymous edge flow satisfies weak/strong flow conservation until ξ as well. The other direction clearly does not hold, i.e. if we only require weak/strong flow conservation for the associated anonymous edge flow, this still allows particles to switch their commodity while travelling along an edge.

Definition 3.31. We say that *flow of commodity i enters edge e at time* θ if $F_{e,i}^+$ is strictly increasing at θ, i.e. if for all $\theta' < \theta$ we have $F_{e,i}^+(\theta') < F_{e,i}^+(\theta)$.

Definition 3.32. For any edge flow $(f_{e,\cdot}^+, f_{e,\cdot}^-)$ we define the *commodity specific experienced current travel time* of commodity i on this edge for all times θ at which flow of commodity i enters edge e by

$$\hat{C}_{e,i}(\theta) := \inf \{ \zeta \geq -\theta \mid F^-_{e,i}(\theta + \zeta) \geq F^+_{e,i}(\theta) \}.$$

Observation 3.33. Whenever flow of some commodity $i \in I$ enters edge e at time θ then flow enters edge e at time θ (i.e. in the sense of the definition for anonymous edge flows). Conversely, whenever flow enters edge e then there must be some commodity such that flow of this commodity enters at that time.

A new aspect of multi-commodity flows is that particles of different commodities could experience congestion differently. For example it could be that in case of congestion particles of a specific commodity get priority over particles of other commodities. In particular, for a multi-commodity edge flow operating at capacity its (commodity-specific) outflow rates are not necessarily uniquely determined by its inflow rates (in contrast to the anonymous outflow rate which is uniquely determined according to Corollary 3.21). This is true, even if we additionally require weak flow conservation for every commodity (i.e. prevent particles from switching their commodity while on an edge):

Example 3.34. Consider an edge with capacity and free flow travel time 1. Then inflow rates of $f^+_{e,1} = f^+_{e,2} = \mathbb{1}_{[0,1]}$ and outflow rates of $f^-_{e,1} = \mathbb{1}_{[1,2]}$ and $f^+_{e,2} = \mathbb{1}_{[2,3]}$ (cf. Fig. 3.4) define a 2-commodity edge flow $(f^+_{e,\cdot}, f^-_{e,\cdot})$ wherein the particles of commodity 1 overtake those of commodity 2 while waiting in the queue, while still satisfying weak flow conservation for each of the two commodities and its queue operating at capacity.

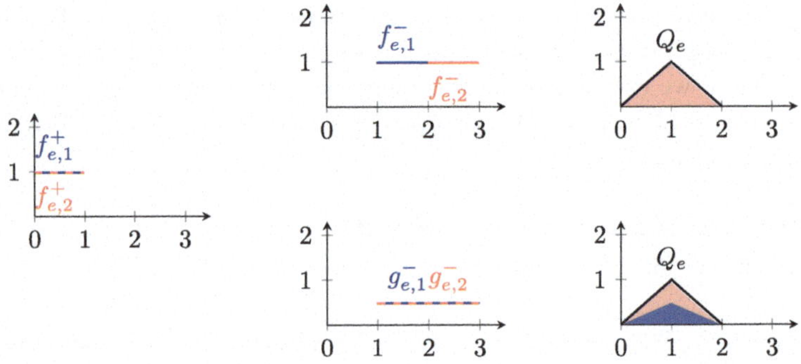

Fig. 3.4 A 2-commodity flow with overtaking (top) and without overtaking (bottom)

Defining outflow rates of $g_{e,1}^- = g_{e,2}^- = \frac{1}{2} \cdot \mathbb{1}_{[1,3]}$ gives us an alternative edge flow $(f_{e,\cdot}^+, g_{e,\cdot}^-)$ without such overtaking but also satisfying weak flow conservation and with its queue operating at capacity.

Therefore, we will impose a stronger condition on the queues, namely that they operate fairly. That is, whenever particles enter an edge in a certain proportion between different commodities, they will also leave the queue (or equivalently the edge) in this same proportion. Intuitively, we would then like to write this fairness condition as

$$f_{e,i}^-(\theta) = \frac{f_{e,i}^+(\vartheta)}{f_e^+(\vartheta)} \cdot f_e^-(\theta),$$

where ϑ is "the" time where flow entered edge e in order to leave at time θ. This definition, however, has two problems: First, even for edge flows with queue operating at capacity, there might be multiple times leading to the same exit time (e.g. Example 3.17) and, second, the above constraint is only well defined if $f_e^+(\vartheta) \neq 0$ holds. The former problem can be solved by choosing a more specific entrance time ϑ, e.g. the earliest or latest time with exit-time θ. A priori, these choices seem to lead to different fairness definitions: However, as noted by Markl in [Mar21, Remark 3.3], for flows respecting capacity these definitions do turn out to be equivalent for almost all $\theta \in \mathbb{R}_{\geq 0}$.

Proposition 3.35. *If an edge flow respects the capacity and satisfies weak flow conservation until ξ, then for almost all $\theta \in [\tau_e, \xi + \tau_e)$ we have*

$$\begin{aligned}\min\{\,\vartheta \mid T_e(\vartheta) = \theta\,\} &= \min\{\,\vartheta \mid T_e(\vartheta) \geq \theta\,\} \\ &= \max\{\,\vartheta \leq \xi \mid T_e(\vartheta) \leq \theta\,\} = \max\{\,\vartheta \leq \xi \mid T_e(\vartheta) = \theta\,\}.\end{aligned}$$

In particular, for almost all $\theta \in [\tau_e, \xi + \tau_e)$ there exists a unique time $\vartheta < \xi$ with $T_e(\vartheta) = \theta$.

Proof. Weak flow conservation gives us $T_e(0) = \tau_e$ and $T_e(\xi) \geq \xi + \tau_e$. Since T_e is continuous, this implies $[\tau_e, \xi + \tau_e) \subseteq T_e([0, \xi))$. Thus, every time $\theta \in [\tau_e, \xi + \tau_e)$ has at least one preimage in $[0, \xi)$ and all maxima and minima in the above statement are attained.

Furthermore, T_e is non-decreasing on $[0, \xi)$ by Proposition 3.16a) (as it respects capacity until ξ). Thus, Proposition 2.10b) implies that almost all $\theta \in [\tau_e, \xi + \tau_e)$ have a unique preimage in $[0, \xi)$. □

For our second problem, it turns out that, if the queue operates at capacity, then for almost all times θ with $f_e^+(\vartheta) = 0$ we already have $f_e^-(\theta) = 0$ as well. Thus, we may just set $f_{e,i}^-(\theta) = 0$ for all these times and still have $\sum_{i \in I} f_{e,i}^-(\theta) = f_e^-(\theta)$ for almost all θ.

Proposition 3.36. *Let f_e be an edge flow with queue operating at capacity until some time $\xi \in \tilde{\mathbb{R}}_{\geq 0}$. Then for almost all $\theta \in [\tau_e, \xi + \tau_e)$ the edge flow satisfies*

$$(\exists \vartheta : T_e(\vartheta) = \theta \wedge f_e^+(\vartheta) = 0) \implies f_e^-(\theta) = 0.$$

Proof. Define by

$$N := \{\, \vartheta \in [0, \xi) \mid T_e(\vartheta) < \xi + \tau_e,\ f_e^+(\vartheta) = 0 \neq f_e^-(T_e(\vartheta)) \,\}$$

the set of all preimages of points where the statement of the proposition does not hold. Since T_e is non-decreasing on $[0, \xi)$ (by Proposition 3.16a)) and F_e^- is absolutely continuous, we can apply the chain rule (Proposition 2.51) to (3.9) to get

$$f_e^-(T_e(\vartheta)) \cdot \partial T_e(\vartheta) = (\partial F_e^-)(T_e(\vartheta)) \cdot \partial T_e(\vartheta) = \partial (F_e^- \circ T_e)(\vartheta) = \partial F_e^+(\vartheta) = f_e^+(\vartheta)$$

which implies that we have $\partial T_e(\vartheta) = 0$ for almost all $\vartheta \in N$. Sard's Theorem (Proposition 2.10a)) and the fact that T_e maps null sets to null sets (cf. Proposition 2.47b)) then imply that the set

$$T_e(N) = \{\, \theta \in [\tau_e, \xi + \tau_e) \mid \exists \vartheta \in [0, \xi) : T_e(\vartheta) = \theta \text{ and } f_e^+(\vartheta) = 0 \neq f_e^-(\theta) \,\}$$

is a null set. Together with the fact that almost all times θ have a unique preimage under T_e (Proposition 3.35) this shows the proposition. □

We can now define our fairness condition for multi-commodity edge flows as follows:

Definition 3.37. Let $(f_{e,i}^+, f_{e,i}^-)_{i \in I}$ be an edge flow such that its associated anonymous edge flow respects the capacity and satisfies weak flow conservation until $\xi \in \tilde{\mathbb{R}}_{\geq 0}$. We then say that the queue on edge e *operates fair until* ξ if for all commodities $i \in I$ we have

$$f_{e,i}^-(\theta+\tau_e) = \begin{cases} f_e^-(\theta+\tau_e)\cdot\frac{f_{e,i}^+(\vartheta)}{f_e^+(\vartheta)}, & \text{if } f_e^+(\vartheta)>0, \\ 0, & \text{else} \end{cases} \quad \text{for almost all } \theta<\xi, \tag{3.13}$$

where ϑ is chosen such that $T_e(\vartheta) = \theta + \tau_e$.

Note, that this constraint is really well defined: Proposition 3.35 ensures that it does not depend on the choice of ϑ while the fact that T_e is an absolutely continuous function and, therefore, maps null sets to null sets (cf. Proposition 2.47b)), implies that it is unaffected by any change of $(f_{e,i}^+, f_{e,i}^-)_{i\in I}$ on some set of measure zero.

Definition 3.38. An edge flow $(f_{e,i}^+, f_{e,i}^-)_{i\in I}$ is a *Vickrey edge flow until* $\xi \in \tilde{\mathbb{R}}_{\geq 0}$ if its queue operates at capacity and fair until ξ.

Observation 3.39. Proposition 3.36 implies that the queue of a single-commodity edge flow automatically operates fairly if it operates at capacity.

We provide two alternative characterisations for when a queue operates at capacity and fair.

Proposition 3.40. *For any edge flow* $(f_{e,i}^+, f_{e,i}^-)_{i\in I}$ *and time* $\xi \in \tilde{\mathbb{R}}_{\geq 0}$ *the following properties are equivalent:*

a) $(f_{e,i}^+, f_{e,i}^-)_{i\in I}$ *is a Vickrey edge flow until* ξ.
b) The associated anonymous flow respects the capacity and satisfies weak flow conservation until ξ *and we have*

$$F_{e,i}^-(T_e(\theta)) = F_{e,i}^+(\theta) \text{ for all } \theta<\xi \text{ with } T_e(\theta)<\xi+\tau_e \text{ and all } i\in I. \tag{3.14}$$

c) The associated anonymous flow respects the capacity and satisfies weak flow conservation until ξ *and for every* $\theta<\xi$ *with* $T_e(\theta)<\xi+\tau_e$ *where* $\hat{C}_{e,i}(\theta)$ *is defined we have* $\hat{C}_{e,i}(\theta) = C_e(\theta)$.

Proof. **a) $\Longrightarrow$ b):** Since the queue operates at capacity until ξ, Proposition 3.19 already shows that the associated anonymous flow respects capacity and satisfies weak flow conservation until ξ.

As T_e is continuous and non-decreasing on $[0, \xi]$ by Proposition 3.16a) and satisfies both $T_e(0) = \tau_e$ and $T_e(\xi) \geq \xi + \tau_e$ (since the flow satisfies weak flow conservation), there exists some $\bar{\xi} := \min\{\theta \leq \xi \mid T_e(\theta) = \xi + \tau_e\}$. We now have to show (3.14) for all $\theta < \bar{\xi}$. For that, we fix some representative of f_e and define new outflow functions $g^-_{e,i}$ by setting

$$g^-_{e,i}(T_e(\vartheta)) := \begin{cases} f^-_e(T_e(\vartheta)) \cdot \frac{f^+_{e,i}(\vartheta)}{f^+_e(\vartheta)}, & \text{if } f^+_e(\vartheta) > 0 \text{ and } \left|T^{-1}_e(T_e(\vartheta))\right| = 1 \\ 0, & \text{else} \end{cases} \tag{3.15}$$

for $\vartheta < \bar{\xi}$ and $g^-_{e,i}(\theta) := 0$ for all other times.

Claim 3.3. *This functions satisfies $g^-_{i,e}(\theta) = f^-_{e,i}(\theta)$ for almost all $\theta \in [0, \bar{\xi} + \tau_e)$ and all $i \in I$.*

Proof. As the queue starts empty and T_e is non-decreasing on $[0, \xi)$, we have $T_e(\vartheta) \geq \tau_e$ for all $\vartheta < \xi$ and, therefore, $g^-_{e,i}(\theta) = 0$ for all $\theta < \tau_e$. At the same time the queue starting empty also implies $f^-_{i,e}(\theta) = 0$ for almost all $\theta < \tau_e$. For the interval $[\tau_e, \bar{\xi} + \tau_e)$ Proposition 3.35 implies that it suffices to only consider those times $\theta \in [\tau_e, \bar{\xi} + \tau_e)$ with a unique time $\bar{\vartheta} < \xi$ satisfying $T_e(\vartheta) = \theta$. For almost all those θ we indeed have

$$\begin{aligned} f^-_{e,i}(\theta) &\overset{(3.13)}{=} \begin{cases} f^-_e(\theta) \cdot \frac{f^+_{e,i}(\vartheta)}{f^+_e(\vartheta)} = f^-_e(T_e(\vartheta)) \cdot \frac{f^+_{e,i}(\vartheta)}{f^+_e(\vartheta)}, & \text{if } f^+_e(\vartheta) > 0 \\ 0, & \text{if } f^+_e(\vartheta) = 0 \end{cases} \\ &\overset{(3.15)}{=} g^-_{e,i}(T_e(\vartheta)) = g^-_{e,i}(\theta) \end{aligned}$$

since the queue operates fair until ξ. ∎

With this claim we now get

$$\begin{aligned} F^-_{e,i}(T_e(\theta)) &= \int_0^{T_e(\theta)} f^-_{e,i}(\zeta)\mathrm{d}\zeta \overset{\text{Cl. 3.3}}{=} \int_0^{T_e(\theta)} g^-_{e,i}(\zeta)\mathrm{d}\zeta = \int_{\tau_e}^{T_e(\theta)} g^-_{e,i}(\zeta)\mathrm{d}\zeta \\ &\overset{(*)}{=} \int_0^{\theta} g^-_{e,i}(T_e(\zeta)) \cdot \partial T_e(\zeta)\mathrm{d}\zeta \overset{(\#)}{=} \int_0^{\theta} f^+_{e,i}(\zeta)\mathrm{d}\zeta = F^+_{e,i}(\theta) \end{aligned}$$

for all $\theta < \bar{\xi}$. At (∗) we used the change of variable formula (Proposition 2.53) while for (#) it remains to show that

$$g_{e,i}^{-}(T_e(\zeta)) \cdot \partial T_e(\zeta) = f_{e,i}^{+}(\zeta) \tag{3.16}$$

holds for almost all $\zeta \in [0, \theta]$. To do that we partition the interval $[0, \theta]$ into the following four sets:

- $N := \{ \zeta \in [0, \theta] \mid F_e^+$ or T_e not differentiable at ζ or $\partial F_e^+(\zeta) \neq f_e^+(\zeta) \}$,
- $M_1 := \{ \zeta \in [0, \theta] \setminus N \mid |T_e^{-1}(T_e(\zeta))| > 1 \}$,
- $M_2 := \{ \zeta \in [0, \theta] \setminus N \mid T_e^{-1}(T_e(\zeta)) = \{ \zeta \}$ and $\partial F_e^+(\zeta) = f_e^+(\zeta) > 0 \}$,
- $M_2 := \{ \zeta \in [0, \theta] \setminus N \mid T_e^{-1}(T_e(\zeta)) = \{ \zeta \}$ and $\partial F_e^+(\zeta) = f_e^+(\zeta) = 0 \}$.

Since F_e^+ and T_e and absolutely continuous, the set N has measure zero. Thus, it suffices to show (3.16) for almost all $\zeta \in M_1 \cup M_2 \cup M_3$.

1. **Case: $\zeta \in M_1$:** For those times ζ we must have $\partial T_e(\zeta) = 0$ (since T_e is non-decreasing). According to Proposition 3.16b) this then implies $f_e^+(\zeta) = 0$ (almost always) and, thus, we have

$$g_{e,i}^{-}(T_e(\zeta)) \cdot \partial T_e(\zeta) = g_{e,i}^{-}(T_e(\zeta)) \cdot 0 = 0 = f_{e,i}^{+}(\zeta).$$

2. **Case: $\zeta \in M_2$:** For (almost all) such times we have

$$\begin{aligned} g_{e,i}^{-}(T_e(\zeta)) \cdot \partial T_e(\zeta) &\overset{(3.15)}{=} f_e^{-}(T_e(\zeta)) \cdot \frac{f_{e,i}^{+}(\zeta)}{f_e^{+}(\zeta)} \cdot \partial T_e(\zeta) \\ &\overset{(\Delta)}{=} f_e^{+}(\zeta) \cdot \frac{f_{e,i}^{+}(\zeta)}{f_e^{+}(\zeta)} = f_{e,i}^{+}(\zeta) \end{aligned}$$

where (Δ) holds since differentiating (3.9) together with the chain rule gives us $f_e^-(T_e(\zeta)) \cdot \partial T_e(\zeta) = f_e^+(\zeta)$.

3. **Case: $\zeta \in M_3$:** For (almost) all such times we have

$$g_{e,i}^{-}(T_e(\zeta)) \cdot \partial T_e(\zeta) \overset{(3.15)}{=} 0 = f_e^{+}(\zeta) \geq f_{e,i}^{+}(\zeta) \geq 0$$

and, therefore $g_{e,i}^{-}(T_e(\zeta)) \cdot \partial T_e(\zeta) = 0 = f_{e,i}^{+}(\zeta)$.

b) $\Longrightarrow$ a): We first observe that (3.14) implies (3.9) for the associated anonymous flow and, thus, the queue operates at capacity until ξ by Proposition 3.19. Moreover, Proposition 3.36 then implies that for almost all $\theta \in [0, \xi)$ and all ϑ with $T_e(\vartheta) = \theta + \tau_e$ and $f_e^+(\vartheta) = 0$ we have $f_{e,i}^-(\theta + \tau_e) \leq f_e^-(\theta + \tau_e) = 0$. In particular, (3.13) holds for almost all such θ.

Thus, it suffices to consider only those θ with a corresponding ϑ such that $f_e^+(\vartheta) > 0$ (and $T_e(\vartheta) = \theta + \tau_e$).

We define the set

$$N := \left\{ \vartheta < \xi \;\middle|\; T_e \text{ is not differentiable at } \vartheta \text{ or } \exists i \in I : \partial F_{e,i}^+(\vartheta) \neq f_{e,i}^+(\vartheta) \right\}$$

which has measure zero. Since T_e is absolutely continuous, the set $T_e(N)$ also has measure zero. The same is then true for the set

$$N' := T_e(N) \cup \left\{ \theta \;\middle|\; \exists i \in I : \partial F_{e,i}^-(\theta + \tau_e) \neq f_{e,i}^-(\theta + \tau_e) \right\}$$

Thus, it suffices to only consider θ which additionally are not in N'. For such θ we can differentiate (3.9) to obtain

$$f_e^-(T_e(\vartheta)) \cdot \partial T_e(\vartheta) = f_e^+(\vartheta).$$

Since the right hand side is strictly positive, all terms are non-zero and we can rearrange this to get

$$\partial T_e(\vartheta) = \frac{f_e^+(\vartheta)}{f_e^-(T_e(\vartheta))}. \tag{3.17}$$

Similarly, differentiating (3.14) gives

$$f_{e,i}^-(T_e(\vartheta)) \cdot \partial T_e(\vartheta) = f_{e,i}^+(\vartheta)$$

which, together with (3.17), implies

$$f_{e,i}^-(T_e(\vartheta)) \cdot \frac{f_e^+(\vartheta)}{f_e^-(T_e(\vartheta))} = f_{e,i}^+(\vartheta).$$

Using $f_e^+(\vartheta) > 0$ we can rearrange this one last time to obtain

$$f_{e,i}^-(\theta + \tau_e) = f_{e,i}^-(T_e(\vartheta)) = f_{e,i}^+(\vartheta) \cdot \frac{f_e^-(T_e(\vartheta))}{f_e^+(\vartheta)} = f_e^-(\theta + \tau_e)) \cdot \frac{f_{e,i}^+(\vartheta)}{f_e^+(\vartheta)}$$

which is exactly what we had to show.

b) $\Longrightarrow$ c): We can show this in exactly the same way as Proposition 3.19e) $\Longrightarrow$ f): Take any time $0 < \theta < \xi$ with $T_e(\theta) < \xi + \tau_e$ such that flow of commodity i enters at time θ (i.e. $F_{e,i}^+$ is strictly increasing at θ). Then, (3.14) directly implies $\hat{C}_{e,i}(\theta) \leq C_e(\theta)$. Now, we assume for contradiction, that we have $\hat{C}_{e,i}(\theta) < C_e(\theta)$. Since C_e is continuous there exists some $\theta' < \theta$ with $\theta + \hat{C}_{e,i}(\theta) \leq \theta' + C_e(\theta')$. As $F_{e,i}^-$ is non-decreasing this gives us

$$F_{e,i}^+(\theta') \overset{(3.14)}{=} F_{e,i}^-(\theta' + C_e(\theta')) \geq F_{e,i}^-(\theta + \hat{C}_{e,i}(\theta)) \geq F_{e,i}^+(\theta).$$

This now implies $f_{e,i}^+(\zeta) = 0$ for almost all $\zeta \in [\theta', \theta]$ which, according to Proposition 2.54a), is a contradiction to flow of commodity i entering at time θ.

c) $\Longrightarrow$ b): By way of contradiction we assume that (3.14) does not hold, i.e. there exists some commodity i and some $\theta < \xi$ with $T_e(\theta) < \xi + \tau_e$ such that either $F_{e,i}^-(T_e(\theta)) < F_{e,i}^+(\theta)$ or $F_{e,i}^-(T_e(\theta)) > F_{e,i}^+(\theta)$.

1. Case: $F_{e,i}^-(T_e(\theta)) < F_{e,i}^+(\theta)$: Since both sides of the inequality are continuous in θ and equality holds for $\theta = 0$ due to weak flow conservation, there must be some proper interval $(a, b) \subseteq [0, \xi)$ such that $F_{e,i}^-(T_e(a)) = F_{e,i}^+(a)$ and $F_{e,i}^-(T_e(\theta)) < F_{e,i}^+(\theta)$ for all $\theta \in (a, b)$. This implies that $F_{e,i}^+$ is constant on (a, b) as otherwise Proposition 2.54d) would guarantee the existence of some $\theta \in (a, b)$ at which flow enters the edge and, thus c) would imply

$$F_{e,i}^-(T_e(\theta)) = F_{e,i}^-(\theta + C_e(\theta)) = F_{e,i}^-(\theta + \hat{C}_{e,i}(\theta)) \geq F_{e,i}^+(\theta).$$

Thus, for any $\theta \in (a, b)$ we have

$$F_{e,i}^-(T_e(\theta)) < F_{e,i}^+(\theta) = F_{e,i}^+(a) = F_{e,i}^-(T_e(a)).$$

As $F_{e,i}^-$ is non-decreasing, this implies

$$\theta + \frac{Q_e(\theta)}{\nu_e} + \tau_e = T_e(\theta) < T_e(a) = a + \frac{Q_e(a)}{\nu_e} + \tau_e \tag{3.18}$$

and, thus, using the fact that F_e^+ is constant on (a, b) and the (aggregated) flow respects the capacity we get

$$\begin{aligned}\nu_e(\theta - a) &\overset{(3.18)}{<} Q_e(a) - Q_e(\theta)\\ &= F_e^+(a) - F_e^+(\theta) + F_e^-(\theta + \tau_e) - F_e^-(a + \tau_e)\\ &= 0 + \int_a^\theta f_e^-(\zeta + \zeta)d\zeta \leq \nu_e(\theta - a),\end{aligned}$$

which is a contradiction.

2. Case: $F_{e,i}^-(T_e(\theta)) > F_{e,i}^+(\theta)$: For the second case we can now assume that $F_{e,i}^-(T_e(\theta)) \geq F_{e,i}^+(\theta)$ holds for all commodities i and all times $\theta < \xi$. If then, additionally, we have $F_{e,i}^-(T_e(\theta)) > F_{e,i}^+(\theta)$ for some $i \in I$ and $\theta < \xi$ with $T_e(\theta) < \xi + \tau_e$, we must also have $F_e^-(T_e(\theta)) > F_e^+(\theta)$. Since the flow satisfies weak flow conservation and respects the capacity, this is a contradiction to Proposition 3.16c). □

Remark 3.41. We note that the alternative outflow rates $g_{e,i}^-$ constructed in the proof of Proposition 3.40a) $\Longrightarrow$ b) satisfy the following 'stronger' version of our fairness condition (3.13):

$$f_{e,i}^-(T_e(\theta)) = \begin{cases} f_e^-(T_e(\theta)) \cdot \frac{f_{e,i}^+(\theta)}{f_e^+(\theta)}, & \text{if } f_e^+(\theta) > 0 \\ 0, & \text{else} \end{cases} \quad \text{for almost all } \theta. \tag{3.19}$$

In [HFY13, (4.19)] this equation is used as the defining constraint for queues operating fairly. Note, however, that, in contrast to constraint (3.13), constraint (3.19) is not indifferent to changes on a set of measure zero. Namely, if there is a proper interval $[a, b]$ such that T_e is constant on $[a, b]$ then changing only the single value of $f_{e,i}^-(T_e(a))$ leads to a violation of constraint (3.19).

Nevertheless, we observe the following connection between the two constraints (3.13) and (3.19): Given functions (not equivalence classes!) $f_{e,i}^+$ and $f_{e,i}^-$. If they satisfy constraint (3.19) then they also satisfy constraint (3.13). If they satisfy constraint (3.13), then there exist alternative representatives $g_{e,i}^-$ of $f_{e,i}^-$ (i.e. functions which are equal almost everywhere) that satisfy constraint (3.19) together with $f_{e,i}^+$.

We now use Propositions 3.19 and 3.40 to deduce several useful properties of Vickrey edge flows. We start by observing that every Vickrey edge flow automatically satisfies weak flow conservation for every commodity:

Corollary 3.42. *Let $(f^+_{e,\bullet}, f^-_{e,\bullet})$ be a Vickrey edge flow until ξ. Then $(f^+_{e,\bullet}, f^-_{e,\bullet})$ satisfies weak flow conservation for every commodity until ξ.*

Proof. Take any commodity $i \in I$ and time $\theta < \xi$. Since the associated anonymous flow satisfies weak flow conservation, there exists some $\vartheta \leq \theta$ with $T_e(\vartheta) = \theta + \tau_e < \xi + \tau_e$. Propositions 3.40b) then implies

$$F^-_{e,i}(\theta + \tau_e) = F^-_{e,i}(T_e(\vartheta)) \overset{(3.14)}{=} F^+_{e,i}(\vartheta) \leq F^+_{e,i}(\theta).$$

□

Next, we show that – as in the case of anonymous edge flows – the commodity-specific inflow rates completely determine the commodity-specific outflow rates. This is formalised in the following corollary which is the multi-commodity analogue to Corollar 3.21:

Corollary 3.43. *For any tuple of locally integrable functions $(g^+_{e,i} : \mathbb{R}_{\geq 0} \to \mathbb{R}_{\geq 0})_{i \in I}$ there exists a tuple of locally integrable functions $(g^-_{e,i} : \mathbb{R}_{\geq 0} \to \mathbb{R}_{\geq 0})_{i \in I}$ such that $(g^+_{e,i}, g^-_{e,i})$ is a Vickrey edge flow.*

Furthermore, if $(f^+_{e,i}, f^-_{e,i})$ and $(g^+_{e,i}, g^-_{e,i})$ are two Vickrey edge flows until $\xi \in \bar{\mathbb{R}}_{\geq 0}$ and $\xi' \leq \xi$ some time with $T^f_e(\xi') \geq \xi + \tau_e$ then

$$\left(\forall i \in I : f^+_{e,i}\big|_{[0,\xi')} =_{a.e.} g^+_{e,i}\big|_{[0,\xi')}\right) \implies \left(\forall i \in I : f^-_{e,i}\big|_{[0,\xi+\tau_e)} =_{a.e.} g^-_{e,i}\big|_{[0,\xi+\tau_e)}\right).$$

Proof. Since $g^+_e := \sum_{i \in I} g^+_{e,i}$ is also locally integrable, Corollary 3.21 already implies that there exists a (locally integrable) function $g^-_e : \mathbb{R}_{\geq 0} \to \mathbb{R}_{\geq 0}$ such that (g^+_e, g^-_e) is an anonymous edge flow operating at capacity. Then, Proposition 3.35 implies that for the corresponding exit time function T_e and almost all times $\theta \in [\tau_e, \infty)$ there is a unique time $\vartheta \in \mathbb{R}_{\geq 0}$ with $T_e(\vartheta) = \theta$. Thus, we can define $g^-_{e,i}$ using (3.13) (and setting $g^-_{e,i}(\theta) = 0$ for all $\theta < \tau_e$) to obtain a Vickrey edge flow $(g^+_{e,i}, g^-_{e,i})$.

For uniqueness we already have $f^-_e\big|_{[0,\xi+\tau_e)} =_{\text{a.e.}} g^-_e\big|_{[0,\xi+\tau_e)}$ from Corollary 3.21. This, in particular, implies $T^f_e(\vartheta) = T^g_e(\vartheta)$ for all $\vartheta < \xi'$ with $T^f_e(\vartheta) < \xi + \tau_e$. Now, for almost every $\theta \in [\tau_e, \xi + \tau_e)$ Proposition 3.35 implies that there exists a unique $\vartheta \in [0, \xi)$ with $\theta = T^f_e(\vartheta)$. Since $T^f_e(\xi') \geq \xi + \tau_e > \theta = T^f_e(\vartheta)$ and T^f_e is non-decreasing on $[0, \xi)$ by Proposition 3.16a), we have $\vartheta < \xi'$ and, therefore, $T_e^g(\vartheta) = T_e^f(\vartheta)$. Thus, Proposition 3.40b) gives us

$$F^-_{e,i}(\theta) = F^-_{e,i}(T_e^f(\vartheta)) = F^+_{e,i}(\vartheta) = G^+_{e,i}(\vartheta) = G^-_{e,i}(T_e^g(\vartheta)) = G^-_{e,i}(\theta)$$

for all such $\theta \in [\tau_e, \xi + \tau_e)$. Since both $F^-_{e,i}$ and $G^-_{e,i}$ are continuous, this already implies that they are equal on $[\tau_e, \xi + \tau_e)$. Equality on $[0, \tau_e)$ follows directly from the definition of cumulative outflow functions. □

For the case of right-constant inflow rates we can, again, describe the corresponding outflow rates of a Vickrey edge flow more explicitly:

Corollary 3.44. *Let $(f^+_{e,i}, f^-_{e,i})$ be a Vickrey edge flow until ξ and $0 \leq a < b \leq \xi$ times with $T_e(b) \leq \xi + \tau_e$. If all $f^+_{e,i}$ are constant on $[a, b)$ then all $f^-_{e,i}$ are right-constant on $[T_e(a), T_e(b))$ with at most one (common) jump. More precisely,*

- *If $Q_e(a) = 0$ or $f_e^+(a) \geq \nu_e$ then*

$$f^-_{e,i}(\theta) = \frac{f^+_{e,i}(a)\nu_e}{\max\{\, f_e^+(a), \nu_e \,\}} \text{ for all } \theta \in [T_e(a), T_e(b)).$$

- *If $Q_e(a) > 0$ and $f_e^+(a) < \nu_e$ then*

$$f_e^-(\theta) = \begin{cases} 0, & \text{for all } \theta \in [T_e(a), T_e(b)) \text{ if } f_e^+(a) = 0 \\ \frac{f^+_{e,i}(a)\nu_e}{f_e^+(a)} & \text{for all } \theta \in [T_e(a), T_e(c)) \text{ if } f_e^+(a) > 0 \\ f^+_{e,i}(a) & \text{for all } \theta \in [T_e(c), T_e(b)) \text{ if } f_e^+(a) > 0 \end{cases}$$

where $c := \min\{\, a + \frac{Q_e(a)}{\nu_e - f_e^+(a)}, b \,\}$.

Proof. First, note that for any time $\theta \in [T_e(a), T_e(b))$ there exists some time $\vartheta \in [a, b)$ with $T_e(\vartheta) = \theta$. We now continue with a case distinction:

1. **Case: $f_e^+(a)$=0:** Here, (3.13) already ensures that $f^-_{e,i}(\theta) = 0$ for almost all $\theta \in [T_e(a), T_e(b))$. Since we also have $f^+_{e,i}(a) \leq f_e^+(a) = 0$ for all $i \in I$, this shows that the outflow has the form stated in the corollary.
2. **Case: $Q_e(a) > 0$ and $0 < f_e^+(a) < \nu_e$:** Using Proposition 3.22 and (3.13) we get

$$f_{e,i}^-(\theta) \overset{(3.13)}{=} f_e^-(\theta) \cdot \frac{f_{e,i}^+(\vartheta)}{f_e^+(\vartheta)} \overset{\text{Prop. 3.22}}{=} \nu_e \cdot \frac{f_{e,i}^+(a)}{f_e^+(a)}$$

for almost all $\theta \in [T_e(a), T_e(c))$ and

$$f_{e,i}^-(\theta) \overset{(3.13)}{=} f_e^-(\theta) \cdot \frac{f_{e,i}^+(\vartheta)}{f_e^+(\vartheta)} \overset{\text{Prop. 3.22}}{=} f_e^+(a) \cdot \frac{f_{e,i}^+(a)}{f_e^+(a)} = f_{e,i}^+(a)$$

for almost all $\theta \in [T_e(c), T_e(b))$.

3. **Case: $f_e^+(a) > 0$ and $Q_e(a) = 0$ or $f_e^+(a) \geq \nu_e$:** Again, using (3.13) and Proposition 3.22 gives us

$$\begin{aligned} f_{e,i}^-(\theta) &\overset{(3.13)}{=} f_e^-(\theta) \cdot \frac{f_{e,i}^+(\vartheta)}{f_e^+(\vartheta)} \overset{\text{Prop. 3.22}}{=} \min\{\, f_e^+(a), \nu_e \,\} \cdot \frac{f_{e,i}^+(a)}{f_e^+(a)} \\ &= \min\left\{ f_{e,i}^+(a), \frac{f_{e,i}^+(a)\nu_e}{f_e^+(a)} \right\} = \min\left\{ \frac{f_{e,i}^+(a)\nu_e}{\nu_e}, \frac{f_{e,i}^+(a)\nu_e}{f_e^+(a)} \right\} \\ &= \frac{f_{e,i}^+(a)\nu_e}{\max\{\, \nu_e, f_e^+(a) \,\}} \end{aligned}$$

for almost all $\theta \in [T_e(a), T_e(b))$. □

Next, we show two continuity properties of Vickrey edge flows: Namely, that the mapping from edge flow to current travel time is weak-strong continuous and the mapping from inflow to outflow rates is weak-strong continuous, provided that the inflow rates are locally p-integrable (instead of just locally integrable):

Corollary 3.45. *For any $p > 1$ and any (finite!) $\xi \in \mathbb{R}_{\geq 0}$ the mapping*

$$L^p([0, \xi], \mathbb{R}_{\geq 0})^I \times L^p([0, \xi + \tau_e], \mathbb{R}_{\geq 0})^I \to C([0, \xi]), (f_{e,\cdot}^+, f_{e,\cdot}^-) \mapsto C_e^f$$

is sequentially weak-strong continuous, i.e. it maps weakly convergent sequences to uniformly converging sequences. Here, we use the natural extension of the definition of the current travel time function C_e^f to "flows" which are only defined up to time ξ.

Proof. Let $(f_{e,\cdot}^{(n),+}, f_{e,\cdot}^{(n),-})_{n \in \mathbb{N}^*}$ be a weakly convergent sequence in $L^p([0, \xi], \mathbb{R}_{\geq 0})^I \times L^p([0, \xi + \tau_e], \mathbb{R}_{\geq 0})^I$. According to Proposition 2.50 the sequence $(F_{e,\bullet}^{(n),+}, F_{e,\bullet}^{(n),-})_{n\in\mathbb{N}^*}$ then converges uniformly. The corresponding sequence of current travel times $(C_e^{f^{(n)}})_{n\in\mathbb{N}^*}$ then converges uniformly as well as linear combination of uniformly convergent sequences. □

Corollary 3.46. *For any $p > 1$ and any $\xi \in \mathbb{R}_{\geq 0}$ the mapping*

$$\begin{aligned}\Phi_e^\xi : L^p([0,\xi],\mathbb{R}_{\geq 0})^I &\to L^p([0,\xi+\tau_e],\mathbb{R}_{\geq 0})^I, \\ f_{e,\bullet}^+ &\mapsto f_{e,\bullet}^- \quad \text{s.th. } (f_{e,\bullet}^+, f_{e,\bullet}^-) \text{ extended by } 0 \\ &\qquad\qquad \text{is a Vickrey edge flow until } \xi\end{aligned}$$

is sequentially weak-weak continuous, i.e. it maps weakly convergent sequences to weakly convergent sequences.

Proof. We first note, that this mapping is well-defined by Corollary 3.43, i.e. there exists a unique family of outflow rates for any given family of inflow rates. Moreover, these outflow rates are p-integrable as they are measurable and bounded (cf. Proposition 2.14).

Now, in order to show that Φ_e^ξ is sequentially weak-weak continuous, take any weakly convergent sequence $(f_{e,\bullet}^{(n),+})_{n\in\mathbb{N}^*} \in (L^p([0,\xi],\mathbb{R}_{\geq 0})^I)^{\mathbb{N}^*}$ with limit point $f_{e,\bullet}^+ \in L^p([0,\xi],\mathbb{R}_{\geq 0})^I$ and let

$$f_{e,\bullet}^{(n),-} := \Phi_e^\xi(f_{e,\bullet}^{(n),+}) \in \left\{ f' \in L^p([0,\xi+\tau_e],\mathbb{R}_{\geq 0})^I \;\middle|\; \sum_{i\in I} f_i' \leq_{\text{a.e.}} \nu_e \right\}$$

be the corresponding sequence of outflow rates. We have to show that this sequence of outflow rates then weakly converges to $f_{e,\bullet}^- := \Phi_e^\xi(f_{e,\bullet}^+)$. Since the set on the right is clearly convex, bounded and strongly closed, it is also sequentially weakly compact (by Propositions 2.41, 2.35 and 2.40). Thus, we can apply Propositions 2.25, i.e. it suffices to show that any weakly convergent subsequence of the given sequence of outflow rates converges to $f_{e,\bullet}^-$.

So, let $(f_{e,\bullet}^{(n_k),-})_{n_k\in\mathbb{N}^*}$ be such a subsequence and $f_{e,\bullet}^{*,-}$ its limit point. We then have

$$\left(f_{e,\bullet}^{(n_k),+}, (f_{e,\bullet}^{(n_k),-}\right) \xrightarrow[k\to\infty]{w} \left(f_{e,\bullet}^+, f_{e,\bullet}^{*,-}\right).$$

We will now show that $\left(f_{e,\bullet}^+, f_{e,\bullet}^{*,-}\right)$ is a Vickrey flow until ξ which then implies $f_{e,\bullet}^{*,-} = \Phi_e^\xi(f_{e,\bullet}^+) = f_{e,\bullet}^-$.

We first note that Propositions 2.37 ensure that $\left(f_{e,\cdot}^{+}, f_{e,\cdot}^{*,-}\right)$ respects the edge capacity until ξ since all $\left(f_{e,\cdot}^{(n_k),+}, (f_{e,\cdot}^{(n_k),-}\right)$ do so as well. By Propositions 2.50 the sequences $F_{e,\cdot}^{(n_k),+}$ and $F_{e,\cdot}^{(n_k),-}$ converge uniformly to $F_{e,\cdot}^{+}$ and $F_{e,\cdot}^{*,-}$ on $[0, \xi]$ and $[0, \xi + \tau_e]$, respectively. Hence, weak flow conservation until ξ for the elements of the sequence implies weak flow conservation until ξ for the limit point.

Furthermore, the sequence $T_e^{f^{(n_k)}}$ also converges uniformly to $T_e^{f^*}$ on $[0, \xi]$ by Corollary 3.45. This, in turn, implies that $F^{(n_k),-} \circ T_e^{f^{(n_k)}}$ converges pointwise to $F^{*,-} \circ T_e^{f^*}$ by Proposition 2.38. Thus, the fact that (3.14) holds for all elements of the sequence guarantees that it holds for the limit point as well.

All together, Proposition 3.40b) now implies that $\left(f_{e,\cdot}^{+}, f_{e,\cdot}^{*,-}\right)$ is a Vickrey flow until ξ and, therefore, $f_{e,\cdot}^{*,-} =_{\text{a.e.}} f_{e,\cdot}^{-}$ by Corollary 3.43. Thus, we have

$$f_{e,\cdot}^{(n),-} \xrightarrow{w} f_{e,\cdot}^{-}$$

which concludes the proof. □

Finally, we note that while it is also possible to show multi-commodity versions of Corollaries 3.23 and 3.24 (i.e. monotonicity and no idling property), those will be much weaker statement than their anonymous counter parts. The following example demonstrates why this is the case with regards to monotonicity:

Example 3.47. Let e be an edge with free flow travel time 1 and capacity 2 and consider the 2-commodity Vickrey flows given by the following two sets of edge inflow rates:

- f_e is given by $f_{e,1}^{+} = f_{e,2}^{+} = \mathbb{1}_{[0,2]}$
- g_e is given by $g_{e,1}^{+} = \mathbb{1}_{[0,2]}$ and $g_{e,2}^{+} = 2 \cdot \mathbb{1}_{[1,2]}$.

The cumulative flows then clearly satisfy $F_{e,1}^{+}(\theta) \leq G_{e,1}^{+}(\theta)$ and $F_{e,2}^{+}(\theta) \leq G_{e,2}^{+}(\theta)$ for all $\theta \in \mathbb{R}_{\geq 0}$. Nevertheless, we also have $F_{e,1}^{-}(3) = 2 > 1 + \frac{2}{3} = G_{e,1}^{-}(3)$ (cf. Fig. 3.5).

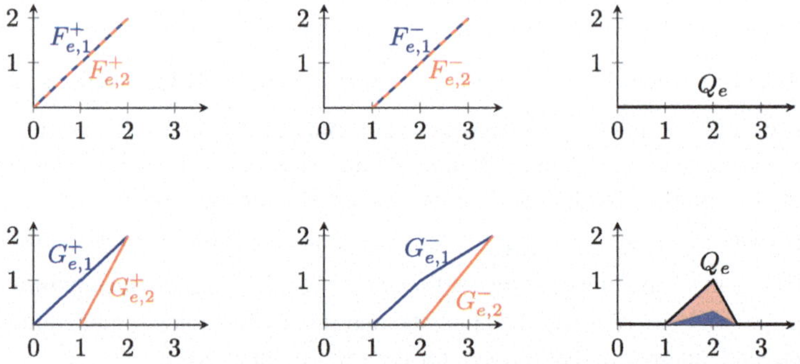

Fig. 3.5 A counter-example to a naive monotonicity property for multi-commodity flows. The blue commodity has the same (cumulative) inflow both in f_e and g_e whereas the red commodity has smaller cumulative inflow in g_e compared to f_e. Nevertheless, for certain times the cumulative outflow for the blue commodity is strictly smaller in g_e compared to f_e

3.1.3 Network Flows

We are now going to extend the concept of dynamic flows from individual edges to whole networks.

Definition 3.48. A *network* $\mathcal{N} = (G, (\tau_e), (\nu_e), I, (u_{v,i}), (T_i))$ consists of

- a directed graph $G = (V, E)$,
- a non-negative *free flow travel time* $\tau_e \in \mathbb{R}_{\geq 0}$ for every edge $e \in E$,
- a positive *capacity* $\nu_e \in \mathbb{R}_{>0}$ for every edge $e \in E$,
- a finite set of *commodities* I,
- a *network inflow rate* $u_{v,i} \in L^1_{\text{loc}}(\mathbb{R}_{\geq 0}, \mathbb{R}_{\geq 0})$ for every commodity $i \in I$ and every node $v \in V$ and
- a non-empty set of *sink nodes* $T_i \subseteq V$ for every commodity $i \in I$.

As for the flow rates we also define the *cumulative network inflow* by

$$U_{v,i} : \mathbb{R}_{\geq 0} \to \mathbb{R}_{\geq 0}, \theta \mapsto \int_0^\theta u_{v,i}(\zeta)\mathrm{d}\zeta.$$

Furthermore, we denote by

$$\hat{\theta}_i := \sup \{ \theta \in \mathbb{R}_{\geq 0} \mid U_{v,i} \text{ is strictly increasing at } \theta \text{ for some } v \in V \}$$

the last time at which additional particles of commodity i enter the network and by $\hat{\theta} := \max \{ \hat{\theta}_i \mid i \in I \}$ the last time where any particles enter the network. We say that commodity i (the network) has finitely *lasting network inflow rates* if $\hat{\theta}_i < \infty$ (if $\hat{\theta} < \infty$).

We call a node $v \in V$ a *dead-end node* of commodity i if there exists no v, T_i-path and denote by $V_i^\dagger \subseteq V$ the set of all dead-end nodes of commodity i. We call a node $v \in V$ a *source node* for commodity $i \in I$ if $u_{v,i} \neq_{\text{a.e.}} 0$ and denote by $S_i \subseteq V$ the set of all source nodes of commodity i.

Finally, we say that $\mathcal{N}$ is *feasible* if for every commodity $i \in I$ and every source node $s \in S_i$ there exists some sink $t \in T_i$ such that there is some s, t-path in G or, equivalently, if $V_i^\dagger \cap S_i = \emptyset$ for all $i \in I$.

Definition 3.49. A *(dynamic) flow* in a network $\mathcal{N}$ is a tuple $f = (f_e)_{e \in E} \in L^1_{\text{loc}}(\mathbb{R}_{\geq 0}, \mathbb{R}_{\geq 0})^{E \times I} \times L^1_{\text{loc}}(\mathbb{R}_{\geq 0}, \mathbb{R}_{\geq 0})^{E \times I}$. We define $\mathcal{F}(\mathcal{N})$ as the set of all dynamic flows in $\mathcal{N}$.

In other words, a dynamic flow f in a network consists of an edge flow $(f^+_{e,i}, f^-_{e,i})_{i \in I}$ for every edge e of that network. Thus, all the properties previously defined for edge flows can also be applied to dynamic flows (both separately to individual edges of the network or to all edges at once). Additionally, we can now also define flow conservation properties between the edge, i.e. at the nodes:

Definition 3.50. For any dynamic flow f we denote by

$$B_{v,i}(\theta) := U_{v,i}(\theta) + \sum_{e \in \delta^-(v)} F^-_{e,i}(\theta) - \sum_{e \in \delta^+(v)} F^+_{e,i}(\theta)$$

the *flow balance of commodity i at node v at time θ* and by

$$F_i^\Delta(\theta) := \sum_{e \in E} F^\Delta_{e,i}(\theta) + \sum_{v \in V \setminus T_i} B_{v,i}(\theta)$$

the *network load of commodity i at time θ*. Finally, $F^\Delta(\theta) := \sum_{i \in I} F_i^\Delta(\theta)$ denotes the *network load at time θ*.

Definition 3.51. We say that a flow f satisfies *weak flow conservation for commodity $i \in I$ at node $v \in V$ until $\xi \in \tilde{\mathbb{R}}_{\geq 0}$* if

$$B_{v,i}(\theta) \geq 0 \text{ for all } \theta < \xi.$$

It satisfies *strong flow conservation for commodity $i \in I$ at a node $v \in V \setminus T_i$ until ξ* if

$$B_{v,i}(\theta) = 0 \text{ for all } \theta < \xi$$

and at a sink node $t \in T_i$ if $B_{v,i}$ is non-decreasing on $[0, \xi)$.

We say that a flow satisfies *weak/strong flow conservation at all nodes until ξ* if it satisfies weak/strong flow conservation until ξ for all commodities at all nodes.

Intuitive explanation: Weak flow conservation allows for temporary storage of flow volume at nodes whereas strong flow conservation requires all flow arriving a node to immediately enter another edge leaving that node (or leave the network entirely if it is a sink node). Thus, weak flow conservation (at nodes) is used in dynamic flows whenever one wants to model congestion on nodes (like for example in [FF58] or [Sku09]) whereas strong flow conservation at nodes is used when congestion is entirely modelled on edges (like in this thesis).

Under strong flow conservation $B_{t,i}(\theta)$ denotes the amount of flow of commodity i which has already arrived at some sink $t \in T_i$ by time θ.

Proposition 3.52. *A flow f satisfies strong flow conservation for commodity i at node v until ξ if and only if it satisfies*

$$u_{i,v}(\theta) + \sum_{e \in \delta^-(v)} f^-_{e,i}(\theta) - \sum_{e \in \delta^+(v)} f^+_{e,i}(\theta) \begin{cases} = 0, & if\ v \in V \setminus T_i \\ \geq 0, & if\ v \in T_i \end{cases} \quad for\ almost\ all\ \theta < \xi. \tag{3.20}$$

Proof. Define by

$$b_{v,i}(\theta) := u_{i,v}(\theta) + \sum_{e\in\delta^-(v)} f^-_{e,i}(\theta) - \sum_{e\in\delta^+(v)} f^+_{e,i}(\theta)$$

the *netto node-inflow rate*. Then we get directly from the definition that $b_{v,i} =_{\text{a.e.}} \partial B_{v,i}$ and $B_{v,i} = \int_0 b_{v,i}(\zeta)\mathrm{d}\zeta$. The equivalence then follows immediately by using the connection between locally integrable and absolutely continuous functions (Proposition 2.49). □

Proposition 3.53. *Let f be a dynamic flow. Then for any subset of nodes $W \subseteq V$ and any time $\theta \in \mathbb{R}_{\geq 0}$ we have*

$$\sum_{e\in E[W]} F^{\Delta}_{e,i}(\theta) = \sum_{v\in W} U_{v,i}(\theta) + \sum_{e\in\delta^-(W)} F^-_{e,i}(\theta) - \sum_{e\in\delta^+(W)} F^+_{e,i}(\theta) - \sum_{v\in W} B_{v,i}(\theta).$$

In particular, if f satisfies strong flow conservation at all nodes until θ, then we have

$$F^{\Delta}_i(\theta) = U_i(\theta) - Z_i(\theta),$$

where $U_i(\theta) := \sum_{v\in V} U_{v,i}(\theta)$ and $Z_i(\theta) := \sum_{t\in T_i} B_{t,i}(\theta)$ denote commodity i's total cumulative network in- and outflow, respectively.

Proof. We can show this via a direct computation:

$$\begin{aligned}
\sum_{e\in E[W]} F^{\Delta}_{e,i}(\theta) &= \sum_{e\in E[W]} \left(F^+_{e,i}(\theta) - F^-_{e,i}(\theta)\right) \\
&= \sum_{v\in W}\sum_{e\in\delta^+(v)} F^+_{e,i}(\theta) - \sum_{e\in\delta^+(W)} F^+_{e,i}(\theta) - \sum_{v\in W}\sum_{e\in\delta^-(v)} F^-_{e,i}(\theta) + \sum_{e\in\delta^-(W)} F^-_{e,i}(\theta) \\
&= \sum_{v\in W}\Big(\sum_{e\in\delta^+(v)} F^+_{e,i}(\theta) - \sum_{e\in\delta^-(v)} F^-_{e,i}(\theta)\Big) + \sum_{e\in\delta^-(W)} F^-_{e,i}(\theta) - \sum_{e\in\delta^+(W)} F^+_{e,i}(\theta) \\
&= \sum_{v\in W} \left(U_{v,i}(\theta) - B_{v,i}(\theta)\right) + \sum_{e\in\delta^-(W)} F^-_{e,i}(\theta) - \sum_{e\in\delta^+(W)} F^+_{e,i}(\theta).
\end{aligned}$$

The second part of the proposition now follows directly from the first by choosing $W = V$. □

Corollary 3.54. *Let f be a dynamic flow and $i \in I$ some commodity such that f satisfies strong flow conservation at all sink nodes of commodity i. Then, the network load of that commodity is non-increasing after $\hat{\theta}_i$, i.e.*

$$F_i^{\Delta}(a) \geq F_i^{\Delta}(b) for\ all\ \hat{\theta}_i \leq a \leq b.$$

Proof. This follows directly from Proposition 3.53 since all $U_{v,i}$ are constant after $\hat{\theta}$ while the functions $B_{t,i}$ are non-decreasing for $t \in T_i$. □

Definition 3.55. We call a flow f a *feasible flow until* $\xi \in \tilde{\mathbb{R}}_{\geq 0}$ if it satisfies weak flow conservation for all commodities at all nodes and edges and respects the capacities of all edges until ξ.

We call f a *Vickrey flow until* $\xi \in \tilde{\mathbb{R}}_{\geq 0}$ if it satisfies strong flow conservation at all nodes until ξ and all queues operate at capacity and fair until ξ (Fig. 3.6).

A helpful property of Vickrey flows is that, as long as the network inflow rates are (essentially) bounded, the cumulative in- and outflows as well as queue length and, most importantly, the current travel times are not just absolutely continuous but even Lipschitz-continuous:

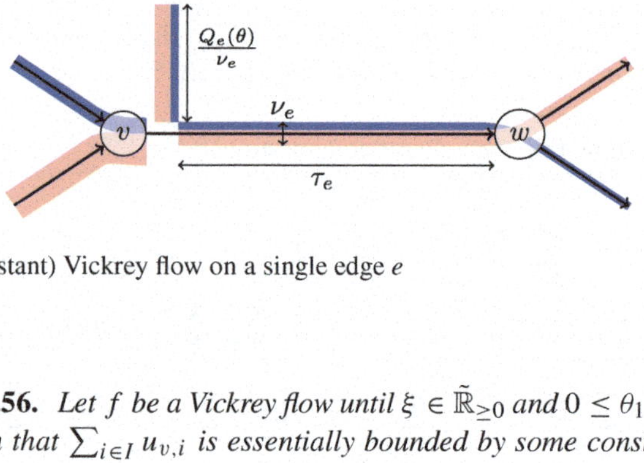

Fig. 3.6 A (constant) Vickrey flow on a single edge e

Proposition 3.56. *Let f be a Vickrey flow until $\xi \in \tilde{\mathbb{R}}_{\geq 0}$ and $0 \leq \theta_1 \leq \theta_2 \leq \xi$ any two times such that $\sum_{i \in I} u_{v,i}$ is essentially bounded by some constant M during $[\theta_1, \theta_2)$. Then for any edge $e = vw \in E$ we have*

$$C_e(\theta_1) - (\theta_2 - \theta_1) \leq C_e(\theta_2) \leq C_e(\theta_1) + (\theta_2 - \theta_1) \cdot \frac{\left(\sum_{e' \in \delta^-(v)} \nu_{e'} + M\right)}{\nu_e}.$$

Proof. We show the following analogous bound for the queue length function from which the proposition then follows immediately:

Claim 3.4. $Q_e(\theta_1) - (\theta_2 - \theta_1)\nu_e \leq Q_e(\theta_2) \leq Q_e(\theta_1) + (\theta_2 - \theta_1) \cdot \Big(\sum_{e' \in \delta^-(v)} \nu_{e'} + M \Big)$.

Proof. For the first inequality we only need the fact that f respects the capacity on edge e:

$$Q_e(\theta_2) = Q_e(\theta_1) + \int_{\theta_1}^{\theta_2} \underbrace{f_e^+(\vartheta)\,\mathrm{d}\vartheta}_{\geq 0} - \int_{\theta_1}^{\theta_2} \underbrace{f_e^-(\vartheta + \tau_e)\,\mathrm{d}\vartheta}_{\leq \nu_e} \geq Q_e(\theta_1) - \nu_e(\theta_2 - \theta_1).$$

For the second inequality we additionally use Proposition 3.52, i.e. strong flow conservation at node v as well as the essential bound M on the network inflow rate at this node:

$$\begin{aligned} Q_e(\theta_2) &\leq Q_e(\theta_1) + \int_{\theta_1}^{\theta_2} f_e^+(\vartheta)\mathrm{d}\vartheta \\ &\overset{(3.20)}{\leq} Q_e(\theta_1) + \sum_{e' \in \delta^-(v)} \int_{\theta_1}^{\theta_2} \underbrace{f_{e'}^-(\vartheta)\,\mathrm{d}\vartheta}_{\leq \nu_{e'}} + \underbrace{\int_{\theta_1}^{\theta_2} \sum_{i \in I} u_{v,i}(\vartheta)\,\mathrm{d}\vartheta}_{\leq M} \\ &\leq Q_e(\theta_1) + (\theta_2 - \theta_1) \cdot \Big(\sum_{e' \in \delta^-(v)} \nu_{e'} + M \Big) \end{aligned}$$

■

Applying the definition $C_e(\theta) = \frac{Q_e(\theta)}{\nu_e} + \tau_e$ now directly yields the desired bounds. □

We conclude our discussion of dynamic network flows by introducing one additional notion: That of restricting a dynamic flow to a subnetwork.

Definition 3.57. Let $\mathcal{N} = (G, \tau, \nu, I, u, T)$ and $\mathcal{N}' = (G', \tau', \nu', I', u', T')$ be two networks. We call $\mathcal{N}'$ a *subnetwork* of $\mathcal{N}$ if

- G' is a subgraph of G, i.e. $G' \subseteq G$,
- the free flow travel times and capacities coincide on the common edges, i.e. $\tau' = \tau|_{E[G']}$ and $\nu' = \nu|_{E[G']}$ and
- the two networks have the same set of commodities, i.e. $I' = I$.

Definition 3.58. Let $\mathcal{N} = (G, \tau, \nu, I, u, T)$ be a network and $\mathcal{N}' = (G', \tau', \nu', I', u', T')$ a subnetwork. Then we define the *restriction mapping*

$$|_{\mathcal{N}'} : \mathcal{F}(\mathcal{N}) \to \mathcal{F}(\mathcal{N}'),\ f \mapsto f|_{\{\pm\}\times E[G']\times I}$$

which restricts any given flow in $\mathcal{N}$ to a flow in $\mathcal{N}'$.

A first simple application of this notion is the following proposition which essentially states that when considering Vickrey flows we may assume without loss of generality that every commodity has exactly one sink node:

Proposition 3.59. *Let $\mathcal{N}$ be a network with $S_i \cap T_i = \emptyset$ for all commodities $i \in I$. Then there exists a larger network $\tilde{\mathcal{N}}$ such that*

- *$\mathcal{N}$ is a subnetwork of $\tilde{\mathcal{N}}$,*
- *in $\tilde{\mathcal{N}}$ every commodity has exactly one sink node and*
- *the restriction mapping $|_{\mathcal{N}}$ is a one to one correspondence between Vickrey flows in $\tilde{\mathcal{N}}$ and in $\mathcal{N}$.*

Proof. We construct $\tilde{\mathcal{N}}$ as follows: For every commodity $i \in I$ we add one new node t_i and edges vt_i for every $v \in T_i$ with free flow travel time 1 and capacity $\tilde{\nu}_{vt_i} := \sum_{e\in\delta^-(v)} \nu_e$. We then define $\tilde{T}_i := \{\, t_i \,\}$ and $\tilde{u}_{t_i,i'} := 0$ for all $i' \in I$ while keeping the network inflow rates the same for all other nodes.

It is now easy to see that $|_{\mathcal{N}}$ maps Vickrey flows to Vickrey flows. Moreover, it is one-to-one since every Vickrey flow in $\mathcal{N}$ can be uniquely extended to a Vickrey flow in $\tilde{\mathcal{N}}$ by sending all excess flow at nodes $v \in T_i$ onto the direct edge vt_i. □

3.2 Behavioural Model

In this section we introduce and discuss the behavioural aspects of our model. We start by defining what dynamic flows we consider to be equilibrium flows in our current information setting using a node label-based approach similar to the one used in [KS11, CCL15] for the full information setting. We then define two quality measures that we will later use to judge the quality of equilibrium flows in our model.

3.2.1 Instantaneous Dynamic Equilibria

As already discussed in the introduction (Chap. 1) we want to study a model wherein individual agents/particles only have access to information on the *current* state of the network (as opposed to the full trajectory of the future flow evolution like in the full information model considered e.g. in [KS11, CCL15]). We then assume that at every time every agent chooses a route with (seemingly) shortest travel time based on this current information. More precisely, we assume that whenever an agent arrives at a node they get the expected current travel times of all edges, compute a shortest path from their current location to one of their sink nodes (based on this information) and then enter the first edge of such a path. A Vickrey flow wherein each particle only travels along such currently shortest paths is then the exact equilibrium concept we will consider from here on out.

In order to formalise this equilibrium concept we will use time dependent node labels denoting for every node, commodity and time the current shortest distance from this node to a sink of that commodity at that time.

Definition 3.60. Given a dynamic flow f we define for any commodity $i \in I$ time dependent node labels denoting for any node $v \in V$ the *(expected) current distance* to the closest sink, i.e.

$$L_{v,i} : \mathbb{R}_{\geq 0} \to \tilde{\mathbb{R}}_{\geq 0}, \theta \mapsto L_{v,i}(\theta) := \inf \left\{ C_p(\theta) \mid p \text{ a } v,t\text{-path for some } t \in T_i \right\}, \tag{3.21}$$

where $C_p(\theta) := \sum_{e \in p} C_e(\theta)$ denotes the current expected travel time along a path p.

We say that a v, T_i-path p is *active* for commodity i at time θ if $C_p(\theta) = L_{v,i}(\theta)$ and denote the set of all active v, T_i-path at time θ by

$$P_{v,i}(\theta) := \{ p \text{ a } v, T_i\text{-path} \mid p \text{ active for } i \text{ at } \theta \}.$$

We say that an edge $e = vw \in E$ is *active* for commodity i at time θ whenever

$$L_{v,i}(\theta) \geq C_e(\theta) + L_{w,i}(\theta)$$

and denote by $E_i(\theta) := \{e \in E \mid e \text{ active for } i \text{ at } \theta\}$ the set of active edges of commodity i at time θ.

In single commodity networks we will drop all indices i.

Remark 3.61. For any fixed time θ and commodity i the vector $(L_{v,i}(\theta))_v$ is a vector of distance labels with respect to the edge costs $(C_e(\theta))_e$. Thus, these labels always satisfy properties a) to e) from Proposition 2.67. If the flow satisfies weak flow conservation on all edges at time θ, then the node labels at that time also satisfy f) to m).

If, additionally, every cycle has a strictly positive free flow time, then the node labels also satisfy properties n) and o) from that proposition. In particular, for such flows Bellman's equations (2.1) can be used as an alternative (recursive) definition of the current distances (cf. e.g. [GHS20, GHKM23]):

$$L_{v,i}(\theta) = \begin{cases} 0, & \text{if } v \in T_i \\ \inf\{\, C_e(\theta) + L_{w,i}(\theta) \mid e = vw \in \delta^+(v) \,\}, & \text{else} \end{cases} \tag{3.22}$$

for all $v \in V, i \in I, \theta \in \mathbb{R}_{\geq 0}$.

In addition to all these properties inherited from Proposition 2.67 we will also make use of two continuity properties of the current distances: First, every such node label is continuous as a function in time and, second, the mapping from current travel times to current distances is continuous as mapping between two function spaces. Together with Corollaries 3.46 and 3.45 the latter then also implies that for Vickrey flows the mapping from inflow rates to current distances is weak-strong continuous which will be an important ingredient for our existence proof in Sect. 4.2.

Proposition 3.62. *For every commodity $i \in I$ and any node $v \in V \setminus V_i^\dagger$ the function $L_{v,i} : \mathbb{R}_{\geq 0} \to \mathbb{R}$ is absolutely continuous.*

Proof. For a node $v \notin V_i^\dagger$ the node label is always finite by Proposition 2.67a). The node label functions are then absolutely continuous as pointwise minima over a finite set of sums of the absolutely continuous functions C_e. □

Proposition 3.63. *For every commodity $i \in I$ and node $v \in V \setminus V_i^\dagger$ the mapping*

$$C(\mathbb{R}_{\geq 0})^E \to C(\mathbb{R}_{\geq 0}), C_\bullet \mapsto L_{v,i}$$

is sequentially (strong-strong) continuous, i.e. it maps uniformly convergent sequences to uniformly convergent sequences.

Proof. Given a uniformly convergent sequence $(C_{\cdot}^{(n)})_n$ the sequences $(C_p^{(n)})_n$ converge uniformly as well as linear combinations of uniformly convergent sequences. This then implies the uniform convergence of the sequence $(L_{v,i}^{(n)})_n$ as minimum of uniformly convergent sequences. □

These node labels now allow us to formally define our main object of study for the rest of this thesis: (Vickrey) flows in which particles always choose an optimal route according to the current travel times. A flow in which (almost) all particles behave this way is then an equilibrium in our setting:

Definition 3.64. A Vickrey flow f until ξ is an *instantaneous dynamic equilibrium (IDE) until* ξ if it satisfies

$$f_{e,i}^{+}(\theta) > 0 \implies e \in E_i(\theta) \text{ for almost all } \theta < \xi \tag{3.23}$$

and for all $i \in I$ and $e \in E$.

With this definition we can now check that the flow seen in Example 1.1 in the introduction is in fact an IDE (for suitably chosen network inflow rates and edge capacities):

Example 3.65. Consider the 2-commodity network depicted in the top left of Fig. 3.7 with edge labels denoting free flow travel time and capacity (in that order), a network inflow rate of 2 at node s_1 during $[0, 1]$ for commodity 1 (blue) and a network inflow rate of 4 at node s_2 during $[1, 2]$ for commodity 2 (red). The flows depicted in Figs. 3.7 and 3.8 are then two possible IDE in this network.

The flow in Fig. 3.8 is also a full information equilibrium flow (and, in fact, also a system optimum flow with respect to both the quality measures we will define in Subsection 3.2.2). Note that the latter two statements would still be true if we were to slightly increase the free flow travel time of the direct edge s_1t while the IDE depicted in Fig. 3.8 would not be an IDE then anymore.

As we can already see in this example, IDE are, not very surprisingly, not necessarily optimal in hindsight – neither from a system wide perspective nor from the point of view an individual particles (as it would be the case in a full information equilibrium). We will see exactly how much worse an IDE can be compared to an optimal flow in Sect. 6.3. In fact, it will turn out that it is possible in certain cases

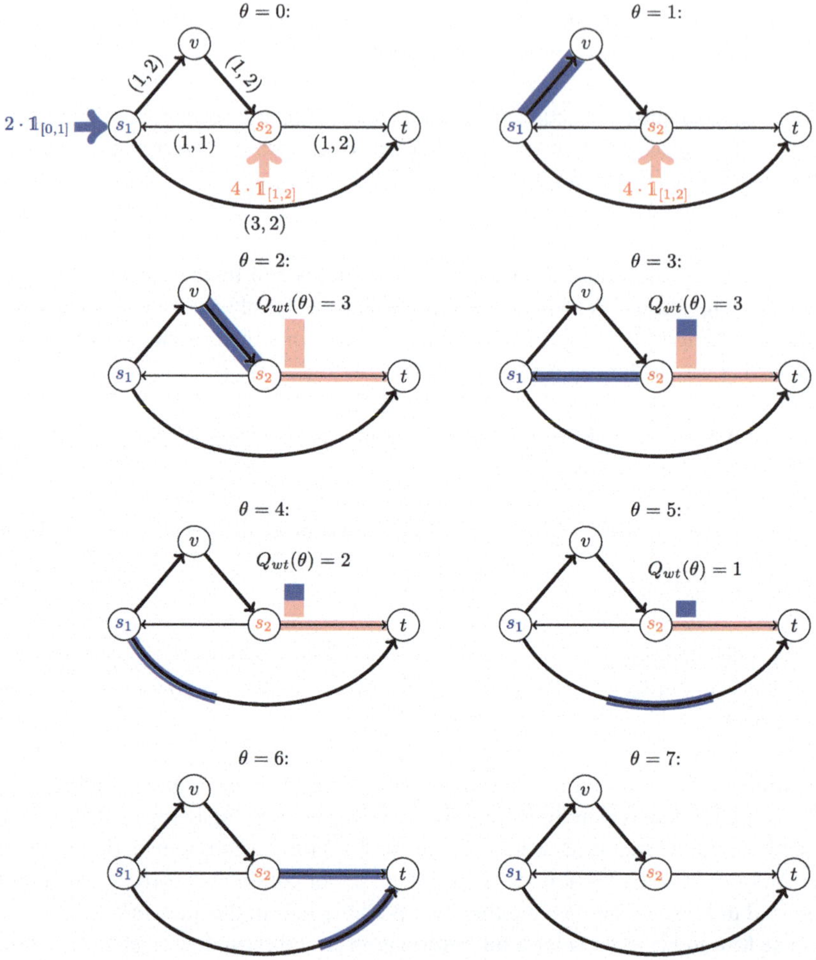

Fig. 3.7 The 2-commodity network with a possible IDE already discussed in Example 1.1. In the first image (top left) the edge labels denote the free flow travel time and the capacity of each respective edge (in this order). Particles of the blue commodity enter the network at node s_1 and have node t as their destination. Particles of the red commodity enter the network at node s_2 and also have node t as their destination

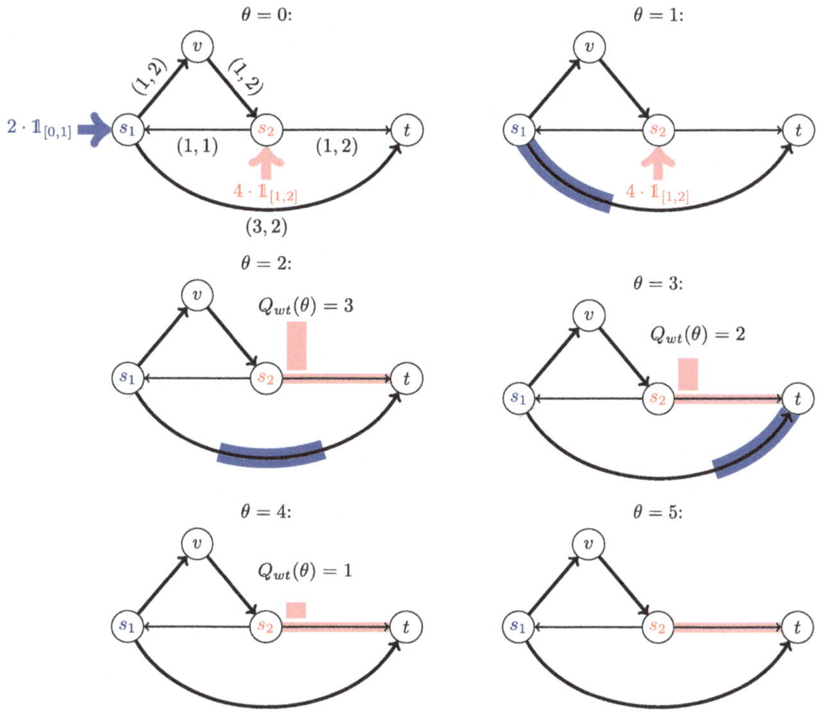

Fig. 3.8 An alternative IDE for the 2-commodity network discussed in Example 1.1

for particles to never reach a sink and travel around in cycles forever (see Theorem 6.18).

Nevertheless, we would at least expect that in an IDE particles never get completely stuck, i.e. arrive at some non-sink node from which there is no way left to go to. This is particularly important for constructing IDE by repeatedly extending IDE up to certain point which we will do in the following chapter. Ideally, we would like to say that in an IDE no particle ever arrives at a dead-end node. This is, in fact true, for feasible networks with strictly positive free flow travel times. For networks with cycles of free flow travel time zero, however, our model technically allows for "ghost particles" to appear at some node in such a cycle, travel around that cycle and then disappear again at their starting node (at the same time at which they appeared to ensure flow conservation at nodes!). This, in particular, is even possible at dead-end nodes. Thus, we have to content ourselves with the following, more technical version of our desired statement which will still be enough to ensure that we will never have to deal with any "stuck" particles when constructing IDE.

Proposition 3.66. *Let f be an IDE until ξ in a feasible network $\mathcal{N}$ and $e = vw \in E$ an edge such that $(f^+_{e,\bullet}, f^-_{e,\bullet})$ is a Vickrey edge flow until $\xi + \frac{Q_e(\xi)}{\nu_e}$.*

If $v \notin V_i^\dagger$ and $w \in V_i^\dagger$ then

$$f^-_{e,i}(\theta) = 0 \; for\; almost\; all\; \theta < T_e(\xi). \tag{3.24}$$

If $v, w \in V_i^\dagger$ then

$$C_e(\theta) > 0 \implies f^-_{e,i}(\theta) = 0 \; for\; almost\; all\; \theta < T_e(\xi). \tag{3.25}$$

Proof. If $v \notin V_i^\dagger$ and $w \in V_i^\dagger$, then $e = vw$ can never be active for commodity i as we always have $L_{v,i}(\theta) < \infty$ and $L_{w,i}(\theta) = \infty$ by Proposition 2.67a). Thus, in an IDE until ξ no flow of commodity i can enter edge e until that time which, in turn, implies

$$F^-_{e,i}(T_e(\xi)) \overset{(3.14)}{=} F^+_{e,i}(\xi)) = 0$$

and, hence, $f^-_{e,i}(\theta) = 0$ for almost all $\theta < T_e(\xi)$.

For edges between dead-end nodes we use Proposition 3.53 to show that for any $\theta < \xi$ we have

$$\begin{aligned} 0 \leq \sum_{e \in E[V_i^\dagger]} F^\Delta_{e,i}(\theta) &\overset{\text{Prop. 3.53}}{=} \sum_{v \in V_i^\dagger} U_{v,i}(\theta) + \sum_{e \in \delta^-(V_i^\dagger)} F^-_{e,i}(\theta) - \sum_{e \in \delta^+(V_i^\dagger)} F^+_{e,i}(\theta) - \sum_{v \in V_i^\dagger} B_{v,i}(\theta) \\ &\overset{(*)}{\leq} \sum_{v \in V_i^\dagger} U_{v,i}(\theta) + \sum_{e \in \delta^-(V_i^\dagger)} F^-_{e,i}(\theta) - 0 \overset{(\#)}{=} 0 + \sum_{e \in \delta^-(V_i^\dagger)} F^-_{e,i}(\theta) \overset{(3.24)}{=} 0, \end{aligned}$$

where $(*)$ holds due to weak flow conservation on edges and nodes and (#) because in a feasible network there can be no network inflow at dead-end nodes. Thus, for any edge $e \in E[V_i^\dagger]$ such that $(f^+_{e,\bullet}, f^-_{e,\bullet})$ is a Vickrey edge flow until $\xi + \frac{Q_e(\xi)}{\nu_e}$ we have

$$\int_\theta^{T_e(\theta)} f^-_{e,i}(\zeta)\mathrm{d}\zeta = F^-_{e,i}(T_e(\theta)) - F^-_{e,i}(\theta) \overset{(3.14)}{=} F^+_{e,i}(\theta) - F^-_{e,i}(\theta) = F^\Delta_{e,i}(\theta) = 0$$

for all times $\theta < \xi$ and, therefore, $f^-_{e,i}(\zeta) = 0$ for almost all $\zeta \in [\theta, T_e(\theta)) = [\theta, \theta + C_e(\theta))$. From this, we can now deduce (3.25) by observing that

$$\{\,\theta \in [0, T_e(\xi)) \mid C_e(\theta) > 0\,\} \subseteq \bigcup_{\theta \in [0,\xi)} [\theta, T_e(\theta))$$

and applying Proposition 2.15. □

3.2.2 Quality Measures for Dynamic Flows

In order to measure the quality of dynamic flows (and, in particular, the instantaneous dynamic equilibria) we need to define some objective that the flow are supposed to achieve. In this thesis we will consider two such quality measures: The arrival time of the last particle at its sink and the "sum" of the travel times of all particles:

Definition 3.67. Let f be a dynamic flow and $i \in I$ a commodity with finitely lasting network inflow rates. We then define

- the *makespan of i in f* as

$$\Psi_i(f) := \inf\left\{\,\theta \geq \hat{\theta}_i \;\middle|\; F^\Delta_i(\theta) = 0\,\right\}.$$

- the *total travel times of i in f* is defined as

$$\Xi_i(f) := \sum_{e \in E} \int_0^\infty C_e(\zeta) f^+_{e,i}(\zeta) \mathrm{d}\zeta.$$

If all commodities have finitely lasting network inflow rates, we additionally define by

$$\Psi(f) := \max\{\,\Psi_i(f) \mid i \in I\,\} \text{ and } \Xi(f) := \sum_{i \in I} \Xi_i(f)$$

the *makespan of f* and *total travel time of f*, respectively.

Definition 3.68. We call a terminating feasible flow f an *optimal flow with respect to makespan/total travel time* if it minimizes makespan/total travel time among all feasible flows.

For the case of single-commodity networks it is known that there always exist so called earliest arrival flows (which can even be computed efficiently – see [BS09]). These flows maximize the amount of flow which has already reached a sink for all times simultaneously. They, therefore, also minimize both makespan and total travel time.

Remark 3.69. For single-commodity networks restricting optimal flows to only Vickrey flows (instead of any feasible flow) would not impact the achievable makespan/total travel time as the monotonicity of the edge dynamics of Vickrey flows (see Corollary 3.23) guarantees that waiting at nodes (or even on edges) is never beneficial. This is, however, not true any more for multi-commodity flows as can be seen in the following example.

Example 3.70. To see why waiting can be beneficial in multi-commodity networks we consider the two commodity network depicted in Fig. 3.9. Since there is only one possible path for each commodity, this instance has a unique Vickrey flow. In this flow the edge vw is first traversed by all flow of the red commodity (travelling from s_2 to t_2 and then by the blue commodity (travelling from s_1 to t_1). Thus, the flow particles of the blue commodity have to wait in a queue on edge vw for one time unit. Hence, the makespan of this flow is 7 (the arrival time of the last particle of the blue commodity at its sink t_1).

On the other hand, in a flow with node storage, the flow of the blue commodity can overtake the second half of the red commodity at node v and immediately enter the middle edge vw. Thus, such a flow can achieve a makespan of 6 (the simultaneous arrival time of the last particle of the blue and the red commodity at their respective sinks).

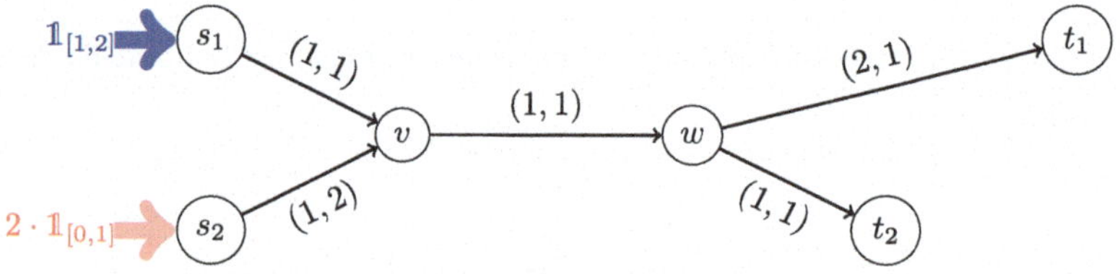

Fig. 3.9 A two-commodity network in which the optimal achievable makespan is different for Vickrey flows and flows which are allowed to wait at nodes

Definition 3.71. Let f be a flow satisfying weak flow conservation at all nodes and edges and strong flow conservation at all sink nodes.

We then say that f *terminates (by time* $\Psi(f)$*)* if $\Psi(f) < \infty$.

The following proposition justifies this definition of terminating flows:

Proposition 3.72. *Let f be a dynamic flow and $i \in I$ some commodity such that f satisfies weak flow conservation for commodity i on all edges and nodes as well as strong flow conservation at all sink nodes of commodity i. Then, after $\Psi_i(f)$ there will never be any flow of commodity i in the network, i.e.*

$$F_{e,i}^{\Delta}(\theta) = 0 \, and \, B_{v,i}(\theta) = 0 \, for all \; e \in E, v \in V \setminus T_i, \theta \geq \Psi(f).$$

Proof. We observe that weak flow conservation on edges implies that the edge load is non-negative as

$$0 \overset{(3.1)}{\leq} F_{e,i}^{+}(\theta) - F_{e,i}^{-}(\theta + \tau_e) \leq F_{e,i}^{+}(\theta) - F_{e,i}^{-}(\theta) = F_{e,i}^{\Delta}(\theta)$$

for all edges $e \in E$. Additionally, Corollary 3.54 implies

$$\sum_{e \in E} F_{e,i}^{\Delta}(\theta) + \sum_{v \in V \setminus T_i} B_{v,i}(\theta) = F_i^{\Delta}(\theta) \leq F_i^{\Delta}(\Psi_i(f)) = 0$$

for all times $\theta \geq \Psi(f)$. As all summands are non-negative, this is only possible if they are all zero. □

For edges with non-zero free flow travel time it is easy to see that the above proposition also implies that the in- and outflow rate (of commodity i) into any such edge is also zero (almost always) after $\Psi_i(f)$. For edges with free flow travel time zero, however, there is again the complicating possibility of "ghost particles" travelling around such cycles (as in the previous subsection in the context of Proposition 3.66) without adding anything to the respective edge load (and, thus, to the network load) of the respective commodity.

Importantly, though, in a Vickrey flow such "ghost particles" may only use edges with current travel time zero and, in particular, cannot build up any queues or contribute to the total travel time of that commodity.

Proposition 3.73. *Let f be a dynamic flow with queues operating fair and at capacity on all edges and $i \in I$ a commodity with finitely lasting inflow rates, strong flow conservation at all sink nodes and weak flow conservation at all other nodes. Then for every time $\theta \geq \Psi_i(f)$ we have*

$$f^+_{e,i}(\zeta) = f^-_{e,i}(\zeta) = 0 \text{ for almost all } \zeta \in [\theta, T_e(\theta)].$$

In particular, we have $f^+_{e,i}(\theta) = 0$ for almost all $\theta \geq \Psi_i(f)$ with $C_e(\theta) > 0$.

Proof. Using Proposition 3.72 and Proposition 3.40b) we have

$$0 = F^{\Delta}_{e,i}(\theta) = F^+_{e,i}(\theta) - F^-_{e,i}(\theta) \overset{(3.14)}{=} F^-_{e,i}(T_e(\theta)) - F^-_{e,i}(\theta) = \int_{\theta}^{T_e(\theta)} f^-_{e,i}(\zeta)\mathrm{d}\zeta$$

as well as

$$\begin{aligned} 0 = F^{\Delta}_{e,i}(T_e(\theta)) &= F^+_{e,i}(T_e(\theta)) - F^-_{e,i}(T_e(\theta)) \\ &\overset{(3.14)}{=} F^+_{e,i}(T_e(\theta)) - F^+_{e,i}(\theta) = \int_{\theta}^{T_e(\theta)} f^+_{e,i}(\zeta)\mathrm{d}\zeta \end{aligned}$$

for all edges $e \in E$. Since both in- and outflow rate are non-negative, this implies the corollary.

From this we can now deduce the second part of the proposition in the same way as in the proof of Proposition 3.66 by observing that

$$\{\theta \geq \Psi_i(f) \mid C_e(\theta) > 0\} \subseteq \bigcup_{\theta \geq \Psi_i(f)} [\theta, T_e(\theta))$$

and applying Proposition 2.15. □

With this proposition we can now show that for Vickrey flows our definition of total travel times coincides with the one used in [GHKM23, Section 6.1]. We start by showing that for a single edge we can compute the total travel time incurred on this edge by particles entering before some time θ in two different ways: Either we "add" for every such particle the expected current travel time at the time of entrance (left side) or we "add" for every particle the difference between its entrance and exit time (right side). Intuitively, the fact that the queue operates fair and at capacity should guarantee that these two measures are the same. The following lemma shows that this is indeed the case. Using a telescope sum argument we will then be able to lift this equality to the whole network.

Lemma 3.74. *Let $(f^+_{e,i}, f^-_{e,i})_{i\in I}$ be a Vickrey edge flow until $\xi \in \tilde{\mathbb{R}}_{\geq 0}$. Then for every time $\theta < \xi$ with $T_e(\theta) < \xi + \tau_e$ and every commodity $i \in I$ we have*

$$\int_0^\theta C_e(\zeta) f^+_{e,i}(\zeta)\mathrm{d}\zeta = \int_0^{T_e(\theta)} \zeta f^-_{e,i}(\zeta)\mathrm{d}\zeta - \int_0^\theta \zeta f^+_{e,i}(\zeta)\mathrm{d}\zeta.$$

Proof. This proof is essentially a straightforward computation:

$$\begin{aligned}
&\int_0^{T_e(\theta)} \zeta f^-_{e,i}(\zeta)\mathrm{d}\zeta - \int_0^\theta \zeta f^+_{e,i}(\zeta)\mathrm{d}\zeta \\
&\overset{(*)}{=} \Big[\zeta F^-_{e,i}(\zeta)\Big]_0^{T_e(\theta)} - \int_0^{T_e(\theta)} F^-_{e,i}(\zeta)\mathrm{d}\zeta - \Big[\zeta F^+_{e,i}(\zeta)\Big]_0^\theta + \int_0^\theta F^+_{e,i}(\zeta)\mathrm{d}\zeta \\
&\overset{(\Delta)}{=} T_e(\theta) F^-_{e,i}(T_e(\theta)) - \int_{\tau_e}^{T_e(\theta)} F^-_{e,i}(\zeta)\mathrm{d}\zeta - \theta F^+_{e,i}(\theta) + \int_0^\theta F^+_{e,i}(\zeta)\mathrm{d}\zeta \\
&\overset{(\#)}{=} T_e(\theta) F^-_{e,i}(T_e(\theta)) - \int_0^\theta F^-_{e,i}(T_e(\zeta))\partial T_e(\zeta)\mathrm{d}\zeta - \theta F^+_{e,i}(\theta) + \int_0^\theta F^+_{e,i}(\zeta)\mathrm{d}\zeta \\
&\overset{(3.14)}{=} T_e(\theta) F^+_{e,i}(\theta) - \int_0^\theta F^+_{e,i}(\zeta)\left(1 + \frac{\partial Q_e(\zeta)}{\nu_e}\right)\mathrm{d}\zeta - \theta F^+_{e,i}(\theta) + \int_0^\theta F^+_{e,i}(\zeta)\mathrm{d}\zeta \\
&= (T_e(\theta) - \theta) F^+_{e,i}(\theta) - \int_0^\theta F^+_{e,i}(\zeta)\frac{\partial Q_e(\zeta)}{\nu_e}\mathrm{d}\zeta \\
&\overset{(*)}{=} (T_e(\theta) - \theta) F^+_{e,i}(\theta) - \left[F^+_{e,i}(\zeta)\frac{Q_e(\zeta)}{\nu_e}\right]_0^\theta + \int_0^\theta f^+_{e,i}(\zeta)\frac{Q_e(\zeta)}{\nu_e}\mathrm{d}\zeta \\
&= \left(\frac{Q_e(\theta)}{\nu_e} + \tau_e\right) F^+_{e,i}(\theta) - F^+_{e,i}(\theta)\frac{Q_e(\theta)}{\nu_e} + \int_0^\theta f^+_{e,i}(\zeta)\frac{Q_e(\zeta)}{\nu_e}\mathrm{d}\zeta \\
&= \int_0^\theta \tau_e f^+_{e,i}(\zeta)\mathrm{d}\zeta + \int_0^\theta f^+_{e,i}(\zeta)\frac{Q_e(\zeta)}{\nu_e}\mathrm{d}\zeta = \int_0^\theta C_e(\zeta) f^+_{e,i}(\zeta)\mathrm{d}\zeta,
\end{aligned}$$

where we use integration by parts (Proposition 2.52) for every equality marked by $(*)$, the change of variable formula (Proposition 2.53) at (#), weak flow conservation at (Δ) and the fact that the queue operates fair and at capacity to be able to use (3.14) from Proposition 3.40b). □

Proposition 3.75. *Let f be a Vickrey flow and $i \in I$ a commodity with finite makespan. Then the following holds*

$$\begin{aligned}\Xi_i(f) &= \int_0^{\Psi_i(f)} \zeta \cdot \partial Z_i(\zeta)\mathrm{d}\zeta - \sum_{v\in V}\int_0^{\hat{\theta}_i} \zeta \cdot u_{v,i}(\zeta)\mathrm{d}\zeta \\ &= \int_0^{\Psi_i(f)} F_i^{\Delta}(\zeta)\mathrm{d}\zeta = \int_0^{\Psi_i(f)} U_i(\zeta)\mathrm{d}\zeta - \int_0^{\Psi_i(f)} Z_i(\zeta)\mathrm{d}\zeta.\end{aligned}$$

Proof. The first equality follows from Lemma 3.74 using a telescope sum argument and strong flow conservation at the nodes:

$$\begin{aligned}\Xi_i(f) &= \sum_{e\in E}\int_0^{\infty} C_e(\zeta) f_{e,i}^{+}(\zeta)\mathrm{d}\zeta \overset{\text{Prop. 3.73}}{=} \sum_{e\in E}\int_0^{\Psi_i(f)} C_e(\zeta) f_{e,i}^{+}(\zeta)\mathrm{d}\zeta \\
&\overset{\text{Lem. 3.74}}{=} \sum_{e\in E}\int_0^{T_e(\Psi_i(f))} \zeta f_{e,i}^{-}(\zeta)\mathrm{d}\zeta - \sum_{e\in E}\int_0^{\Psi_i(f)} \zeta f_{e,i}^{+}(\zeta)\mathrm{d}\zeta \\
&\overset{\text{Prop. 3.73}}{=} \sum_{e\in E}\int_0^{\Psi_i(f)} \zeta f_{e,i}^{-}(\zeta)\mathrm{d}\zeta - \sum_{e\in E}\int_0^{\Psi_i(f)} \zeta f_{e,i}^{+}(\zeta)\mathrm{d}\zeta \\
&= \sum_{v\in V}\sum_{e\in\delta^-(v)}\int_0^{\Psi_i(f)} \zeta f_{e,i}^{-}(\zeta)\mathrm{d}\zeta - \sum_{v\in V}\sum_{e\in\delta^+(v)}\int_0^{\Psi_i(f)} \zeta f_{e,i}^{+}(\zeta)\mathrm{d}\zeta \\
&= \sum_{v\in V}\int_0^{\Psi_i(f)} \zeta\Big(\sum_{e\in\delta^-(v)} f_{e,i}^{-}(\zeta) - \sum_{e\in\delta^+(v)} f_{e,i}^{+}(\zeta)\Big)\mathrm{d}\zeta \\
&= \sum_{v\in V}\int_0^{\Psi_i(f)} \zeta\Big(\partial B_{v,i}(\zeta) - u_{v,i}(\zeta)\Big)\mathrm{d}\zeta \\
&\overset{(\Delta)}{=} \sum_{v\in T_i}\int_0^{\Psi_i(f)} \zeta\partial B_{v,i}(\zeta)\mathrm{d}\zeta - \sum_{v\in V}\int_0^{\Psi_i(f)} \zeta u_{v,i}(\zeta)\mathrm{d}\zeta \\
&= \int_0^{\Psi_i(f)} \zeta\partial Z_i(\zeta)\mathrm{d}\zeta - \sum_{v\in V}\int_0^{\hat{\theta}} \zeta u_{v,i}(\zeta)\mathrm{d}\zeta,\end{aligned}$$

where we use strong flow conservation at the nodes at (Δ).

The second equality from the proposition's statement can be shown using integration by parts:

$$\int_0^{\Psi_i(f)} \zeta \partial Z_i(\zeta) \mathrm{d}\zeta - \sum_{v \in V} \int_0^{\hat{\theta}} \zeta u_{v,i}(\zeta) \mathrm{d}\zeta = \int_0^{\Psi_i(f)} \zeta \partial Z_i(\zeta) \mathrm{d}\zeta - \int_0^{\Psi_i(f)} \zeta \partial U_i(\zeta) \mathrm{d}\zeta$$
$$= \int_0^{\Psi_i(f)} \zeta \cdot \partial(Z_i - U_i)(\zeta) \mathrm{d}\zeta \overset{\text{Prop. 3.53}}{=} - \int_0^{\Psi_i(f)} \zeta \partial F_i^{\Delta}(\zeta) \mathrm{d}\zeta$$
$$\overset{(*)}{=} - \left[\zeta F_i^{\Delta}(\zeta)\right]_0^{\Psi_i(f)} + \int_0^{\Psi_i(f)} F_i^{\Delta}(\zeta) \mathrm{d}\zeta = -\Psi_i(f) F_i^{\Delta}(\Psi_i(f)) + \int_0^{\Psi_i(f)} F_i^{\Delta}(\zeta) \mathrm{d}\zeta$$
$$= \int_0^{\Psi_i(f)} F_i^{\Delta}(\zeta) \mathrm{d}\zeta,$$

where we use integration by parts at $(*)$ and the definition of $\Psi_i(f)$ at the last equality.

Finally, the proposition's third equality now follows directly by applying Proposition 3.53). □

Using this proposition we get the following relation between our two quality measures:

Corollary 3.76. *For any Vickrey flow f and commodity i with finite makespan under f we have*

$$\Xi_i(f) \leq U_i(\hat{\theta}_i) \cdot \Psi_i(f).$$

Proof. This follows directly from Proposition 3.75:

$$\Xi_i(f) \overset{\text{Prop. 3.75}}{=} \int_0^{\Psi_i(f)} U_i(\zeta) \mathrm{d}\zeta - \int_0^{\Psi_i(f)} Z_i(\zeta) \mathrm{d}\zeta$$
$$\leq \int_0^{\Psi_i(f)} U_i(\zeta) \mathrm{d}\zeta \leq U_i(\hat{\theta}_i) \cdot \Psi_i(f).$$

□

3.3 Model Summary

For easier reference we repeat here the complete definition of the model introduced over the course of this chapter:

Network: A *feasible network* $\mathcal{N}$ is given by a tuple $(G, (\tau_e), (\nu_e), I, (u_{v,i}), (T_i))$ consisting of a directed graph $G = (V, E)$, free flow travel times $\tau_e \in \mathbb{R}_{\geq 0}$

and capacities $\nu_e \in \mathbb{R}_{>0}$ for all edges $e \in E$, a finite set I of commodities and for each commodity $i \in I$ a non-empty set $T_i \subseteq V$ of sink nodes and a locally integrable network inflow rate $u_{v,i} : \mathbb{R}_{\geq 0} \to \mathbb{R}_{\geq 0}$ for every node $v \in V$. Moreover, we assume that for every commodity $i \in I$ and every node $s \in V$ with $u_{s,i} \neq_{\text{a.e.}} 0$ there exists some sink node $t \in T_i$ which is reachable from s.

Dynamic flow: A *dynamic flow* f in such a network is a vector $(f^+_{e,i}, f^-_{e,i}) \in L^1_{\text{loc}}(\mathbb{R}_{\geq 0}, \mathbb{R}_{\geq 0})^{E \times I \times \{\pm\}}$. Here, $f^+_{e,i}(\theta)$ denotes the inflow rate of particles of commodity i into edge e at time θ and, analogously, $f^-_{e,i}(\theta)$ denotes the (edge) outflow rate. For any such dynamic flow we define the cumulative inflow (outflow) of commodity i into (from) edge e by $F^+_{e,i}(\theta) := \int_0^\theta f^+_{e,i}(\zeta)\mathrm{d}\zeta$ ($F^-_{e,i}(\theta) := \int_0^\theta f^-_{e,i}(\zeta)\mathrm{d}\zeta$). We drop the index i to refer to the associated anonymous flow (rates), e.g. we define $f^+_e := \sum_{i \in I} f^+_{e,i}$. Finally, we denote the queue length on edge e at time θ by $Q_e(\theta) := F^+_e(\theta) - F^-_e(\theta + \tau_e)$, its current travel time by $C_e(\theta) := \frac{Q_e(\theta)}{\nu_e} + \tau_e$ and its current exit time by $T_e(\theta) := \theta + C_e(\theta)$.

Vickrey flow: A dynamic flow f is a *Vickrey flow* until some time $\xi \in \tilde{\mathbb{R}}_{\geq 0}$ if it satisfies the following three properties:

- The queues of all edges operate at capacity until ξ, i.e. for all $e \in E$ we have

$$f^-_e(\theta + \tau_e) = \begin{cases} \nu_e, & \text{if } Q_e(\theta) > 0 \\ \min\{\, f^+_e(\theta), \nu_e \,\}, & \text{else} \end{cases} \quad \text{f.a.a. } \theta < \xi \tag{3.4}$$

  as well as $Q_e(0) = 0$ (if $\xi > 0$).

- The queues of all edges operate fair until ξ, i.e. for all $i \in I$ and $e \in E$ we have

$$f^-_{e,i}(\theta + \tau_e) = \begin{cases} f^-_e(\theta + \tau_e) \cdot \frac{f^+_{e,i}(\vartheta)}{f^+_e(\vartheta)}, & \text{if } f^+_e(\vartheta) > 0, \\ 0, & \text{else} \end{cases} \quad \text{f.a.a. } \theta < \xi, \tag{3.13}$$

  where ϑ is chosen such that $T_e(\vartheta) = \theta + \tau_e$.

- The flow satisfies strong flow conservation at all nodes until ξ, i.e. for all $i \in I$ and $v \in V$ we have

$$u_{i,v}(\theta) + \sum_{e \in \delta^-(v)} f^-_{e,i}(\theta) - \sum_{e \in \delta^+(v)} f^+_{e,i}(\theta) \begin{cases} = 0, & \text{if } v \in V \setminus T_i \\ \geq 0, & \text{if } v \in T_i \end{cases} \quad \text{f.a.a. } \theta < \xi. \tag{3.20}$$

There are also several equivalent characterisations for queues operating at capacity or fair given in Propositions 3.19 and 3.40, respectively.

IDE: A Vickrey flow until ξ is an *instantaneous dynamic equilibrium (IDE)* until ξ if it additionally satisfies the following IDE-property for all $i \in I$ and $e \in E$:

$$f^+_{e,i}(\theta) > 0 \implies e \in E_i(\theta) \text{ for almost all } \theta < \xi \qquad (3.23)$$

where we denote by $E_i(\theta) := \{\, e = vw \in E \mid L_{v,i}(\theta) \geq C_e(\theta) + L_{w,i}(\theta) \,\}$ the set of active edges for commodity i at time θ and the current distance from node v to commodity i's closest sink node by

$$L_{v,i}(\theta) := \inf \{\, \sum_{e \in p} C_e(\theta) \mid p \text{ a } v, t\text{-path for some } t \in T_i \,\}.$$

3.4 Bibliographic Notes and Open Questions

The flow model introduced in Sect. 3.1 is the same which was also used in [GHS20, GH22, GH23] which in turn is mostly based on the flow model used by Koch and Skutella in [KS11] and Cominetti, Correa and Larré in [CCL15]. The alternative characterisations in Propositions 3.19 and 3.40 are essentially a collection of several different ways the edge dynamics are defined in the literature. The properties of the flow model deduced from these characterisation are also mostly well known (so much in fact that they are often used without even explicitly stating them).

Our definition of IDE presented here was first introduced in [GH19] (which is a joint work with Tobias Harks) and is based on the equilibrium definition in the full information model used e.g. in [KS11, CCL15]. Note, that in [GH19] the node labels are directly defined by the recursive Bellman equations (3.22) (which is possible there since only strictly positive free flow travel times are allowed there).

We note that the physical and the behavioural model (i.e. Sects. 3.1 and 3.2) are largely independent of each other. This is what allowed us to reuse many properties of the physical model originally shown in the full information setting to our model. At the same time, this also means that one could easily replace the deterministic queuing model by another physical model while keeping our behavioural model in order to define IDE with different flow dynamics. As long as the chosen physical model is reasonable well-behaved (i.e. satisfies all or at least some of the fundamental properties stated at the beginning of Sect. 3.1), it seems quite likely that several of our results in the coming chapters could then be transferred to such a related model.

As a final remark we would also like to point out that even though the deterministic queuing model is generally well understood there are also still some open questions related to it (cf. [Gai+22, Section 4.6]). One such example is the computational complexity of its network-loading procedure (i.e. the network-wide analogue to the edge-loading function Φ_e^{ξ} discussed in Corollary 3.46).

Existence of IDE

4

In this chapter we will give several proofs for the existence of IDE in various different settings. The proofs are ordered by decreasing generality (of instances for which they apply). As will become important in the next chapter (Chap. 5), this coincides with increasing levels of constructiveness.

One common aspect of all proofs is the following general idea: To show the existence of an IDE it suffices to show that, when given an IDE flow up to some point in time, we can always extend this to an IDE up to some later point in time. Then, starting with the trivial IDE up to time 0, a simple limit argument guarantees existence of an IDE for all times.

For dynamic equilibria in the full information setting this same approach (only for a different notion of equilibrium flows up to a certain point in time) together with the nice combinatorial structure of the individual extensions (so called "thin flows") found by Koch and Skutella in [KS11] was and still is one of the main sources for the recent progress in better understanding these equilibria in the last two decades. As we will see in this chapter, IDE exhibit a very similar structure.

4.1 A Meta-Theorem on IDE-Existence

We start by formalising the concepts of partial IDE and IDE-extensions and then showing a meta-theorem which basically states that in order to show existence of IDE it suffices to show that partial IDE can always be extended.

L. Graf, *Dynamic Network Flows with Adaptive Route Choice based on Current Information*, Mathematische Optimierung und Wirtschaftsmathematik | Mathematical Optimization and Economathematics,
https://doi.org/10.1007/978-3-658-44948-3_4

Definition 4.1. For any network $\mathcal{N}$ we denote by

$$\mathcal{F}_{pa}(\mathcal{N}) := \mathcal{F}(\mathcal{N}) \times \tilde{\mathbb{R}}_{\geq 0}$$

the set of all *partial flows* in $\mathcal{N}$. We call a partial flow $(f, \xi) \in \mathcal{F}_{pa}(\mathcal{N})$ a *partial Vickrey flow* if f is a Vickrey flow until ξ and a *partial IDE* if f is an IDE until ξ. We denote the set of all partial IDE in $\mathcal{N}$ by $\mathcal{F}_{pa}^{\text{IDE}}(\mathcal{N}) \subseteq \mathcal{F}_{pa}(\mathcal{N})$.

For any two partial flows $(f, \xi), (f', \xi') \in \mathcal{F}_{pa}(\mathcal{N})$ we then say that (f', ξ') is an *extension of* (f, ξ) *(until* ξ'*)* if $\xi' \geq \xi$ and we have

$$f^+_{e,i}\big|_{[0,\xi)} =_{\text{a.e.}} f'^+_{e,i}\big|_{[0,\xi)} \text{ and } f^-_{e,i}\big|_{[0,\xi+\tau_e)} =_{\text{a.e.}} f'^-_{e,i}\big|_{[0,\xi+\tau_e)} \text{ for all } e \in E, i \in I.$$

We denote this by $(f, \xi) \preceq (f', \xi')$. Finally, we consider two partial flows to be equal as partial flows if both $(f, \xi) \preceq (f', \xi')$ and $(f', \xi') \preceq (f, \xi)$ hold and denote this by $(f, \xi) \approx (f', \xi')$.

Observation 4.2. For any two partial flows (f, ξ) and (f', ξ') we have

$$(f, \xi) \approx (f', \xi') \iff (f, \xi) \preceq (f', \xi') \text{ and } \xi = \xi'.$$

Observation 4.3. For any two equivalent partial flows $(f, \xi) \approx (f', \xi')$ we have that (f, ξ) is a partial Vickrey flow/IDE if and only (f', ξ') is a partial Vickrey flow/IDE. This is because the constraints defining Vickrey flows/IDE up to some time ξ only depend on the edge inflow rates until ξ and the edge outflow rates until $\xi + \tau_e$.

Observation 4.4. If (f, ξ) and (f', ξ') are two partial flows which are Vickrey edge flows until ξ on all edges, then we have

$$(f, \xi) \preceq (f', \xi') \iff \xi \leq \xi' \text{ and } f^+_{e,i}\Big|_{[0,\xi)} =_{\text{a.e.}} f'^+_{e,i}\Big|_{[0,\xi)} \quad \text{for all } e \in E, i \in I.$$

This follows directly from Corollary 3.43, i.e. the fact that in a Vickrey edge flow the edge outflow rates until $\xi + \tau_e$ are uniquely determined by the edge inflow rates until ξ.

Observation 4.5. The relation $\preceq$ is a preorder on $\mathcal{F}_{pa}(\mathcal{N})$.

This now allows us to state the following meta-theorem: Whenever we have some non-empty set of partial flows which contains

a) a *proper* extension for every individual element of that set and
b) a *common* extension for every sequence of elements in that set,

then this set also contains a flow for all time. In particular, this meta-theorem reduces the challenge of proving existence of IDE (potentially with some additional property) to showing a) an extension-lemma for such IDE and b) that limits of such IDE are again an IDE (still satisfying any required additional property).

Theorem 4.6. *Let $\mathcal{N}$ be any network and $\mathfrak{F} \subseteq \mathcal{F}_{pa}(\mathcal{N})$ be some subset of all partial flows in $\mathcal{N}$ satisfying the following properties:*

1. *$\mathfrak{F}$ is non-empty.*
2. *For any $(f, \xi) \in \mathfrak{F}$ with $\xi < \infty$ there exists some $(f', \xi + \varepsilon) \in \mathfrak{F}$ with $\varepsilon > 0$ and $(f, \xi) \preceq (f', \xi + \varepsilon)$.*
3. *For any (non-trivial) sequence $(f^{(1)}, \xi_1) \preceq (f^{(2)}, \xi_2) \preceq \dots$ of partial flows in $\mathfrak{F}$, there exists some $(f, \xi) \in \mathfrak{F}$ with $(f^{(k)}, \xi_k) \preceq (f, \xi)$ for all $k = 1, 2, \dots$.*

Then there exists a dynamic flow $\tilde{f}$ such that $(\tilde{f}, \infty) \in \mathfrak{F}$.

Proof idea: Properties 1 and 3 are essentially just the conditions of Zorn's Lemma specified for our situation. Thus, the set $\mathfrak{F}$ must contain a maximal element $(\tilde{f}, \tilde{\xi})$ which, by property 2, must satisfy $\tilde{\xi} = \infty$.

Proof. By Observation 4.5 $(\mathfrak{F}, \preceq)$ is a preordered set. Now, let $\mathfrak{J} \subseteq \mathfrak{F}$ be a chain in $\mathfrak{F}$. If it is the trivial chain (i.e. $\mathfrak{J} = \emptyset$) then any element of $\mathfrak{F}$ is an upper bound—and such an element exists by property 1.

Otherwise, we define $\hat{\xi} := \sup\{\, \xi \mid \exists (f, \xi) \in \mathfrak{J} \,\}$ and find an upper bound to $\mathfrak{J}$ as follows:

1. Case: $\exists(\hat{f}, \hat{\xi}) \in \mathfrak{J}$: Then this is an upper bound to $\mathfrak{J}$: Since $\mathfrak{J}$ is totally ordered we have $(f, \xi) \preceq (\hat{f}, \hat{\xi})$ or $(f, \xi) \succeq (\hat{f}, \hat{\xi})$ for every $(f, \xi) \in \mathfrak{J}$. In the latter case we then have $\xi \geq \hat{\xi}$ but also $\xi \leq \hat{\xi}$ by the definition of $\hat{\xi}$. Thus, we have $\xi = \hat{\xi}$ and, therefore, $(f, \xi) \preceq (f, \hat{\xi})$ by Observation 4.2.

2. Case: $\nexists(\hat{f}, \hat{\xi}) \in \mathfrak{J}$**:** Then we have $\xi < \hat{\xi}$ for all $(f, \xi) \in \mathfrak{J}$ and we can fix some sequence $(f^{(k)}, \xi_k)_{k \in \mathbb{N}^*}$ such that the sequence $(\xi_k)_{k \in \mathbb{N}^*}$ is non-descending and converges to $\hat{\xi}$. Since $\mathfrak{J}$ is totally ordered, we have $(f^{(k)}, \xi_k) \preceq (f^{(k+1)}, \xi_{k+1})$ or $(f^{(k)}, \xi_k) \succeq (f^{(k+1)}, \xi_{k+1})$ for every k and as (ξ_k) is a non-decreasing the latter also implies the former by Observation 4.2. Thus, by property 3, there exists some $(f', \xi') \in \mathfrak{F}$ with $(f^{(k)}, \xi_k) \preceq (f', \xi')$ for all $k \in \mathbb{N}^*$. This (f', ξ') is then also an upper bound to $\mathfrak{J}$: For any $(f, \xi) \in \mathfrak{J}$ there exists some $k \in \mathbb{N}^*$ with $\xi < \xi_k$ and, therefore, $(f, \xi) \preceq (f^{(k)}, \xi_k) \preceq (f', \xi')$.

Thus, we can apply Zorn's Lemma (Lemma 2.58) to obtain a maximal element $(f, \xi) \in \mathfrak{F}$. If $\xi = \infty$, then we are done. So, assume for contradiction that $\xi < \infty$. Then, by property 2, we also have some $(f', \xi + \varepsilon) \in \mathfrak{F}$ with $\varepsilon > 0$ and $(f, \xi) \preceq (f', \xi + \varepsilon)$—a contradiction to the maximality of (f, ξ). □

Remark 4.7. It is possible to avoid Zorn's Lemma in the proof above if one assumes additionally, that the set $\mathfrak{F}$ is closed under restriction (i.e. if $(f, \xi) \in \mathfrak{F}$ and $\xi' < \xi$ then $(f, \xi') \in \mathfrak{F}$ as well) which will be the case in all our applications of this theorem. See [GHS20, Theorem 3.4] for such a proof. However, this alternative proof is still non-constructive as it still uses a limit argument together with a proof by contradiction.

If there is some general lower bound on the ε in property 2, then existence follows by a simple inductive argument and, thus, the proof becomes constructive—provided that the extension lemma ensuring that property 2 holds is constructive itself.

In the following we will use Theorem 4.6 to show existence of IDE by applying it to certain subsets $\mathfrak{F} \subseteq \mathcal{F}_{pa}^{\text{IDE}}(\mathcal{N})$. The first property will always be trivially satisfied (e.g. by the zero-flow which is an IDE until time 0). Ensuring the second property will be the main part of the proof and will be done by appropriate "extension-lemmas" which we will see in the following sections (Lemmas 4.14 and 4.34 and Corollary 4.30). The third property will again be automatically satisfied since we can just take the limit of any given chain of partial IDE to obtain the required upper bound to such a chain. This can be formalised as follows:

Definition 4.8. Let $(f^{(1)}, \xi_1) \preceq (f^{(2)}, \xi_2) \preceq \ldots$ be an ascending sequence of partial flows in some fixed network $\mathcal{N}$. Then we define the *limit of this sequence of partial flows* by

$$\lim_k(f^{(k)}, \xi_k) := (\hat{f}, \hat{\xi}) \text{ where}$$

$$\hat{\xi} := \limsup_k \xi_k$$

$$f^+_{e,i}(\theta) := \begin{cases} f^{(k),+}_{e,i}(\theta), & \text{if } \theta < \xi_k \text{ for some } k \\ 0, & \text{else} \end{cases}$$

$$f^-_{e,i}(\theta) := \begin{cases} f^{(k),-}_{e,i}(\theta), & \text{if } \theta < \xi_k + \tau_e \text{ for some } k \\ 0, & \text{else} \end{cases}.$$

In order to show that this definition is not only well defined but also retains all relevant properties, we need the following lemma:

Lemma 4.9. *A dynamic flow f is a Vickrey flow/IDE until some $\xi \in \tilde{\mathbb{R}}_{\geq 0}$ if and only if it is a Vickrey flow/IDE until ξ' for every $\xi' < \xi$.*

Proof. The 'only if'-part is clear. For the 'if'-part let $N \subseteq [0, \xi)$ be the set of all times where f violates at least one of the constraints of being a Vickrey flow (an IDE), i.e. one of the constraints (3.4), (3.13) and (3.20) (and (3.23)). Since (f, ξ') is a partial Vickrey flow (an IDE) for all $\xi' < \xi$, the set $N \cap [0, \xi')$ has measure zero for all those ξ'. But this then implies that N itself has measure zero. Thus, (f, ξ) is a partial Vickrey flow (an IDE) as well. □

Proposition 4.10. *If, for every edge commodity $i \in I$ and edge $e \in E$ there exist some locally integrable functions $g_{e,i} : \mathbb{R}_{\geq 0} \to \mathbb{R}_{\geq 0}$ such that $f^+_{e,i}\big|_{[0,\xi_k)} \leq_{a.e.} g_{e,i}$ and $f^-_{e,i}\big|_{[0,\xi_k+\tau_e)} \leq_{a.e.} g_{e,i}$, then $\lim_k(f^{(k)}, \xi_k)$ is a well-defined dynamic flow with $(f^{(k)}, \xi_k) \preceq \lim_k(f^{(k)}, \xi_k)$ for all $k \in \mathbb{N}_0$. Furthermore,*

a) it is a partial flow with locally p-integrable flow rates if all $g_{e,i}$ are locally p-integrable,
b) it is a partial flow with right-constant flow rates if all $(f^{(k)}, \xi_k)$ have right-constant flow rates,
c) it is a partial Vickrey flow if all $(f^{(k)}, \xi_k)$ are partial Vickrey flows and
d) it is a partial IDE if all $(f^{(k)}, \xi_k)$ are partial IDE.

Proof. We first show that $\hat{f}$ is a dynamic flow. Take any edge $e \in E$ and commodity $i \in I$ and observe that for every $\theta < \hat{\xi}$ there exists some $K \in \mathbb{N}^*$ with $\xi_K > \theta$ and therefore all (countably many) $f^{(k)+}_{e,i}$ for $k \geq K$ coincide on $[0, \theta]$ almost everywhere. Thus, $\hat{f}^+_{e,i}$ is well-defined. Furthermore, the functions $f^{(k)+}_{e,i} \cdot \mathbb{1}_{[0,\xi_k)}$ are measurable and converge pointwise almost everywhere to $\hat{f}^+_{e,i}$. Thus, $\hat{f}^+_{e,i}$ is measurable as well by Proposition 2.8. Finally, $\hat{f}^+_{e,i}$ is essentially bounded by the locally integrable function $g_{e,i}$ and, hence, Proposition 2.13 implies that it is locally integrable. The same is true for $\hat{f}^-_{e,i}$ just with all intervals extended by τ_e. Since all this is true for all edges and commodities, $\hat{f}$ is indeed a dynamic flow.

That $\lim_k(f^{(k)}, \xi_k)$ is an extension of each of the $(f^{(k)}, \xi_k)$ now follows immediately from the definition of $\lim_k(f^{(k)}, \xi_k)$.

We now show the additional properties:

a) Since we have $\hat{f}^+_{e,i}, \hat{f}^-_{e,i} \leq g_{e,i}$ this follows directly from Proposition 2.13

b) For any $\theta \in [0, \hat{\xi})$ there exists some $k \in \mathbb{N}_0$ such that we have $\theta < \xi_k$. Since $f^{(k)}$ has right-constant flow rates, there exists some $0 < \varepsilon \leq \xi_k - \theta$ such that all edge inflow rates of $f^{(k)}$ are constant on $[\theta, \theta + \varepsilon)$. The same is then true for $\hat{f}$ as we have $\hat{f}^+_{e,i}\Big|_{[0,\xi_k)} =_{\text{a.e.}} f^{(k),+}_{e,i}\Big|_{[0,\xi_k)}$ for all $e \in E$ and $i \in I$. An analogous argument for shows that the same is true for the edge outflow rates.

c),d) These two properties follow from Lemma 4.9: For any $\xi' < \hat{\xi}$ there exists some $k \in \mathbb{N}^*$ such that $\xi_k \geq \xi'$. We then have $(\hat{f}, \xi') \approx (f^{(k)}, \xi')$ and $(f^{(k)}, \xi')$ is a partial Vickrey flow/IDE (by trivial direction of Lemma 4.9). Thus, $\hat{f}$ is a Vickrey flow/IDE until ξ' as well by Observation 4.3. □

Note that, for general dynamic flows, the assumptions on the existence of the bounding functions $g_{e,i}$ is necessary since there are functions like $x \mapsto \frac{1}{1-x}$ which is (locally) integrable on any interval $[0, \xi]$ with $\xi < 1$ but not on $[0, 1]$. However, for Vickrey flows this assumption can be dropped as respecting capacity, strong flow conservation at nodes and local integrability of network inflow rates together automatically guarantee the existence of such bounds:

Corollary 4.11. *Let $(f^{(1)}, \xi_1) \preceq (f^{(2)}, \xi_2) \preceq \dots$ be an ascending sequence of Vickrey flows/IDE in some fixed network $\mathcal{N}$. Then $\lim_k(f^{(k)}, \xi_k)$ is a partial Vickrey flow/IDE extending all $(f^{(k)}, \xi_k)$. Moreover, if the network inflow rates are locally p-integrable for some $p \geq 1$, then $\lim_k(f^{(k)}, \xi_k)$ has locally p-integrable flow rates as well.*

Proof. For any edge $e = vw \in E$ and commodity $i \in I$ we define a function $g_{e,i} : \mathbb{R}_{\geq 0} \to \mathbb{R}_{\geq 0}$ by setting

$$g_{e,i}(\theta) := \max\{\, \nu_e, u_{v,i}(\theta) + \sum_{e' \in \delta^-(v)} \nu_{e'} \,\}.$$

This function is clearly locally integrable (and even locally p-integrable if $u_{v,i}$ is so as well). Moreover, for every $k \in \mathbb{N}^*$ this function satisfies

$$f^+_{e,i}(\theta) \overset{(3.20)}{\leq} u_{i,v}(\theta) + \sum_{e' \in \delta^-(v)} f^{(k),-}_{e',i}(\theta) \overset{(3.2)}{\leq} u_{i,v}(\theta) + \sum_{e' \in \delta^-(v)} \nu_{e'} \leq g_{e,i}(\theta)$$

for almost all $\theta \in [0, \xi)$ and

$$f^{(k),-}_{e,i}(\theta) \overset{(3.2)}{\leq} \nu_e \leq g_{e,i}(\theta)$$

for almost all $\theta \in [0, \xi_k + \tau_e)$. Thus, we can apply Proposition 4.10 to show the corollary. □

4.2 Extension Lemma for General Inflow Rates

In this section we want to show a general existence result for IDE in networks with locally p-integrable network inflow rates for some $p > 1$. We will do so by showing that any partial IDE with locally p-integrable flow rates can be extended for any additional *finite* time interval. Our meta-existence theorem from the previous section then implies the existence of IDE for all time. Even more, we will be able to deduce that any partial IDE (with locally p-integrable flow rates) can be extended to an IDE for all time.

Note, that we restrict ourselves to $p > 1$ since then the space $L^p([a,b])^d$ is a reflexive Banach space for any $[a,b] \subseteq \mathbb{R}$ and any $d \in \mathbb{N}^*$ (cf. Propositions 2.40 and 2.43).

So, let $\mathcal{N}$ be some feasible network with locally p-integrable network inflow rates and (f, ξ) a partial IDE with $\xi < \infty$ and locally p-integrable flow rates. Furthermore, let $\varepsilon > 0$ be any (finite) extension period. Our goal is then to find a partial IDE $(g, \xi + \varepsilon)$ with $(f, \xi) \preceq (g, \xi + \varepsilon)$.

Proof idea: In order to show the existence of such an extension we will start with a set of candidates

$$\left\{ g = (g^+, g^-) \,\middle|\, \begin{array}{l} (f, \xi) \preceq (g, \xi + \varepsilon), \quad g \text{ respects capacity and} \\ \text{satisfies strong flow conservation at all nodes until } \xi + \varepsilon \end{array} \right\}.$$

Clearly, any actual extension of (f, ξ) is contained in this set. Moreover, it is easy to see that this set is non-empty. However, this set of course also contains many flows which are not IDE-extensions as we are missing both the IDE-property and the constraints for the flow dynamics on the edges. Thus, we will additionally define a mapping Γ which for any such candidate g gives us the set of candidates h such that

- the inflow rates of h satisfy the IDE-property with respect to the current travel times induced by g and
- the outflow rates of h satisfy the Vickrey edge flow dynamics together with the edge inflow rates of g.

It is again easy to see that there always exists at least one such candidate h as we can just derive h^- from g^+ using the Vickrey edge flow dynamics and then send all this flow further on into some edge which is active for g.

If now, we can find some candidate g which gets mapped to itself by Γ, then we have found the desired extension. To get such a fixed point of the mapping Γ we will then use the Kakutani–Fan–Glicksberg fixed point theorem.

Before we can start our formal proof we define the set $K \subseteq L^p(\mathbb{R}_{\geq 0}, \mathbb{R}_{\geq 0})^{E \times I \times \{\pm\}}$ by

$$K := \left\{ g \,\middle|\, \begin{array}{l} (f, \xi) \approx (g, \xi), \;\; g \text{ respects the edge capacities and} \\ \text{satisfies strong flow conservation at all nodes until } \xi + \varepsilon, \\ g^+_{e,i}\big|_{[\xi+\varepsilon,\infty)} =_{\text{a.e.}} 0, \; g^-_{e,i}\big|_{[\xi+\varepsilon+\tau_e,\infty)} =_{\text{a.e.}} 0 \text{ f.a. } e \in E, i \in I, \\ g^+_{e,i}\big|_{[\xi,\xi+\varepsilon)} =_{\text{a.e.}} 0 \text{ for all } e = vw \in E, i \in I \text{ with } w \in V_i^\dagger \end{array} \right\} \tag{4.1}$$

as well as a correspondence $\Gamma : K \to 2^K$ by setting

$$\Gamma(g) := \left\{ h \in K \;\middle|\; \begin{array}{l} (g^+_{e,\bullet}, h^-_{e,\bullet}) \text{ is a Vickrey flow until } \xi + \varepsilon \text{ for all } e \in E \text{ and} \\ h^+_{e,i}(\theta) = 0 \text{ for all } e = vw \in E, i \in I \text{ and almost all } \theta \in \mathbb{R}_{\geq 0} \\ \text{with } L^g_{v,i}(\theta) < L^g_{w,i}(\theta) + C^g_e(\theta) \end{array} \right\} \tag{4.2}$$

for any $g \in K$. The additional constraints in the definition of set K (compared to the set of candidates from our proof idea) are there to ensure that the set K is (weakly) compact and that $\Gamma(g)$ is non-empty for any $g \in K$. It is easy to see that a fixed point of Γ is a partial IDE until ξ. So the main part of the proof is to show that K and Γ satisfy the conditions of the Kakutani–Fan–Glicksberg fixed point theorem (Theorem 2.56) which then guarantees the existence of such a fixed point.

Lemma 4.12. *For any feasible network with locally p-integrable network inflow rates and any partial IDE with locally p-integrable flow rates the set K is non-empty, convex, weakly and sequentially weakly compact (i.e. (sequentially) compact with respect to the weak topology on $L^p(\mathbb{R}_{\geq 0})$).*

Proof. Convexity of K is clear since all constraints defining this set are linear. The set is non-empty as well, as it contains at least the following flow: For any commodity $i \in I$ and any node $v \in V \setminus (V_i^\dagger \cup T_i)$ we choose some edge $e_{v,i} \in \delta^+(v) \setminus \delta^-(V_i^\dagger)$ (such an edge must exist as otherwise v would be a dead-end node for commodity i). We then define g by setting

$$g^-_{e,i}(\theta) := \begin{cases} f^-_{e,i}(\theta), & \text{if } \theta < \xi + \tau_e \\ 0, & \text{else} \end{cases}$$

and

$$g^+_{e,i}(\theta) := \begin{cases} f^+_{e,i}(\theta), & \text{if } \theta < \xi \\ u_{v,i}(\theta) + \sum_{e' \in \delta^-(v)} g^-_{e,i}(\theta), & \text{if } \theta \in [\xi, \xi + \varepsilon) \text{ and } e = e_{v,i} \\ 0, & \text{else} \end{cases}$$

for all $e \in E$, $i \in I$ and $\theta \in \mathbb{R}_{\geq 0}$. Since $f^-_{e,i}$, $f^+_{e,i}$ and $u_{v,i}$ are all locally p-integrable, $g^-_{e,i}$ and $g^+_{e,i}$ are so as well. As they have bounded support we even have $g^-_{e,i}, g^+_{e,i} \in L^p(\mathbb{R}_{\geq 0})$. Moreover, this flow respects the capacities of all edges since f does so until ξ. For the same reason it also satisfies strong flow conservation at all nodes until ξ. For $[\xi, \xi + \varepsilon)$ we show this flow conservation by distinguishing three cases:

1. Case: $v \in V \setminus (V_i^\dagger \cup T_i)$: Here, we have

$$u_{v,i}(\theta) + \sum_{e \in \delta^-(v)} g_{e,i}^-(\theta) = g_{e_{v,i},i}^+(\theta) = \sum_{e \in \delta^+(v)} g_{e,i}^+(\theta)$$

for all $\theta \in [\xi, \xi + \varepsilon)$.

2. Case: $v \in T_i$: In this case we have $g_{e,i}^+(\theta) = 0$ for all outgoing edges from v during $[\xi, \xi + \varepsilon)$. Hence, strong flow conservation holds at v since v is a sink.

3. Case: $v \in V_i^\dagger$: Here, we also have $g_{e,i}^+(\theta) = 0$ for all $e \in \delta^+(v)$ during $[\xi, \xi + \varepsilon)$. Thus, we have to show that we also have $u_{v,i}(\theta) =_{\text{a.e.}} 0$ and $g_{e,i}^-(\theta) =_{\text{a.e.}} 0$ for all $e \in \delta^-(v)$ during that time. The first part holds since the network is feasible. The second part follows directly from our definition of g^- for $\theta \geq \xi + \tau_e$, so we only have to consider times $\theta \in [\xi, \xi + \tau_e)$ here and, in particular, only those edges with $\tau_e > 0$. For those edges Proposition 3.66 ensures that we have $g_{e,i}^-(\theta) = f_{e,i}^-(\theta) = 0$ for all $\theta < \xi + \tau_e$ (note that there must be some time $\xi' \leq \xi$ with $T_e^f(\xi') = \xi + \tau_e$ and then f is an IDE until ξ' and $(f_{e,\cdot}^+, f_{e,\cdot}^-)$ a Vickrey edge flow until $\xi' + \frac{Q_e(\xi')}{\nu_e}$).

The remaining constraints follow directly from the definition of g. Thus, we have $g \in K$.

Finally, K is weakly compact since it is convex, bounded (both network-inflow and edge outflow rates are bounded—thus, the edge inflow rates are bounded as well due to strong flow conservation at the nodes) and closed (with respect to the strong topology) subset of a reflexive space (cf. Proposition 2.41). Sequential weak compactness then follows as well since K is a subset of a normed space (cf. Proposition 2.35). □

Lemma 4.13. *For any feasible network with locally p-integrable network inflow rates and any partial IDE with locally p-integrable flow rates the mapping Γ has a closed graph (with respect to the weak topology) and non-empty convex images.*

Proof. We first note that for any $g \in K$ and $h, h' \in \Gamma(g)$ we have $h^- =_{\text{a.e.}} h'^-$. This is, because for Vickrey edge flows the outflow rates are uniquely determined by the edge inflow rates (cf. Corollary 3.43). Convexity of $\Gamma(g)$ is then obvious as the other constraint defining $\Gamma(g)$ is just a linear constraint. To show $\Gamma(g) \neq \emptyset$ we can find an element in $\Gamma(g)$ by defining all $h_{e,i}^-$ on $[0, \xi + \varepsilon + \tau_e]$ such that all $(g_{e,\cdot}^+, h_{e,\cdot}^-)$ are Vickrey edge flows until $\xi + \varepsilon$ (this is always possible by Corollary 3.43) and set them to zero on $[\xi+\varepsilon+\tau_e, \infty)$. This, in particular, implies $h^-_{e,i}\big|_{[0,\xi+\tau_e)} =_{\text{a.e.}} f^-_{e,i}\big|_{[0,\xi+\tau_e)}$ since we have $g^+_{e,i}\big|_{[0,\xi)} =_{\text{a.e.}} f^+_{e,i}\big|_{[0,\xi)}$ and f is a Vickrey flow until ξ.

In order to define $h^+_{e,i}$ we first observe that for any commodity i and any node $v \in V \setminus (V_i^\dagger \cup T_i)$ there will always be at least one outgoing active edge with respect to g (cf. Proposition 2.67e)). Thus, for any such node v we can partition $[\xi, \xi+\varepsilon)$ into measurable subsets $(M_{v,e})_{e\in\delta^+(v)}$ such that for every time $\theta \in [\xi, \xi+\varepsilon)$ we have

$$\theta \in M_{v,e} \implies e \in E_i^g(\theta).$$

We now define

$$h^+_{e,i}(\theta) := \begin{cases} f^+_{e,i}(\theta), & \text{if } \theta < \xi \\ u_{v,i}(\theta) + \sum_{e'\in\delta^-(v)} h^-_{e,i}(\theta), & \text{if } \theta \in M_{v,e}, e = vw, v \in V \setminus (V_i^\dagger \cup T_i) \\ 0, & \text{else} \end{cases}$$

for all $e \in E$, $i \in I$ and $\theta \in \mathbb{R}_{\geq 0}$. This gives us locally p-integrable flow rates such that h^+ only uses active edges (with respect to g) until $\xi+\varepsilon$ and is zero afterwards. Strong flow conservation at all nodes can be shown in essentially the same way as in the proof of Lemma 4.13: It holds until ξ since we have $(h, \xi) \approx (f, \xi)$ and the latter is a partial Vickrey flow. Afterwards, it holds at all nodes $v \in V \setminus V_i^\dagger$ by our definition and at all dead-end nodes $v \in V_i^\dagger$ provided that we have $h^-_{e,i}(\theta) = 0$ for all $e \in \delta^-(v)$ and almost all $\theta \in [\xi, \xi+\varepsilon)$ (note that, again, we have $u_{v,i} =_{\text{a.e.}} 0$ for all such nodes v as the network is feasible). We show this by distinguishing two cases:

1. **Case: $\xi+\varepsilon < \tau_e$:** In this case we immediately get $h^-_{e,i}(\theta) = 0$ for almost all $\theta < \xi+\varepsilon$ since $(g^+_{e,\cdot}, h^-_{e,\cdot})$ is a Vickrey edge flow until $\xi+\varepsilon > 0$ and, therefore, satisfies $h^-_{e,i}(\theta) = 0$ for almost all $\theta < \tau_e$.
2. **Case: $\xi+\varepsilon \geq \tau_e$:** In this case we define $\vartheta := \max\{\theta \leq \xi \mid T_e(\theta) \leq \xi+\varepsilon\}$. By Proposition 3.66 we then have $h^-_{e,i}(\theta) = 0$ for almost all $\theta \in [\xi, T_e(\vartheta))$. If $T_e(\vartheta) = \xi+\varepsilon$, then we are done. Otherwise, we have $\vartheta = \xi$ and

$$H^-_{e,i}(\xi+\varepsilon) \overset{(\#)}{\leq} G^+_{e,i}(\xi+\varepsilon) \overset{(*)}{=} G^+_{e,i}(\xi) \overset{(\#)}{=} H^-_{e,i}(T_e(\xi)) = H^-_{e,i}(T_e(\vartheta)),$$

where we used $g \in K$ at (*) and the fact that $(g^+_{e,\cdot,\cdot}, h^-_{e,\cdot})$ is a Vickrey edge flow until $\xi+\varepsilon$ at (#). This, now implies $h^-_{e,i}(\theta) = 0$ for almost all $\theta \in [T_e(\vartheta), \xi+\varepsilon)$.

Thus, we have $h \in \Gamma(g)$ and, therefore, $\Gamma(g) \neq \emptyset$.

It now remains to show that Γ has a closed graph. We will accomplish this by showing that the graph of Γ is weakly sequentially compact using the continuity properties of the edge loading as well as the mapping from flows rates to distance labels (Corollaries 3.45 and 3.46 and Proposition 3.63). Since $\left(L^p(\mathbb{R}_{\geq 0})^{E\times I\times\{+,-\}}\right)^2$ is a normed space, this is enough to show that the graph of Γ is weakly compact (cf. Proposition 2.35). Since $\left(L^p(\mathbb{R}_{\geq 0})^{E\times I\times\{+,-\}}\right)^2$ is Hausdorff with respect to the weak topology (Propositions 2.34 and 2.43), this implies that the graph of Γ is also weakly closed.

So, let $(g^{(n)}, h^{(n)}) \in \text{graph}(\Gamma)^{\mathbb{N}^*}$ be a sequence in the graph of Γ. We want to show that it has a subsequence converging in graph(Γ). Since $K \times K$ is weakly sequentially compact by Lemma 4.12 and Proposition 2.43, there exists a subsequence of $(g^{(n)}, h^{(n)})$ converging to some $(g, h) \in K \times K$. By some abuse of notation we will also denote this subsequence by $(g^{(n)}, h^{(n)})$. We now want to show that $(g, h) \in \text{graph}(\Gamma)$ or, equivalently, $h \in \Gamma(g)$.

For any edge $e \in E$ we have $h^{(n),-}_{e,\cdot} = \Psi^{\xi+\varepsilon}_e(g^{(n),+}_{e,\cdot}) \xrightarrow{w} \Psi^{\xi+\varepsilon}_e(g^+_{e,\cdot})$ by Corollary 3.46, where $\Psi^{\xi+\varepsilon}_e$ is the mapping from inflow rates to outflow rates. At the same time we have $h^{(n),-}_{e,\cdot} \xrightarrow{w} h^-_{e,\cdot}$ and, therefore, $\Psi^{\xi+\varepsilon}_e(g^+_{e,\cdot}) = h^-_{e,\cdot}$ as limit points in Hausdorff spaces are unique and $L^p(\mathbb{R}_{\geq 0})^I$ is a Hausdorff space with respect to the weak topology (cf. Proposition 2.34). Thus, h satisfies the first property required of elements in $\Gamma(g)$.

Now, assume that h does not satisfy the second property, i.e. there exist some $\gamma > 0$, $e = vw \in E$, $i \in I$ and a subset $J \subseteq \mathbb{R}_{\geq 0}$ of positive measure such that we have $h^+_{e,i} > 0$ on J and $L^g_{v,i} + \gamma \leq L^g_{w,i} + C^g_e$ on J. As $g^{(n),+}_{e,i}$ converges weakly to $g^+_{e,i}$, the edge flows $(g^{(n),+}_{e,\cdot}, g^{(n),-}_{e,\cdot})$ converges weakly to $(g^+_{e,\cdot}, g^-_{e,\cdot})$ by Corollary 3.46, the travel times $C^{g^{(n)}}_e$ converge uniformly to C^g_e by Corollary 3.45 and the distance labels $L^{g^{(n)}}_{v',i}$ converge uniformly to $L^g_{v',i}$ for all $v' \in V$ by Proposition 3.63. Thus, we have $L^{g^{(n)}}_{v,i} + \frac{\gamma}{2} \leq L^{g^{(n)}}_{w,i} + C^{g^{(n)}}_e$ on J for large enough n. But at the same time we also have $\int_J h^{(n),+}_{e,i}(\zeta)\mathrm{d}\zeta > 0$ for large enough n since $h^{(n),+}_{e,i}$ converges weakly to $h^+_{e,i}$—a contradiction to $(g^{(n)}, h^{(n)}) \in \text{graph}(\Gamma)$.

Thus, we do have $h \in \Gamma(g)$ and, hence, graph(Γ) is weak sequentially compact. This then implies, as described before, that graph(Γ) is weakly closed. □

Lemma 4.14. *Let $\mathcal{N}$ be a feasible network with locally p-integrable network inflow rates and (f, ξ) a partial IDE in $\mathcal{N}$ with p-integrable flow rates. Then, for any $\varepsilon > 0$ there exists a partial IDE $(g, \xi + \varepsilon)$ in $\mathcal{N}$ with p-integrable flow rates and which extends (f, ξ).*

Proof. $L^p(\mathbb{R}_{\geq 0})$ equipped with the weak topology is a locally convex Hausdorff space (cf. Proposition 2.34) and, thus, so is $L^p(\mathbb{R}_{\geq 0})^{E \times I \times \{+,-\}}$ (equipped with the product topology). Lemmas 4.12 and 4.13 then allow us to apply Theorem 2.56 in order to obtain a fixed point $g \in K$. The definition of K then already ensures that g satisfies strong flow conservation at all nodes until $\xi + \varepsilon$ and that it is an extension of (f, ξ), while $g \in \Gamma(g)$ means that g is a Vickrey edge flow until $\xi + \varepsilon$ on all edges and satisfies the IDE property until $\xi + \varepsilon$. Hence, $(g, \xi + \varepsilon)$ is a partial IDE extending (f, ξ). □

With this extension lemma we now immediately get our first existence result for IDE:

Theorem 4.15. *Let $\mathcal{N}$ be a feasible network with locally p-integrable network inflow rates for some $p > 1$. Then any partial IDE with p-integrable flow rates can be extended to an IDE for all times. In particular, there exists an IDE (for all times) in $\mathcal{N}$.*

Proof. We want to apply Theorem 4.6: So, let (f, ξ) be a partial IDE with p-integrable flow rates and define

$$\mathfrak{F} := \{\, (g, \xi') \in \mathcal{F}_{pa}^{\mathrm{IDE}}(\mathcal{N}) \mid (f, \xi) \preceq (g, \xi'),\ g \text{ has locally } p\text{-integrable flow rates} \,\}.$$

This set is clearly not empty (as it contains at least (f, ξ)). Furthermore, it satisfies property 2 due to Lemma 4.14 and property 3 because of Corollary 4.11. Thus, Theorem 4.6 shows the existence of an IDE $(f, \infty) \in \mathfrak{F}$.

The existence of IDE now follows from the observation that the zero-flow is always an IDE until $\xi = 0$. □

Remark 4.16. It is interesting to note that this existence proof is largely independent of both the underlying physical and the behavioural model. More precisely, we essentially only use the following properties of our model:

- The edge dynamics induce a well defined edge loading function Φ_e (Corollary 3.43) which is sequentially weak-weak continuous (Corollary 3.46).
- The distance labels are defined such that the mapping from flow rates to distance labels is sequentially weak-strong continuous (Corollary 3.45 and Proposition 3.63).
- Changing a partial flow (f, ξ) after time ξ does not affect whether it is a partial IDE up to time ξ.

Thus, the existence proof in this section should be easily adaptable to other models satisfying these properties, e.g. flow dynamics induced by the linear edge delay model (cf. e.g. [CM02, Section 4]) or distance labels derived from other predictions depending only on the past flow (e.g. causal predictors as defined in [GHKM23, Definition 4]).

4.3 Extension-Lemmas using IDE-Thin Flows

In this section we consider the case of networks with right-constant network inflow rates and show extension-lemmas for partial flows with right-constant edge inflow rates[1] (by Corollary 3.44 such a flow automatically also has right-constant edge outflow rates). By choosing the extension period small enough we can then ensure that all in- and outflow rates are constant during this period. Thus, we can describe the extension by a single value (instead of a whole function) for every edge-commodity pair.

In other words, the goal is to find a constant extension of a given partial IDE for some proper interval. Such an extension is, therefore, completely determined by the length $\varepsilon > 0$ of the extension interval and for every commodity i and edge e a number $x^+_{e,i} \geq 0$ denoting the constant inflow rate of commodity i into edge e during the extension interval. If (f, ξ) is the given partial IDE, the extension $(g, \xi + \varepsilon)$ is then defined by

$$g^+_{e,i}(\theta) := \begin{cases} f^+_{e,i}(\theta), & \text{if } \theta < \xi \\ x^+_{e,i}, & \text{if } \theta \geq \xi. \end{cases} \tag{4.3}$$

$$g^-_{e,i} \text{ such that } (g^+_{e,\cdot}, g^-_{e,\cdot}) \text{ is a Vickrey edge flow}$$

[1] Recall that for right-constant functions $[f] \in L^p_{\text{loc}}(\mathbb{R}_{\geq 0})$ we always assume that we are using the unique right constant representative of this equivalence class. In particular, we are allowed to evaluate such functions at individual points (cf. Definition 2.20).

where the outflow rates are uniquely defined according to Corollary 3.43. This clearly is an extension of (f, ξ) with queues operating fair and at capacity on all edges until $\xi + \varepsilon$. To ensure that it is also a partial IDE (up to time $\xi + \varepsilon$) we have to verify that g satisfies strong flow conservation at all nodes and only enters active edges between ξ and $\xi + \varepsilon$. Since all edge in- and outflow rates are constant right after ξ and, therefore, all distance labels are linear for some time, it actually suffices to check whether these two conditions hold at ξ and whether they will continue to holds for the immediate future. Then we can just choose ε small enough such that this does not change before $\xi + \varepsilon$.

For flow conservation at the nodes we introduce additional variables $x^-_{e,i} \geq 0$ representing the edge outflow rates for the extension interval $[\xi, \xi + \varepsilon)$. Given the edge inflow rates $x^+_{e,i}$ and the partial flow (f, ξ) we can compute these outflow rates as follows (cf. Corollary 3.44): For edges $e \in E^0(\xi) := \{ e \in E \mid \tau_e = 0, q_e(\xi) = 0 \}$ with a current travel time of zero at time ξ by

$$x^-_{e,i} = \frac{x^+_{e,i} \cdot \nu_e}{\max\{ x^+_e, \nu_e \}}, \tag{4.4}$$

where we use x^+_e as shorthand for $\sum_{j \in I} x^+_{e,j}$, and for all other edges $e \in E \setminus E^0(\xi)$ by

$$x^-_{e,i} = f^-_{e,i}(\xi). \tag{4.5}$$

Note that, for (4.5) we assume that for edges with free flow travel time zero $(f^+_{e,\cdot}, f^-_{e,\cdot})$ is a Vickrey edge until some time $\xi + \beta$ with $\beta > 0$. We can do this without any effect on the sought-after extension g since for edges e with $T_e(\xi) > 0$ the outflow rates of a Vickrey edge flow are already completely determined by the inflow rates until time ξ (Corollary 3.43). Flow conservation at nodes for the extension interval then translates to the condition

$$\sum_{e \in \delta^+(v)} x^+_{e,i} = u_{v,i}(\xi) + \sum_{e \in \delta^-(v)} x^-_{e,i} \tag{4.6}$$

for all $i \in I$, $v \in V \setminus T_i$ and

$$\sum_{e \in \delta^+(t)} x^+_{e,i} \leq u_{t,i}(\xi) + \sum_{e \in \delta^-(t)} x^-_{e,i} \tag{4.7}$$

for all sink nodes $t \in T_i$ (cf. Proposition 3.52).

For the IDE-condition of only entering active edges we first of all require flow to only enter edges which are active at time ξ, i.e.

$$x^+_{e,i} = 0 \text{ for all } e \in E \setminus E_i(\xi). \tag{4.8}$$

However, this is not enough to guarantee for the IDE-condition to hold for any proper interval after ξ as—depending on the edge inflow rates—waiting times may change and edges might become inactive immediately after ξ (note that the opposite, i.e. edges becoming newly active, is not a problem due to the continuity of the travel times). To ensure that we only use active edges which also *stay* active for some proper time interval we introduce another set of variables $a_{v,i} \in \mathbb{R}$ representing the derivative of the node labels during the extension interval.

Note that, by Proposition 2.67a), we have $L_{v,i}(\theta) = \infty$ for some time θ if and only if we have $L_{v,i} \equiv \infty$. Thus, we will define the derivative of such a constantly infinite function to be 0. We can then compute these derivatives from the edge inflow rates (cf. Definition 3.60) by

$$a_{v,i} = \min\left\{ \sum_{e \in p} \psi_e(x^+_e) \;\middle|\; p \in P_{v,i}(\xi) \right\}, \tag{4.9}$$

for all $i \in I$ and $v \in V \setminus V^\dagger_i$ and

$$a_{i,v} = 0 \tag{4.10}$$

for all $i \in I$ and $v \in V^\dagger_i$. Here, ψ_e denotes the change of the waiting time on edge e depending on the aggregated inflow rate x_e (cf. Proposition 3.19b)), i.e.

$$\psi_e : \mathbb{R}_{\geq 0} \to \mathbb{R}, x_e \mapsto \begin{cases} \frac{x_e}{\nu_e} - 1, & \text{if } Q_e(\xi) > 0 \\ \max\left\{ \frac{x_e}{\nu_e} - 1, 0 \right\}, & \text{else} \end{cases}. \tag{4.11}$$

The condition that used edges stay active for some time after ξ is can now be formally stated as:

$$a_{v,i} = \psi_e(x^+_e) + a_{w,i} \text{ for all } e = vw \in E \text{ with } x^+_{e,i} > 0. \tag{4.12}$$

Thus, finding an extension to a partial IDE with right-constant flow rates is equivalent to finding a vector $(x^+, x^-, a) \in \mathbb{R}^{E \times I}_{\geq 0} \times \mathbb{R}^{E \times I}_{\geq 0} \times \mathbb{R}^{V \times I}$ satisfying eqs. (4.4) to (4.10) and (4.12). This motivates the following definition of IDE-thin flows in analogy to the thin flows with resetting for dynamic equilibria introduced in [KS11, Definition 6]:

Definition 4.17. Let (f, ξ) be a partial IDE with right-constant flow rates and a Vickrey edge flow for all edges and all times. Then a vector $(x^+, x^-, a) \in \mathbb{R}_{\geq 0}^{E \times I} \times \mathbb{R}_{\geq 0}^{E \times I} \times \mathbb{R}^{V \times I}$ an *IDE-thin flow* for (f, ξ) if it is a solution to the following system of equations:

$$x_{e,i}^- = \frac{x_{e,i}^+ \cdot \nu_e}{\max\{x_e^+, \nu_e\}} \qquad \text{for all } i \in I, e \in E^0(\xi) \tag{4.4}$$

$$x_{e,i}^- = f_{e,i}^-(\xi) \qquad \text{for all } i \in I, e \in E \setminus E^0(\xi) \tag{4.5}$$

$$\sum_{e \in \delta^+(v)} x_{e,i}^+ = u_{v,i}(\xi) + \sum_{e \in \delta^-(v)} x_{e,i}^- \qquad \text{for all } i \in I, v \in V \setminus T_i \tag{4.6}$$

$$\sum_{e \in \delta^+(t)} x_{e,i}^+ \leq u_{t,i}(\xi) + \sum_{e \in \delta^-(t)} x_{e,i}^- \qquad \text{for all } i \in I, t \in T_i \tag{4.7}$$

$$x_{e,i}^+ = 0 \qquad \text{for all } e \in E \setminus E_i(\xi) \tag{4.8}$$

$$a_{v,i} = \min\left\{ \sum_{e \in p} \psi_e(x_e^+) \;\middle|\; p \in P_{v,i}(\xi) \right\} \qquad \text{for all } i \in I, v \in V \setminus V_i^\dagger \tag{4.9}$$

$$a_{v,i} = 0 \qquad \text{for all } i \in I, v \in V_i^\dagger \tag{4.10}$$

$$a_{v,i} = \psi_e(x_e^+) + a_{w,i} \text{ if } x_{e,i}^+ > 0 \qquad \text{for all } i \in I, v \in V \text{ and } e = vw \in \delta^+(v). \tag{4.12}$$

Remark 4.18. It should be noted that, in contrast to the thin flows with resetting in the full information setting, the values $x_{e,i}^+$ of our IDE-thin flows do not form a static flow in the given network. This is because, with the exception of edges of free flow time zero, there is no direct connection between the inflow and outflow rates of any edge at time ξ.

Our goal for the rest of this section will now be to first formally show that IDE-thin flows can always be used to extend partial IDE with right-constant flow rates and then derive extension-lemmas from this by providing sufficient conditions for the existence of IDE-thin flows. Our first step on this path is to collect several basic properties of IDE-thin flows:

Proposition 4.19. *The property of being an IDE-thin flow is well defined for partial flows, i.e. if $(f, \xi) \approx (f', \xi)$ are two equivalent partial flows, then any vector (x^+, x^-, a) is an IDE-thin flow for (f, ξ) if and only if it is an IDE-thin flow for (f', ξ).*

Proof. This immediately clear for eqs. (4.4), (4.6) to (4.10) and (4.12) since the sets $E^0(\xi)$ and $E_i(\xi)$ as well as the queue length at time ξ are completely determined by the inflow rates on $[0, \xi)$ and the outflow rates $[0, \xi + \tau_e)$. For eq. (4.5) it follows from Corollary 3.43 since for (4.5) we assume that $f^-_{e,i}$ is defined on some slightly larger time interval $[0, \xi + \beta)$ in such a way that $(f^+_{e,\cdot}, f^-_{e,\cdot})$ becomes a Vickrey edge flow until $\xi + \beta > \xi$ in case we have $\tau_e = 0$ (and equivalently so for f'). □

Proposition 4.20. *Let (x^+, x^-, a) be an IDE-thin flow for some partial IDE (f, ξ) and $i \in I$ any commodity. Then $(a_{v,i})_{v \in V \setminus V_i^\dagger}$ are node labels in the active subgraph $G(\xi) := (V \setminus V_i^\dagger, E[V \setminus V_i^\dagger] \cap E_i(\xi))$ with edge costs $(\psi_e(x_e^+))_{e \in E[G(\xi)]}$ satisfying the properties a) to m) in Proposition 2.67.*

Proof. Since f satisfies weak flow conservation until ξ, we get from Proposition 2.67l) (cf. Remark 3.61) that $P_{v,i}(\xi)$ contains exactly those v, T_i-paths consisting exclusively of edges in $E_i(\xi)$. Thus, (4.9) is equivalent to the definition of node labels in $G(\xi)$ with edge costs $(\psi_e(x_e^+))_{e \in E[G(\xi)]}$.

Moreover, for any cycle c in G' we have $C_c(\xi) = 0$ by Proposition 2.67h) and, therefore, $C_e(\xi) = 0$ for all edges $e \in c$. This, in turn, implies $\psi_e(x_e^+) \geq 0$ for all such edges. Similarly, we have $C_p(\xi) = 0$ for any T_i, T_i-path in G' by Proposition 2.67k) and, hence, $C_e(\xi) = 0$ and $\psi_e(x_e^+) \geq 0$ for all edges e on such a path. Thus, all the properties a) to m) in Proposition 2.67 hold for $(a_{v,i})_{v \in V \setminus V_i^\dagger}$ as well. □

Proposition 4.21. *The mapping*

$$\mathbb{R}_{\geq 0}^{E \times I} \to \mathbb{R}^{V \times I}, (x_{e,i}^+)_{e,i} \mapsto (a_{v,i}) \text{ such that } (x^+, a) \text{ satisfy eqs. (4.9) and (4.10)}$$

is well defined and continuous.

Proof. By Proposition 2.67b) the set $P_{v,i}(\xi)$ is non-empty for any node $v \in V \setminus V_i^\dagger$. Thus, the minimum in (4.9) is well defined for any such node as it is taken over a finite non-empty set. This already shows that the given mapping is well defined. Continuity then follows directly from the continuity of the function ψ_e which, in turn, follows directly from its definition (see (4.11)). □

The following proposition now states that there is indeed a one-to-one correspondence between IDE-thin flows and constant extensions of partial IDE with right-constant flow rates.

Proposition 4.22. *Let (f, ξ) be a partial IDE with right-constant in- and outflow rates which is a Vickrey edge flow for all times on all edges in a network with right-constant network inflow rates. A vector $x^+ \in \mathbb{R}_{\geq 0}^{E \times I}$ defines a non-trivial constant extension of (f, ξ) via (4.3) to some partial IDE $(g, \xi + \varepsilon)$ if and only if there exist vectors $x^- \in \mathbb{R}_{\geq 0}^{E \times I}$ and $a \in \mathbb{R}^{V \times I}$ such that (x^+, x^-, a) is an IDE-thin flow for f at ξ.*

In this case we have $g_{e,i}^-(\theta) = x_{e,i}^-$ and $\partial L_{v,i}^g(\theta) = a_{v,i}$ during the extension period.

Proof. **Sufficiency:** Given an IDE-thin flow (x^+, x^-, a) we first want to determine some $\varepsilon > 0$ such that a constant extension is possible for an interval of length ε. First, we need to ensure that the edge outflow rates do not change during the interval:

$$\varepsilon_1 := \min\left\{ \max\left\{ \varepsilon' \leq C_e(\xi) \;\middle|\; f_{e,i}^-\big|_{[\xi,\xi+\varepsilon')} \text{ constant} \right\} \;\middle|\; e \in E \setminus E^0(\xi) \right\}$$

Similarly, the network inflow rates should stay constant for our extension interval:

$$\varepsilon_2 := \max\left\{ \varepsilon' \geq 0 \;\middle|\; u_{v,i}\big|_{[\xi,\xi+\varepsilon')} \text{ is constant for all } i \in I, v \in V \right\}$$

Next, no queues should deplete during the extension period (as this changes the rate at which the travel time along that edge changes):

$$\varepsilon_3 := \sup\left\{ \varepsilon' \geq 0 \mid Q_e(\xi) + \varepsilon'(x_e^+ - \nu_e) \geq 0 \text{ for all } e \in E \text{ with } Q_e(\xi) > 0 \right\}$$

Finally, we have to ensure that no path becomes newly active during the extension interval:

$$\varepsilon_4 := \max\left\{ \varepsilon' \geq 0 \;\middle|\; \begin{array}{c} L_{v,i}(\xi) + \varepsilon' \cdot a_{v,i} \leq C_p(\xi) + \varepsilon' \cdot \left(\sum_{e \in p} \psi_e(x_e^+)\right) \\ \text{for all } i \in I, v \in V \text{ and all } v, T_i\text{-paths } p \\ \text{which are inactive for } i \text{ at time } \xi \end{array} \right\}$$

We now choose $\varepsilon := \min\{\varepsilon_1, \varepsilon_2, \varepsilon_3, \varepsilon_4\}$. This is a strictly positive number since all four ε_k are positive:

- $\varepsilon_1 > 0$ because the edge outflow rates of f are assumed to be right-constant and by the definition of the set $E^0(\xi)$ we have $C_e(\xi) > 0$ for all $e \in E^0(\xi)$,
- $\varepsilon_2 > 0$ because the network-inflow rates are assumed to be right-constant,
- $\varepsilon_3 > 0$ because the left side of the inequality is continuous in ε' and strictly larger than 0 for $\varepsilon' = 0$ and
- $\varepsilon_4 > 0$ because both sides of the inequality are continuous in ε', the inequality is strict for $\varepsilon' = 0$ (by the definition of (non-)active paths).

We will show that extending (f, ξ) by constant edge inflow rates $(x^+_{e,i})$ for $[\xi, \xi + \varepsilon)$ results in a partial IDE $(g, \xi + \varepsilon)$. By construction g is already a Vickrey edge flow on every edge. Furthermore, since we have $(f, \xi) \preceq (g, \xi + \varepsilon)$ (cf. Observation 4.4), we also know that the IDE-property and strong flow conservation at all nodes holds until ξ. Thus, it suffices to check that these two properties also hold during the extension interval. For this, we show that during the extension interval $[\xi, \xi+\varepsilon)$ the variables $a_{v,i}$ and $x^-_{e,i}$ fulfil their role correctly:

Claim 4.1. *For all $i \in I$, $e \in E$, $v \in V$, paths p and $\theta \in [\xi, \xi + \varepsilon)$ we have*

- $g^-_{e,i}(\theta) = x^-_{e,i}$,
- $Q^g_e(\theta) = Q_e(\xi) + (\theta - \xi) \cdot \nu_e \cdot \psi_e(x^+_e)$,
- $C^g_p(\theta) = C_p(\xi) + (\theta - \xi) \sum_{e \in p} \psi_e(x^+_e)$ *and*
- $L^g_{v,i}(\theta) = L_{v,i}(\xi) + (\theta - \xi) \cdot a_{v,i}$.

Proof. **Outflow rates:** For all edges $e \in E^0(\xi)$ we have

$$g^-_{e,i}(\theta) \overset{\text{Cor. 3.44}}{=} \frac{g^+_{e,i}(\xi)\nu_e}{\max\{g^+_e(\xi), \nu_e\}} \overset{(4.3)}{=} \frac{x^+_{e,i}\nu_e}{\max\{x^+_e, \nu_e\}} \overset{(4.4)}{=} x^-_{e,i}.$$

For all other edges $e \in E \setminus E^0(\xi)$ we have

$$g^-_{e,i}(\theta) = f^-_{e,i}(\theta) = f^-_{e,i}(\xi) \overset{(4.5)}{=} x^-_{e,i},$$

where the second equality holds since ε was chosen such that $f^-_{e,i}$ is constant on $[\xi, \xi + \varepsilon)$ and the first equality holds by Corollary 3.43 since we have

$$T^f_e(\xi) = \xi + C^f_e(\xi) \geq \xi + \varepsilon_1 \geq \xi + \varepsilon > \theta$$

by our choice of ε.

Queue lengths: This follows directly from Corollary 3.43 and the definition of ψ_e and the fact that we chose ε such that no queue depletes before time $\xi + \varepsilon$:

$$Q_e^g(\theta) \overset{\text{Cor. 3.43}}{=} \begin{cases} Q_e^g(\xi) + (\theta - \xi)\nu_e \max\{x_e^+ - \nu_e, 0\}, & \text{if } Q_e^g(\xi) = 0 \\ Q_e^g(\xi) + (\theta - \xi)\nu_e(x_e^+ - \nu_e), & \text{if } Q_e^g(\xi) > 0 \end{cases}$$

$$\overset{(4.11)}{=} Q_e(\xi) + (\theta - \xi)\nu_e\psi_e(x_e^+).$$

Path travel times: This follows directly from the the previous result for the queue length functions: For all paths p and times $\theta \in [\xi, \xi + \varepsilon)$ we have:

$$\begin{aligned} C_p^g(\theta) &= \sum_{e \in p} C_e^g(\theta) = \sum_{e \in p} \left(\tau_e + \tfrac{Q_e^g(\theta)}{\nu_e}\right) \\ &= \sum_{e \in p} \left(\tau_e + \tfrac{Q_e(\xi)}{\nu_e} + (\theta - \xi)\psi_e(x_e^+)\right) \\ &= \sum_{e \in p} C_e(\xi) + (\theta - \xi)\sum_{e \in p} \psi_e(x_e^+) = C_p(\xi) + (\theta - \xi)\sum_{e \in p} \psi_e(x_e^+). \end{aligned}$$

Node distances: Here we distinguish between two cases:

1. **Case: $v \in V_i^\dagger$:** Then there are no v, T_i-paths and, thus, we have

$$\begin{aligned} L_{v,i}^g(\theta) &\overset{\text{Prop. 2.67a)}}{=} \infty = \infty + (\theta - \xi) \cdot 0 \\ &\overset{\text{Prop. 2.67a),(4.10)}}{=} L_{v,i}(\xi) + (\theta - \xi) \cdot a_{v,i} \end{aligned}$$

   for all $\theta \in [\xi, \xi + \varepsilon)$.

2. **Case: $v \notin V_i^\dagger$:** Then the set $P_{v,i}(\xi)$ is non-empty by Proposition 2.67b). In particular, there exists a v, T_i-path p with

$$p \in \arg\min\left\{\sum_{e \in p'} \psi_e(x_e^+) \;\middle|\; p' \in P_{v,i}(\xi)\right\}.$$

   Using the previous result on path travel times and the fact that p is active for commodity i at time ξ, this path then satisfies

$$C_p^g(\theta) = C_p(\xi) + (\theta - \xi)\sum_{e\in p} \psi_e(x_e^+) \overset{(4.9)}{=} C_p(\xi) + (\theta - \xi)a_{v,i}$$
$$= L_{v,i}(\xi) + (\theta - \xi)a_{v,i}.$$

We now want to show that p remains active for the whole interval $[\xi, \xi + \varepsilon)$. So, take any other path $p' \in P_{v,i}(\xi)$ and observe that by the choice of p we then have

$$C_{p'}^g(\theta) = C_{p'}(\xi) + (\theta - \xi)\sum_{e\in p'} \psi_e(x_e^+)$$
$$\geq C_p(\xi) + (\theta - \xi)\sum_{e\in p} \psi_e(x_e^+) = C_p^g(\theta)$$

for all $\theta \in [\xi, \xi + \varepsilon)$. On the other hand, for any v, T_i-path p' not active at time ξ the choice of ε_4 guarantees

$$C_{p'}^g(\theta) = C_{p'}(\xi) + (\theta - \xi)\sum_{e\in p'} \psi_e(x_e^+) \geq L_{v,i}(\xi) + (\theta - \xi)a_{v,i} = C_p^g(\theta).$$

Thus, p remains active for the whole interval $[\xi, \xi + \varepsilon)$ and, therefore, we have

$$L_{v,i}^g(\theta) = C_p^g(\theta) = L_{v,i}(\xi) + (\theta - \xi) \cdot a_{v,i}$$

for all $\theta \in [\xi, \xi + \varepsilon)$. ■

With this claim strong flow conservation at the node now follows directly from eqs. (4.6) and (4.7) (note that the choice of ε_3 ensures that all network inflow rates are constant on $[\xi, \xi + \varepsilon)$). Similarly, the IDE-property now follows directly from eq. (4.12) as for any edge $e = vw$ with positive inflow of commodity i during the extension interval we get:

$$L_{v,i}^g(\theta) \overset{\text{Claim 4.1}}{=} L_{v,i}(\xi) + (\theta - \xi)a_{v,i} \overset{(4.12)}{=} L_{v,i}(\xi) + (\theta - \xi)\big(\psi_e(x_e^+) + a_{w,i}\big)$$
$$\overset{\text{Claim 4.1}}{=} L_{v,i}(\xi) + \tfrac{1}{\nu_e} \cdot \big(Q_e^g(\theta) - Q_e(\xi)\big) + L_{w,i}^g(\theta) - L_{w,i}(\xi)$$
$$= L_{v,i}(\xi) - \Big(\tfrac{Q_e(\xi)}{\nu_e} + L_{w,i}(\xi)\Big) + L_{w,i}^g(\theta) + \tfrac{Q_e^g(\theta)}{\nu_e}$$
$$\overset{(4.8)}{=} \tau_e + L_{w,i}^g(\theta) + Q_e^g(\theta) = C_e^g(\theta) + L_{w,i}^g(\theta),$$

for all $\theta \in [\xi, \xi + \varepsilon)$ which shows that e is active for commodity i during the whole extension interval.

Neccessity: Let $(f, \xi) \preceq (g, \xi+\varepsilon)$ be two partial IDE with right-constant flow rates and $\varepsilon > 0$ chosen such that we have $g^+_{e,i}|_{[\xi,\xi+\varepsilon)} = x^+_{e,i}$ for all $e \in E$ and $i \in I$. We define vectors $(x^-_{e,i})$ and $(a_{v,i})$ by setting $x^-_{e,i} := g^-_{e,i}(\xi)$ and $a_{v,i} := \partial_+ L^g_{v,i}(\xi)$ and claim that (x^+, x^-, a) is then an IDE-thin flow for (f, ξ). As in the proof of Proposition 4.19 we observe that $E^f_i(\xi) = E^g_i(\xi)$ and $E^{0,f}(\xi) = E^{0,g}(\xi)$ since f and g coincide until ξ. Thus, we will not distinguish between those sets for f and g in the following proof and drop the corresponding index.

(4.4): Let $e \in E^0(\xi)$ be an edge with zero travel time at ξ and $i \in I$ any commodity. Using the fact that g is a Vickrey flow until $\xi + \varepsilon$ (in particular the queue on edge e operates fair and at capacity) we get

$$x^-_{e,i} = g^-_{e,i}(\xi) \overset{(3.13)}{=} g^-_e(\xi) \cdot \frac{g^+_{e,i}(\xi)}{g^+_e(\xi)} \overset{(3.4)}{=} \min\{g^+_e(\xi), \nu_e\} \cdot \frac{g^+_{e,i}(\xi)}{g^+_e(\xi)}$$

$$= \frac{g^+_{e,i}(\xi)\nu_e}{\max\{g^+_e(\xi), \nu_e\}} = \frac{x^+_{e,i} \cdot \nu_e}{\max\{x^+_e, \nu_e\}}.$$

if $g^+_e(\xi) > 0$ and $x^-_{e,i} = 0 = \frac{x^+_{e,i} \cdot \nu_e}{\max\{x^+_e, \nu_e\}}$ directly from (3.13) if $x^+_e = g^+_e(\xi) = 0$.

(4.5): Let $e \in E \setminus E^0_\xi$ be an edge with non-zero travel time at ξ and $i \in I$ any commodity. Then the edge outflow close to time ξ is not affected by changes of the edge inflow at or after time ξ (cf. Corollary 3.43). Thus, we have

$$x^-_{e,i} = g^-_{e,i}(\xi) = f^-_{e,i}(\xi).$$

(4.6),(4.7): Since g is a Vickrey flow until $\xi + \varepsilon$, (4.6) and (4.7) follow directly from Proposition 3.52.

(4.8): This follows directly from the fact that g is an IDE until $\xi + \varepsilon$.

(4.9): Since g is a Vickrey flow until $\xi + \varepsilon > \xi$ with constant edge inflow rates x^+_e during $[\xi, \xi + \varepsilon)$, Proposition 3.22 guarantees the existence of some $\beta \in (0, \varepsilon]$ such that all queue lengths functions and, consequently, all node label functions are linear on $[\xi, \xi + \beta)$. For any v, T_i-path p we can then compute the current path travel times during this interval by

$$C_p^g(\theta) = \sum_{e \in p} \left(\tau_e + \frac{Q_e^g(\theta)}{\nu_e}\right) \overset{\text{Prop. 3.22}}{=} \sum_{e \in p} \left(\tau_e + \frac{Q_e(\xi)}{\nu_e} + (\theta - \xi)\psi_e(g_e^+(\xi))\right)$$
$$= C_p(\xi) + (\theta - \xi) \sum_{e \in p} \psi_e(x_e^+).$$

Furthermore, we can even choose β small enough such that there is a v, T_i-path $p \in P_{v,i}(\xi)$ which is active for the whole interval $[\xi, \xi + \beta)$. Then we have for any path $p' \in P_{v,i}(\xi)$:

$$C_{p'}(\xi) + (\theta - \xi) \sum_{e \in p'} \psi_e(x_e^+) = C_{p'}^g(\theta)$$
$$\geq C_p^g(\theta) = C_p(\xi) + (\theta - \xi) \sum_{e \in p} \psi_e(x_e^+)$$

and, therefore,

$$\sum_{e \in p'} \psi_e(x_e^+) \geq \sum_{e \in p} \psi_e(x_e^+)$$

Thus, we have

$$a_{v,i} = \partial_+ L_{v,i}^g(\xi) = \partial_+ C_p^g(\xi) = \sum_{e \in p} \psi_e(x_e^+) \leq \sum_{e \in p'} \psi_e(x_e^+)$$

for all $p' \in P_{v,i}(\xi)$.

(4.10): This follows directly from Proposition 2.67a) and our convention that the derivative of a function which is constantly infinite is 0.

(4.12): Let $e = vw$ be any edge with $x_{e,i}^+ > 0$ for some commodity i. Since the IDE-property holds for g until $\xi + \varepsilon$, we must have

$$L_{v,i}^g(\theta) = C_e^g(\theta) + L_{w,i}^g(\theta)$$

for all $\theta \in [\xi, \xi + \varepsilon)$ and, therefore,

$$a_{v,i} = \partial_+ L_{v,i}^g(\xi) = \partial_+ C_e^g(\xi) + \partial_+ L_{w,i}^g(\xi) = \frac{\partial_+ Q_e^g(\xi)}{\nu_e} + \partial_+ L_{w,i}^g(\xi)$$
$$\overset{(3.6)}{=} \psi_e(x_e^+) + a_{w,i}$$

Thus, (x^+, x^-, a) is indeed an IDE-thin flow for f at ξ. □

With this proposition we have now reduced the existence of IDE-extensions for partial IDE with right-constant flow rates to the existence of IDE-thin flows. We will show existence of such IDE-thin flows in Subsections 4.3.2 and 4.3.3. Before that, however, we will introduce and additional framework which will allow us to break down existence (and, later, computation) of IDE-thin flows further down into smaller subproblems.

4.3.1 IDE-Thin Flow Augmentation

The goal of this section is to show how under certain assumption we can split a given network into an "inner" and an "outer" part such that we can augment any IDE-thin flow for the inner part to an IDE-thin flow for the whole network. More precisely, let $\mathcal{N}$ be a network and (f, ξ) be a partial IDE in $\mathcal{N}$ with right-constant flow rates. We then consider a partition $V = \check{V} \mathbin{\dot\cup} \hat{V}$ of the node set into inner nodes $\check{V}$ and outer nodes $\hat{V}$ satisfying the following two assumptions

(i) $\forall i \in I : \delta^+(\check{V}) \cap E_i(\xi) = \emptyset$ and
(ii) $\forall i \in I : \delta^-(\check{V}) \cap E^0(\xi) \cap E_i(\xi) = \emptyset$.

Note that it is explicitly allowed here that one of the two sets may be empty.

Example 4.23. Consider an urban area with a central city (the inner part) and surrounding suburbs (the outer part). If we want to model the morning rush hour traffic it might be a reasonable assumption that there are only three types of travellers: Travellers with origin and destination either both inside or both outside the city and travellers with origin outside and destination inside the city. In this situation it might very well be the case that there are no relevant (=active) routes which leave the city at some point and, thus, assumption (i) would be satisfied.

Another example could be the holiday traffic at the start of the summer holidays in Bavaria where (at least according to folklore) everyone is travelling southwards and, thus, every split of the road network between a northern and a southern part would satisfy assumption (i).

From a more mathematical point of view networks which always allow partitions satisfying assumption (i) are single-sink networks and acyclic networks with strictly positive free flow travel times.

We now define the restricted inner network

$$\check{\mathcal{N}} := (G[\check{V}], \tau|_{E[\check{V}]}, \nu|_{E[\check{V}]}, I, \check{u}, (T_i \cap \check{V}))$$

where $\check{u}_{v,i} := u_{v,i} + \sum_{e \in (\delta^+(\hat{V}) \cap \delta^-(v)) \setminus E^0(\xi)} f^-_{e,i}$ defines the network inflow rates of the inner network.

Assumption (ii) ensures that in any IDE extension of (f, ξ) the outflow rates of all edges in $\delta^-(\check{V})$ remain unchanged for some time after ξ. Thus, the adjusted network inflow rates $\check{u}_{v,i}$ for the inner network ensure that any IDE-thin flow for the inner network already satisfies the flow conservation constraints at nodes when considered as part of an IDE-thin flow for the whole network. Assumption (i) guarantees that at time ξ all active paths starting at an inner node remain within the inner network. This, in particular, allows us to speak of IDE-thin flows for $(f|_{G[\check{V}]}, \xi)$ in the restricted network $\check{\mathcal{N}}$. Additionally, it also allows us to describe the v, T_i-paths for outer nodes v by introducing for any node $v \in \hat{V}$ and commodity $i \in I$ the following two sets:

$$P^+_{v,i}(\xi) := \left\{ (p, w) \;\middle|\; p \text{ a } v, w\text{-path, } w \in \check{V} \setminus V^\dagger_i, \; p \subseteq \hat{E} \cap E_i(\xi) \right\}$$

denotes the set of prefixes of active v, T_i-paths ending with an active edge connecting the outer and the inner part of the network while

$$\hat{P_{v,i}}(\xi) := \left\{ p \in P_{v,i}(\xi) \;\middle|\; \text{all nodes on } p \text{ are in } \hat{V} \right\}$$

denotes the set of active v, T_i-paths completely within the outer part of the network. The following lemma now explains how these two sets are related to the set v, T_i-path (in the full network):

Lemma 4.24. *Let (f, ξ) be a partial flow, $V = \check{V} \dot{\cup} \hat{V}$ a partition satisfying assumptions (i) and (ii), $(\check{x}^+, \check{x}^-, \check{a})$ an IDE-thin flow for $(f|_{G[\check{V}]}, \xi)$ in $\check{\mathcal{N}}$, $\hat{x}^+ \in \mathbb{R}^{\hat{E}}_{\geq 0}$ some vector and $x^+ := \check{x}^- \oplus \hat{x}^+ \in \mathbb{R}^E_{\geq 0}$. Then the following holds for any commodity $i \in I$:*

a) For any $p \in P_{v,i}(\xi) \setminus \hat{P_{v,i}}(\xi)$ there exists a pair $(p', w) \in P^+_{v,i}(\xi)$ such that

$$\sum_{e \in p} \psi_e(x^+_e) \geq \check{a}_{w,i} + \sum_{e \in p'} \psi_e(\hat{x}^+_e).$$

b) *For any pair* $(p', w) \in P^+_{v,i}(\xi)$ *there exists a path* $p \in P_{v,i}(\xi)$ *with*

$$\sum_{e \in p} \psi_e(x^+_e) = \check{a}_{w,i} + \sum_{e \in p'} \psi_e(\hat{x}^+_e).$$

Proof. a): Take any path $p \in P_{v,i}(\xi) \setminus \hat{P_{v,i}}(\xi)$. Due to assumption (i) we can subdivide this path as $p = p', p''$ where p' only uses edges in $E[\hat{V}]$ and p'' only uses edges in $E[\check{V}]$. Furthermore, let $w \in \check{V}$ be the node where the two subpaths connect. Then, we have $p'' \in P_{w,i}(\xi)$ and, thus, $w \notin V_i^{\dagger}$ by Proposition 2.67b). Furthermore, Proposition 2.67l) ensures $p' \subseteq E_i(\xi)$ and, therefore, $(p', w) \in P^+_{v,i}$ as well as

$$\sum_{e \in p} \psi_e(x^+_e) = \sum_{e \in p''} \psi_e(\check{x}^+_e) + \sum_{e \in p'} \psi_e(\hat{x}^+_e) \overset{(4.9)}{\geq} \check{a}_{w,i} + \sum_{e \in p'} \psi_e(\hat{x}^+_e).$$

b): Take any pair $(p', w) \in P^+_{v,i}(\xi)$. Then, due to (4.9), there must be a path $p'' \in P_{w,i}(\xi)$ with $\check{a}_{w,i} = \sum_{e \in p''} \psi_e(\check{x}^+_e)$. Since assumption (i) guarantees that p'' does not leave the inner network, $p := p', p''$ is still a simple path using only active edges and, thus, we have $p \in P_{v,i}(\xi)$ by Proposition 2.67l). Additionally, we have

$$\sum_{e \in p} \psi_e(x^+_e) = \sum_{e \in p''} \psi_e(\check{x}^+_e) + \sum_{e \in p'} \psi_e(\hat{x}^+_e) = \check{a}_{w,i} + \sum_{e \in p'} \psi_e(\hat{x}^+_e).$$

□

We can now formally describe the conditions a vector $(\hat{x}^+, \hat{x}^-, \hat{a}) \in \mathbb{R}^{\hat{E} \times I}_{\geq 0} \times \mathbb{R}^{\hat{E} \times I}_{\geq 0} \times \mathbb{R}^{\hat{V} \times I}$ must satisfy such that we can use it to augment a given IDE-thin flow for the inner part of the network. Here, we denote by $\hat{E} := E \setminus E[\check{V}]$ the set of edges which are not part of the inner network.

Definition 4.25. Let (f, ξ) be a partial IDE with right-constant flow rates in some network $\mathcal{N}$ with right-constant network inflow rates. Furthermore, let $V = \check{V} \dot{\cup} \hat{V}$ be a partition of the node set of $\mathcal{N}$ and $(\check{x}^+, \check{x}^-, \check{a}) \in \mathbb{R}^{E[\check{V}] \times I}_{\geq 0} \times \mathbb{R}^{E[\check{V}] \times I}_{\geq 0} \times \mathbb{R}^{\check{V} \times I}$ an IDE-thin flow for $(f|_{G[\check{V}]}, \xi)$ in $\check{\mathcal{N}}$.

Then a vector $(\hat{x}^+, \hat{x}^-, \hat{a}) \in \mathbb{R}^{\hat{E} \times I}_{\geq 0} \times \mathbb{R}^{\hat{E} \times I}_{\geq 0} \times \mathbb{R}^{\hat{V} \times I}$ is an *IDE-thin flow augmentation* of $(\check{x}^+, \check{x}^-, \check{a})$ if it satisfies the following equations:

$$\hat{x}^-_{e,i} = \frac{\hat{x}^+_{e,i} \cdot \nu_e}{\max\{\hat{x}^+_e, \nu_e\}} \qquad \text{for all } e \in E^0(\xi) \cap \hat{E} \tag{4.4*}$$

$$\hat{x}^-_{e,i} = f^-_{e,i}(\xi) \qquad \text{for all } e \in \hat{E} \setminus E^0(\xi) \tag{4.5*}$$

$$\sum_{e \in \delta^+(v)} \hat{x}^+_{e,i} = u_{v,i}(\xi) + \sum_{e \in \delta^-(v)} \hat{x}^-_{e,i} \qquad \text{for all } i \in I, v \in \hat{V} \setminus T_i \tag{4.6*}$$

$$\sum_{e \in \delta^+(t)} \hat{x}^+_{e,i} \leq u_{t,i}(\xi) + \sum_{e \in \delta^-(t)} \hat{x}^-_{e,i} \qquad \text{for all } i \in I, t \in T_i \cap \hat{V} \tag{4.7*}$$

$$\hat{x}^+_{e,i} = 0 \qquad \text{for all } e \in \hat{E} \setminus E_i(\xi) \tag{4.8*}$$

$$\hat{a}_{v,i} = \min \left(\begin{array}{l} \left\{ \check{a}_{w,i} + \sum_{e \in p} \psi_e(\hat{x}^+_e) \;\middle|\; (p, w) \in P^+_{v,i}(\xi) \right\} \\ \cup \left\{ \sum_{e \in p} \psi_e(\hat{x}^+_e) \;\middle|\; p \in \hat{P_{v,i}}(\xi) \right\} \end{array} \right) \tag{4.9*}$$

$$\text{for all } i \in I, v \in \hat{V} \setminus V^\dagger_i$$

$$\hat{a}_{v,i} = 0 \qquad \text{for all } i \in I, v \in \hat{V} \cap V^\dagger_i \tag{4.10*}$$

$$\hat{a}_{v,i} = \psi_e(\hat{x}^+_e) + \hat{a}_{w,i} \text{ if } \hat{x}^+_{e,i} > 0 \qquad \begin{array}{l} \text{for all } i \in I, v \in \hat{V} \text{ and} \\ e = vw \in \delta^+(v) \cap \delta^-(\check{V}). \end{array} \tag{4.12*a}$$

$$\hat{a}_{v,i} = \psi_e(\hat{x}^+_e) + \check{a}_{w,i} \text{ if } \hat{x}^+_{e,i} > 0 \qquad \begin{array}{l} \text{for all } i \in I, v \in \hat{V} \text{ and} \\ e = vw \in \delta^+(v) \cap \delta^-(\check{V}). \end{array} \tag{4.12*b}$$

Intuitively, the defining constraints for an IDE-thin flow augmentation are just the general IDE-thin flow constraints specialised to nodes in the outer part of the network (under the assumption that $(\check{x}^+, \check{x}^-, \check{a})$ already is an IDE-thin flow for the inner part). Thus, it should not be surprising that combining an IDE-thin flow for the inner network with an IDE-thin flow augmentation gives us an IDE-thin flow for the whole network. The following lemma shows that this is indeed the case:

Lemma 4.26. *Let (f, ξ) be a partial IDE with right-constant flow rates in some network $\mathcal{N}$ with right-constant network inflow rates and $V = \check{V} \,\dot{\cup}\, \hat{V}$ a partition satisfying assumptions (i) and (ii). Let $(\check{x}^+, \check{x}^-, \check{a})$ an IDE-thin flow for $(f|_{G[\check{V}]}, \xi)$ in $\check{\mathcal{N}}$ and $(\hat{x}^+, \hat{x}^-, \hat{a})$ an IDE-thin flow augmentation for $(\check{x}^+, \check{x}^-, \check{a})$.*

Then $(x^+, x^-, a) := (\check{x}^+, \check{x}^-, \check{a}) \oplus (\hat{x}^+, \hat{x}^-, \hat{a})$ is an IDE-thin flow for (f, ξ) in $\mathcal{N}$.

Proof. We first note that, since there are no active edges leaving $G[\check{V}]$ at time ξ (assumption (i)), all active paths from nodes in $\check{V}$ also stay inside $G[\check{V}]$. Thus, the sets of active paths from nodes in the inner network and, consequently, the sets of active edges within the inner network are the same regardless of whether we compute them with respect to (f, ξ) in the full network or with respect to $(f|_{G[\check{V}]}, \xi)$ in the inner network. Therefore, we do not need to distinguish between these two types of sets in the following proof.

(4.4),(4.5) **and** (4.8): For edges in $E[\check{V}]$ these are satisfied because $(\check{x}^+, \check{x}^-, \check{a})$ is an IDE-thin flow in the restricted network. For all other edges they are satisfied because $(\hat{x}^+, \hat{x}^-, \hat{a})$ satisfies the equivalent constraints (4.4*), (4.5*) and (4.8*).

(4.6): For nodes $v \in \hat{V} \setminus T_i$ this constraint is the same as (4.6*) for $(\hat{x}^+, \hat{x}^-, \hat{a})$. For nodes $v \in \check{V} \setminus T_i$ we have

$$\sum_{e\in\delta^+(v)} x^+_{e,i} = \sum_{e\in\delta^+(v)\cap E[\check{V}]} \check{x}^+_{e,i} + \sum_{e\in\delta^+(v)\cap\delta^+(\check{V})} \hat{x}^+_{e,i} \overset{\text{(i),(4.8*)}}{=} \sum_{e\in\delta^-_{G[\check{V}]}(v)} \check{x}^+_{e,i} + 0$$

$$\overset{(4.6)}{=} \check{u}_{v,i}(\xi) + \sum_{e\in\delta^-_{G[\check{V}]}(v)} \check{x}^-_{e,i} = u_{v,i}(\xi) + \sum_{e\in(\delta^+(\hat{V})\cap\delta^-(v))\setminus E^0(\xi)} f^-_{e,i}(\xi) + \sum_{e\in\delta^-(v)\cap E[\check{V}]} \check{x}^-_{e,i}$$

$$\overset{(4.5^*)}{=} u_{v,i}(\xi) + \sum_{e\in(\delta^+(\hat{V})\cap\delta^-(v))\setminus E^0(\xi)} \hat{x}^-_{e,i} + \sum_{e\in\delta^-(v)\cap E[\check{V}]} \check{x}^-_{e,i}$$

$$\stackrel{\text{(ii),(4.8*)}}{=} u_{v,i}(\xi) + \sum_{e \in (\delta^+(\hat{V}) \cap \delta^-(v)) \setminus E^0(\xi)} \hat{x}^-_{e,i} + \sum_{e \in \delta^+(\hat{V}) \cap \delta^-(v) \cap E^0(\xi)} \hat{x}^-_{e,i} + \sum_{e \in \delta^-(v) \cap E[\check{V}]} \check{x}^-_{e,i}$$

$$= u_{v,i}(\xi) + \sum_{e \in \delta^-(v)} x^-_{e,i}.$$

(4.7): This constraint holds for the same reasons as (4.6) (just using (4.7) instead of (4.6) for the inner IDE-thin flow).

(4.9): For nodes $v \in \check{V}$ this is the same as (4.9) for $(\check{x}^+, \check{x}^-, \check{a})$ in the restricted network since, as mentioned at the beginning of the proof, assumption (i) ensures that the set of active v, T_i-path is the same with respect to the inner network and to the whole network. For nodes $v \in \hat{V}$ this follows from (4.9*) using Lemma 4.24a) and b):

$$a_{v,i} = \hat{a}_{v,i}$$

$$\stackrel{(4.9^*)}{=} \min\left(\begin{array}{l}\left\{ \check{a}_{w,i} + \sum_{e \in p} \psi_e(\hat{x}^+_e) \;\middle|\; (p,w) \in P^+_{v,i}(\xi) \right\} \\ \cup \left\{ \sum_{e \in p} \psi_e(\hat{x}^+_e) \;\middle|\; p \in \hat{P_{v,i}}(\xi) \right\}\end{array}\right)$$

$$\stackrel{\text{Lem. 4.24a),b)}}{=} \min\left(\left\{\sum_{e \in p} \psi_e(x^+_e) \;\middle|\; p \in P_{v,i}(\xi)\right\} \cup \left\{\sum_{e \in p} \psi_e(x^+_e) \;\middle|\; p \in \hat{P_{v,i}}(\xi)\right\}\right)$$

$$= \min\left\{\sum_{e \in p} \psi_e(x^+_e) \;\middle|\; p \in P_{v,i}(\xi)\right\}.$$

(4.10): This is just the same as (4.10) for $(\check{x}^+, \check{x}^-, \check{a})$ and (4.10*) for $(\hat{x}^+, \hat{x}^-, \hat{a})$, respectively.

(4.12): This is the same as (4.12) for $(\check{x}^+, \check{x}^-, \check{a})$ and (4.12*a)/(4.12*b) for $(\hat{x}^+, \hat{x}^-, \hat{a})$, respectively. □

Remark 4.27. It is not hard to see that the reverse direction of Lemma 4.26 holds as well, i.e. in the situation of this lemma any IDE-thin flow for (f, ξ) in $\mathcal{N}$ can be split into an IDE-thin flow for $(f|_{G[\check{V}]}, \xi)$ in $\check{\mathcal{N}}$ and an IDE-thin flow augmentation for it. However, we will not need this direction and, therefore, omit a formal proof for it.

Before we can now show the existence of IDE-thin flow augmentations (and, therefore, of IDE-thin flows), we need one more technical lemma:

Lemma 4.28. *Let (f, ξ) be a partial IDE with right-constant flow rates in some network $\mathcal{N}$ with right-constant network inflow rates. Moreover, let $V = \check{V} \dot{\cup} \hat{V}$ a partition satisfying assumptions (i) and (ii), $(\check{x}^+, \check{x}^-, \check{a})$ an IDE-thin flow for $(f|_{G[\check{V}]}, \xi)$ in $\check{\mathcal{N}}$ and $i \in I$ some fixed commodity. Now let, additionally, $(\hat{x}^+, \hat{a}) \in \mathbb{R}_{\geq 0}^{\hat{E}\times I} \times \mathbb{R}^{\hat{V}\times I}$ be any vector satisfying (4.9*) and (4.10*) and define $(x^+, a) := (\check{x}^+, \check{a}) \oplus (\hat{x}^+, \hat{a})$.*

Then for any node $v \in \hat{V} \setminus (T_i \cup V_i^\dagger)$ there exists an edge $e = vw \in \delta^+(v) \cap E_i(\xi)$ with $a_{v,i} = \psi_e(x_e^+) + a_{w,i}$.

Proof. We define a new graph

$$\hat{G} := (V \setminus V_i^\dagger \dot{\cup} \{t\}, E_i(\xi) \setminus E[\check{V}] \dot{\cup} \{vt \mid v \in \check{V} \setminus V_i^\dagger\})$$

together with edge costs

$$\gamma_e := \begin{cases} \psi_e(x_e^+), & \text{for } e \in E_i(\xi) \setminus E[\check{V}] \\ \check{a}_{v,i}, & \text{for } e = vt \end{cases}.$$

Claim 4.2. *$(a_{v,i})_{v\in V\setminus V_i^\dagger} \oplus (0)_{v\in\{t\}}$ is a vector of node labels for $\hat{G}$ with respect to (γ_e) and $T := T_i \cup \{t\}$. Furthermore, $\hat{G}$ does not have any dead-end nodes and every cycle in $\hat{G}$ has non-negative total costs.*

Proof. We show that $(a_{v,i})_{v\in V\setminus V_i^\dagger} \oplus (0)_{v\in\{t\}}$ is a vector of node labels by a case distinction on the node v:

1. **Case: $v = t$:** Since t is a terminal node with no outgoing edges in $\hat{G}$ the empty path is the only t, T-path and, thus, the node label at t must be 0.
2. **Case: $v \in \check{V} \setminus V_i^\dagger$:** Here Assumption (i) ensures that vt is the only edge leaving v in $\hat{G}$ and, hence, also its unique v, T-path. Consequently, the node label of v must be $\gamma_{vt} = \check{a}_{v,i} = a_{v,i}$.
3. **Case: $v \in \hat{V} \setminus V_i^\dagger$:** Let p be any v, T-path in $\hat{G}$. If all nodes on p are from $\hat{V}$, then we have $p \in \hat{P_{v,i}}(\xi)$ and its cost with respect to (γ_e) is $\sum_{e\in p} \psi_e(x_e^+)$. Otherwise, p either ends at some node $w \in \check{V} \cap T_i$ or with an edge wt for some node $w \in \check{V}$. In both cases we have $(p', w) \in P_{v,i}^+(\xi)$ with either $p = p'$

or $p = p', wt$ and the cost of p with respect to (γ_e) is $\check{a}_{w,i} + \sum_{e\in p'} \psi_e(x_e^+)$. For the case of $w \in \check{V} \cap T_i$ we use here that $(\check{x}^+, \check{x}^-, \check{a})$ is an IDE-thin flow for $(f|_{G[\check{V}]}, \xi)$ in $\check{\mathcal{N}}$ and, therefore, we have $\check{a}_{w,i} = 0$ by Proposition 2.67j) (cf. Proposition 4.20). In conclusion, defining $\hat{a}_{v,i}$ by (4.9*) is equivalent to the definition of the node label in $\hat{G}$ with respect to (γ_e).

To see that there are no dead end nodes, we first observe that any node in $\check{V} \cup \{t\}$ has a path towards the terminal node t. For any node $v \in \hat{V} \setminus V_i^\dagger$ there exists at least one path $p \in P_{v,i}(\xi)$ according to Proposition 2.67b). This path then either uses only edges in $\hat{E}$ or it uses at least on edge in $E[\check{V}]$. In the former case p is a v, T-path in $\hat{G}$. In the latter case let p' be a prefix of p ending in some node $w \in \check{V}$. Then p', wt is a v, T-path path in $\hat{G}$.

Finally, let c be a cycle in $\hat{G}$. Since there are no edges leaving t in $\hat{G}$, such a cycle must lie completely in $E_i(\xi)$. Thus, all edges on c have a current travel time of zero by Proposition 2.67h) (applied, again, to G with respect to $(C_e(\xi))$). This, in turn, implies $Q_e(\xi) = 0$ and, therefore, $\psi_e(x_e^+) \geq 0$ for all $e \in c$. Thus, we have $\sum_{e\in c} \gamma_e = \sum_{e\in c} \psi_e(x_e^+) \geq 0$. ■

Now, take any node $v \in \hat{V} \setminus (T_i \cup V_i^\dagger)$. Then, applying Proposition 2.67e) and g) to $\hat{G}$ with respect to (γ_e) (which we are allowed to by Claim 4.2) provides us with the required edge $e = vw \in \delta^+(v) \cap E_i(\xi)$ satisfying $a_{v,i} = \psi_e(x_e^+) + a_{w,i}$. □

4.3.2 IDE-Thin Flows via a Fixed Point Theorem

We will now show the existence of IDE-thin flow (augmentations) in essentially the same way that we showed existence of IDE-extensions for general flow rates in Sect. 4.2: That is, we define a set K of candidates using only the "easy" constraints from the definition of IDE-thin flow augmentations. Next, we define a mapping Γ which gives us for any such candidate $(x^+, x^-) \in K$ the set of candidates $(y^+, y^-) \in K$ such that (x^+, y^-) satisfies the IDE-thin flow constraints missing in the definition of K. Finally, we apply Kakutani's Fixed Point Theorem to deduce existence of a fixed point of Γ which is then our IDE-thin flow (augmentation). Note that for finite dimensional Banach spaces (which is what we are considering here) weak and strong topology coincide (cf. [AB06, Theorem 6.26]) which is why we can just use the standard topology on $\mathbb{R}^n$ here.

Lemma 4.29. *Let (f, ξ) be a partial IDE with right-constant flow rates in a feasible network with right-constant network inflow rates. Moreover, assume that f is a Vickrey edge flow for all times on all edges, let $V = \check{V} \dot\cup \hat{V}$ be a partition satisfying assumptions (i) and (ii) and $(\check{x}^+, \check{x}^-, \check{a})$ an IDE-thin flow for $(f|_{G[\check{V}]}, \xi)$ in $\check{\mathcal{N}}$.*

Then there exists an IDE-thin flow augmentation for $(\check{x}^+, \check{x}^-, \check{a})$.

Proof. We want to show the existence of a vector $(\hat{x}^+, \hat{x}^-, \hat{a}) \in \mathbb{R}^{\hat{E}\times I} \times \mathbb{R}^{\hat{E}\times I} \times \mathbb{R}^{\hat{V}\times I}$ satisfying constraints 4.4* to (4.12*b), i.e. an IDE-thin flow augmentation, by using Kakutani's fixed point theorem (Theorem 2.56). To do so, we first define the set

$$K := \left\{ (\hat{x}^+, \hat{x}^-) \in \mathbb{R}^{I\times\hat{E}}_{\geq 0} \times \mathbb{R}^{I\times\hat{E}}_{\geq 0} \;\middle|\; \begin{array}{l} (\hat{x}^+, \hat{x}^-) \text{ satisfies (4.5*), (4.6*), (4.7*), (4.8*),} \\ \hat{x}^-_{e,i} \leq \nu_e \text{ f.a. } e \in \hat{E}, i \in I \text{ and } \hat{x}^+_{e,i} = \hat{x}^+_{e,i} = 0 \\ \text{for all } i \in I, e = vw \in \hat{E} \text{ with } w \in V_i^\dagger \end{array} \right\}.$$

Since all constraints defining this set are linear, it is clearly convex and closed. Furthermore, it is bounded as all $\hat{x}^-_{e,i}$ are bounded by ν_e and then all $\hat{x}^+_{e,i}$ are bounded by (4.6*) and (4.7*). Finally, to show that K is non-empty we can construct an element of K as follows: We set $\hat{x}^-_{e,i} := f^-_{e,i}(\xi)$ for all $e \in \hat{E} \setminus E^0(\xi)$ and $\hat{x}^-_{e,i} := 0$ for all $e \in \hat{E} \cap E^0(\xi)$. Since f respects capacity, we then have $\hat{x}^-_{e,i} \leq \nu_e$. Moreover, we have $\hat{x}^+_{e,i} = f^-_{e,i}(\xi) = 0$ for all $e = vw \in E \setminus E^0(\xi)$ with $w \in V_i^\dagger$ by Proposition 3.66. Thus, (4.5*) as well as the two additional constraints of K are already satisfied.

For defining $\hat{x}^+_{e,i}$ in such a way that (4.6*), (4.7*) and (4.8*) hold as well, we look at all edges leaving a common node $v \in \hat{V}$ at once and distinguish three cases:

1. If $v \in T_i$, we set $\hat{x}^+_{e,i} := 0$ for all $e \in \delta^+(v)$.
2. If $v \in V_i^\dagger$, we have $u_{v,i}(\xi) = 0$ since the network is feasible and $f^-_{e,i}(\xi) = 0$ for all $e \in \delta^-(v) \setminus E^0(\xi)$ by Proposition 3.66. Thus, we can set $\hat{x}^+_{e,i} := 0$ for all $e \in \delta^-(v)$.
3. If $v \notin T_i \cup V_i^\dagger$, then we must have $\delta^+(v) \neq \emptyset$ and, therefore, Proposition 2.67e) guarantees the existence of least one active edge leaving v. We pick one such edge $e' = vw \in \delta^+(v) \cap E_i(\xi) \subseteq \hat{E}$ and define $\hat{x}^+_{e',i} := u_{v,i}(\xi) + \sum_{e\in\delta^-(v)\setminus E^0(\xi)} f^-_{e,i}(\xi)$ for this edge and $\hat{x}^+_{e,i} := 0$ for all other edges $e \in \delta^+(v) \setminus \{e'\}$. Since $v \notin V_i^\dagger$ and e' is active, we have $w \notin V_i^\dagger$ as well due to Proposition 2.67a).

The vector $(\hat{x}^+, \hat{x}^-)$ defined this way is now clearly an element of K i.e. a witness for $K \neq \emptyset$.

Next, we define a mapping $\Gamma : K \to 2^K$ by setting

$$\Gamma(\hat{x}) := \left\{ \hat{y} \in K \;\middle|\; \begin{array}{l} \hat{y}^-_{e,i} = \frac{\hat{x}^+_{e,i}\nu_e}{\max\{\hat{x}^+_e, \nu_e\}} \text{ for all } e \in E^0(\xi) \cap \hat{E} \text{ and} \\ \hat{y}^+_{e,i} = 0 \text{ for all } e = vw \in \hat{E} \text{ with } a_{v,i} < \psi_e(\hat{x}^+_e) + a_{w,i} \end{array} \right\},$$

where $(a_{v,i})$ is the node labels corresponding to $(\hat{x}^+, \hat{x}^-)$ according to eqs. (4.9*) and (4.10*) for nodes $v \in \hat{V}$ and the node labels $\check{a}_{v,i}$ of $(\check{x}^+, \check{x}^-, \check{a})$ for nodes $v \in \check{V}$. Since the constraints on $(\hat{y}^+, \hat{y}^-)$ are linear (for fixed $(\hat{x}^+, \hat{x}^-)$), these sets are clearly convex. They are also non-empty, as for any given $(\hat{x}^+, \hat{x}^-)$ we can define an elements $(\hat{y}^+, \hat{y}^-) \in \Gamma(\hat{x}^+, \hat{x}^-)$ as follows: First, we set $\hat{y}^-_{e,i} := \frac{\hat{x}^+_{e,i}\nu_e}{\max\{\hat{x}^+_e, \nu_e\}}$ for all $e \in E^0(\xi) \cap \hat{E}$ and $\hat{y}^-_{e,i} := f^-_{e,i}(\xi)$ for all $e \in \hat{E} \setminus E^0(\xi)$. Observe, that, in both cases, these $\hat{y}^-_{e,i}$ then satisfy $\hat{y}^-_{e,i} \leq \nu_e$ (in the former case by construction and in the latter case since f respects capacity) and we have $\hat{y}^-_{e,i} = 0$ whenever $e \in \delta^-(w)$ for some $w \in V_i^\dagger$ (in the first case because then we also have $\hat{x}^+_{e,i} = 0$, in the second case due to Proposition 3.66). For defining $\hat{y}^+_{e,i}$ we again look at all edges leaving some node $v \in \hat{V}$ at once and distinguish four cases:

1. If $v \in T_i$ or $u_{v,i}(\xi) + \sum_{e \in \delta^-(v)} \hat{y}^-_{e,i} = 0$, we set $\hat{y}^+_{e,i} := 0$ for all $e \in \delta^+(v)$.
2. If $v \notin T_i$ and $u_{v,i}(\xi) > 0$, we have $v \notin V_i^\dagger$ since the given network is feasible. Lemma 4.28 then ensures that that there exists an active edge $e' = vw \in \delta^+(v) \cap E_i(\xi)$ with $a_{v,i} = \psi_{e'}(\hat{x}^+_{e'}) + a_{w,i}$. Thus, we can set $\hat{y}^+_{e',i} := u_{v,i}(\xi) + \sum_{e \in \delta^-(v)} \hat{y}^-_{e,i}$ and $\hat{y}^+_{e,i} := 0$ for all other edges $e \in \delta^+(v) \setminus \{e'\}$. Since $v \notin V_i^\dagger$ and e' is active, we have $w \notin V_i^\dagger$ as well due to Proposition 2.67a).
3. If $v \notin T_i$ and $\hat{y}^-_{e,i} > 0$ for some edge $e \in E^0(\xi) \cap \delta^-(v)$, we must have $\hat{x}^+_{e,i} > 0$ as well and, therefore, $v \notin V_i^\dagger$ since $(\hat{x}^+, \hat{x}^-) \in K$. Thus, we can proceed to define $\hat{y}^+_{e,i}$ for the edges $e \in \delta^+(v)$ in the same way as in the previous case.
4. If $v \notin T_i$ and $f^-_{e,i}(\xi) > 0$ for some edge $e \in \delta^-(v) \setminus E^0(\xi)$, we have $v \notin V_i^\dagger$ by Proposition 3.66. Hence, we can again proceed as in the second case.

Clearly, any vector $(\hat{y}^+, \hat{y}^-)$ defined this way is an element of $\Gamma(\hat{x}^+, \hat{x}^-)$ which is, therefore, non-empty.

Finally, we need to show that Γ has a closed graph. Let $(\hat{x}^{+,n}, \hat{x}^{-,n}, \hat{y}^{+,n}, \hat{y}^{-,n})$ be a sequence in $\mathbb{R}^{\hat{E}\times I}_{\geq 0} \times \mathbb{R}^{\hat{E}\times I}_{\geq 0} \times \mathbb{R}^{\hat{E}\times I}_{\geq 0} \times \mathbb{R}^{\hat{E}\times I}_{\geq 0}$ with $(\hat{x}^+, \hat{x}^-) \in K$ and

$(\hat{y}^{+,n}, \hat{y}^{-,n}) \in \Gamma(\hat{x}^{+,n}, \hat{x}^{-,n})$ for all $n \in \mathbb{N}_0$ converging to some vector $(\hat{x}^+, \hat{x}^-, \hat{y}^+, \hat{y}^-)$. As all $(\hat{x}^{+,n}, \hat{x}^{-,n})$ and $(\hat{y}^{+,n}, \hat{y}^{-,n})$ are elements of the compact set K, so are their limit points $(\hat{x}^+, \hat{x}^-)$ and $(\hat{y}^+, \hat{y}^-)$. Let $\hat{a}^n \in \mathbb{R}^{\hat{V} \times I}$ be the sequence of node labels associated to $(\hat{x}^{+,n}, \hat{x}^{-,n})$ via eqs. (4.9*) and (4.10*). Since the mapping $(\hat{x}^+, \hat{x}^-) \mapsto \hat{a}$ is continuous by exactly the same argument as in the proof of Proposition 4.21, we know that $a := \lim_n a^n$ is the vector of node labels associated with $(\hat{x}^+, \hat{x}^-)$. Now, assume for contradiction that $(\hat{y}^+, \hat{y}^-) \notin \Gamma(\hat{x}^+, \hat{x}^-)$. There are then two possible reasons for this—both leading to a contradiction:

- If we have $\hat{y}^-_{e,i} \neq \frac{\hat{x}^+_{e,i} \nu_e}{\max\{\hat{x}^+_e, \nu_e\}}$ for some $e \in E^0(\xi)$, then due to the continuity of both sides, there must be some $n \in \mathbb{N}_0$ such that $\hat{y}^{-,n}_{e,i} \neq \frac{\hat{x}^{+,n}_{e,i} \nu_e}{\max\{\hat{x}^{+,n}_e, \nu_e\}}$, which is a contradiction to $(\hat{y}^{+,n}, \hat{y}^{-,n}) \in \Gamma(\hat{x}^{+,n}, \hat{x}^{-,n})$—note here, that the set $E^0(\xi)$ is independent of n.
- If we have $\hat{y}^+_{e,i} > 0$ for some edge $e = vw$ with $a_{v,i} < \psi_e(\hat{x}^+_e) + a_{w,i}$, then, again because of continuity, we must also have $\hat{y}^{n,+}_{e,i} > 0$ and $a^n_{v,i} < \psi_e(\hat{x}^{+,n}_e) + a^n_{w,i}$ for some $n \in \mathbb{N}_0$ which is a contradiction to $(\hat{y}^{+,n}, \hat{y}^{-,n}) \in \Gamma(\hat{x}^{+,n}, \hat{x}^{-,n})$.

Thus, all conditions of Theorem 2.56 are satisfied and we get the existence of a fixed point $(\hat{x}^+, \hat{x}^-) \in K$ with $(\hat{x}^+, \hat{x}^-) \in \Gamma(\hat{x}^+, \hat{x}^-)$. The vector $(\hat{x}^+, \hat{x}^-, \hat{a})$ with $\hat{a}$ defined by eqs. (4.9*) and (4.10*) is then an IDE-thin flow augmentation for $(\check{x}^+, \check{x}^-, \check{a})$. □

Note, that we explicitly allow $\check{V} = \emptyset$ in the above lemma. In this case $\check{\mathcal{N}}$ is the empty network and, thus, the empty vector is an IDE-thin flow for this restricted network. The above lemma then implies that there always exists an IDE-thin flow for the whole network—i.e. the following existence result for IDE-thin flows is an immediate corollary of the above lemma:

Corollary 4.30. *For any partial IDE (f, ξ) with right-constant flow rates in a feasible network with right-constant network inflow rates there exists an IDE-thin flow (x^+, x^-, a).*

Proof. Choose $\check{V} := \emptyset$ and $\hat{V} := V$ as a partition in Lemma 4.29. This clearly satisfies both assumptions (i) and (ii). Furthermore, the empty vector is then a trivial IDE-thin flow for the inner network. Thus, Lemma 4.29 guarantees the existence of an IDE-thin flow augmentation which, by Lemma 4.26, gives us an IDE-thin flow for (f, ξ). □

This, in turn, now implies the existence of IDE in networks with right-constant network inflow rates:

Theorem 4.31. *Let $\mathcal{N}$ be a feasible network with right-constant network inflow rates. Then there exists an IDE-flow with right-constant flow rates.*

Proof. This follows directly from the above corollary and Theorem 4.6: We define $\mathfrak{F}$ as the set of partial IDE in $\mathcal{N}$ with right-constant flow rates. This set is non-empty as $(0, 0) \in \mathfrak{F}$. By Corollary 4.30 and Proposition 4.22 we can extend any partial IDE $(f, \xi) \in \mathfrak{F}$ for some additional non-zero time interval. Finally, for any chain of right-constant partial IDE their limit (in the sense of Definition 4.8) is again a partial IDE with right-constant flow rates by Proposition 4.10/Corollary 4.11. □

4.3.3 IDE-Thin Flows via Convex Optimization

Looking at the previous subsection one may wonder why we even introduced the concept of IDE-thin flow augmentations as in the proof of Theorem 4.31 we only used Lemma 4.26 to deduce the existence of an IDE-thin flow for the whole network at once. Thus, for this result it would have been enough to directly state and prove Lemma 4.26 as an existence result for IDE-thin flows as it was done with essentially the same proof in [GHS20, Lemma 5.3].

The usefulness of IDE-thin flow augmentations will become clear in this subsection where we will use it to derive existence of IDE-thin flows without needing to refer to some fixed point theorem (though, on the flip side, only for a more restricted class of networks). More precisely, we will consider partitions where the outer part of the network consists of only a single node. This then reduces the set of constraints for IDE-thin flow augmentations to a much simpler set of equations for which we can show the existence of a solution more directly:

Lemma 4.32. *Let (f, ξ) be a partial IDE with right-constant flow rates in a feasible network with right-constant network inflow rates. Moreover, assume that f is a Vickrey edge flow for all times on all edges, let $V = \check{V} \dot{\cup} \hat{V}$ be a partition with $\hat{V} = \{v\}$ and satisfying assumptions (i) and (ii) and $(\check{x}^+, \check{x}^-, \check{a})$ an IDE-thin flow for $(f|_{G[\check{V}]}, \xi)$ in $\check{\mathcal{N}}$.*

Then any solution $(x^+_{e,i})_{e,i} \in \mathbb{R}^{\delta^+(v)\times I}$ *to*

$$\begin{aligned}
\min \sum_{e\in\delta^+(v)} \int_0^{\sum_{i\in I} x^+_{e,i}} \psi_e(z)\mathrm{d}z + \sum_{e=vw\in\delta^+(v), i\in I} x^+_{e,i}\cdot \check{a}_{w,i} & \\
\text{s.t.} \sum_{e\in\delta^+(v)} x^+_{e,i} = u_{v,i}(\xi) + \sum_{e\in\delta^-(v)\setminus E^0(\xi)} f^-_{e,i}(\xi) \quad & \text{f.a. } i\in I \text{ with } v\notin T_i \qquad \text{(COPT)}\\
x^+_{e,i} = 0 \quad & \text{f.a. } i\in I \text{ with } v\in T_i \text{ and } e\in\delta^+(v)\\
x^+_{e,i} \geq 0 \quad & \text{f.a. } i\in I, e\in\delta^+(v)\cap E_i(\xi)\\
x^+_{e,i} = 0 \quad & \text{f.a. } i\in I, e\in\delta^+(v)\setminus E_i(\xi)
\end{aligned}$$

together with $x^+_{e,i} := 0$ *for all* $e\in\delta^-(v)$, $x^-_{e,i}$ *defined by (4.4*) and (4.5*) and* $a_{v,i}$ *defined by (4.9*) and (4.10*) is an IDE-thin flow augmentation for* $(\check{x}^+, \check{x}^-, \check{a})$.

Furthermore, such a solution always exists.

Proof. Let $(x^+_{e,\bullet}, x^-_{e,\bullet})$ be an optimal solution to (COPT). Then the feasibility of $(x^+_{e,\bullet}, x^-_{e,\bullet})$ together with the way all the remaining variables are defined already ensures that (x^+, x^-, a) satisfies 4.4* to (4.10*). Furthermore, (4.12*a) is an empty condition here since $\hat{V}$ contains only a single node.

Thus, we only have to show that (4.12*b) is satisfied as well. Since the objective function of (COPT) is differentiable and $(x^+_{e,\bullet}, x^-_{e,\bullet})$ an optimal solution, there must then exist vectors $(\alpha_i)_i \in \mathbb{R}^I$ and $(\beta_{e,i})_{e,i} \in \mathbb{R}^{\delta^+(v)\times I}$ satisfying the necessary KKT-condition (Proposition 2.69):

$$\psi_e(x^+_e) + \check{a}_{w,i} + \alpha_i - \beta_{e,i} = 0 \text{ for all } i\in I, e=vw\in\delta^+(v) \tag{4.14}$$

$$\beta_{e,i} \geq 0 \text{ for all } i\in I, e\in\delta^+(v)\cap E_i(\xi) \tag{4.15}$$

$$\beta_{e,i}\cdot x^+_{e,i} = 0 \text{ for all } i\in I, e\in\delta^+(v)\cap E_i(\xi) \tag{4.16}$$

Hence, for any commodity i with $v\notin V_i^\dagger \cup T_i$ and edges $e=vw\in\delta^+(v)\cap E_i(\xi)$ we have

$$\psi_e(x^+_e) + \check{a}_{w,i} \overset{(4.14)}{=} -\alpha_i + \beta_{e,i} \overset{(4.15)}{\geq} -\alpha_i,$$

while for any edge $e\in\delta^+(v)\cap E_i(\xi)$ with $x^+_{e,i} > 0$ we get

$$\psi_e(x^+_e) + \check{a}_{w,i} \overset{(4.14),(4.16)}{=} -\alpha_i.$$

Together with (4.9*) this implies $a_{v,i} = -\alpha_i$ whenever there is at least one edge $e \in \delta^+(v)$ with $x^+_{e,i} > 0$, which, in turn, guarantees that (4.12*b) holds for such an edge.

Now, in order to show that (COPT) always has an optimal solution, we observe that the objective function is continuous while the set of feasible solutions is compact. Thus, by the extreme value theorem (Proposition 2.68) it suffices to show that the set of feasible solutions is non-empty. This is, indeed, the case as we can only have $u_{v,i}(\xi) + \sum_{e \in \delta^-(v) \setminus E^0(\xi)} f^-_{e,i}(\xi) > 0$ for commodities with $v \notin V^\dagger_i$ (since the network is feasible and because of Proposition 3.66). Such a node then always has at least one outgoing edge if it is not a sink node. Thus, we we are always able to choose $x^+_{e,i} \geq 0$ such that all feasibility conditions of (COPT) are satisfied. □

Remark 4.33. The problem considered in Lemma 4.32 can also be interpreted as a Wardrop equilibrium problem for static flows in a rather simple network (see Fig. 4.1). The objective function in (COPT) is then exactly the potential function typically used to show existence of Wardrop equilibria (cf. e.g. [NRTV07, Section 18.3.1]). Note that the edge costs of this static flow network might be negative. However, since all source,sink-paths consist of exactly two edges, we can easily transform this network into an equivalent network with non-negative costs by adding

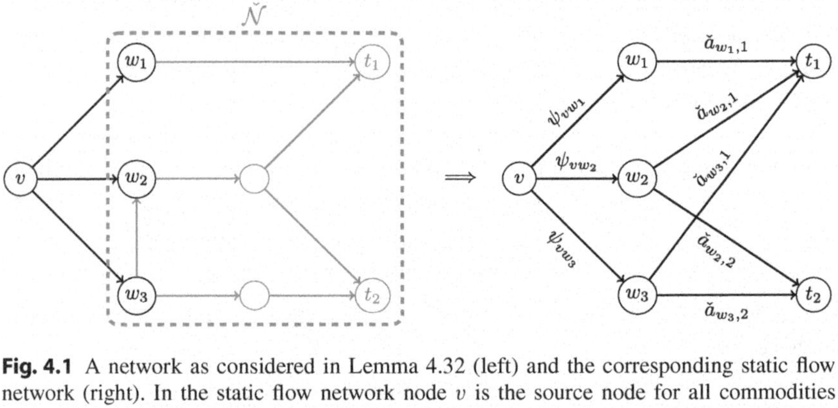

Fig. 4.1 A network as considered in Lemma 4.32 (left) and the corresponding static flow network (right). In the static flow network node v is the source node for all commodities and there is a single sink node t_i for each commodity i. The demand of commodity i is $u_{v,i}(\xi) + \sum_{e \in \delta^-(v) \setminus E^0(\xi)} f^-_{e,i}(\xi)$ if $v \notin T_i$ and 0 otherwise. The edge costs on the first edges are the (flow dependent) functions ψ_{vw_i} while the costs on the second edges are constant. It is easy to see that Wardrop equilibria in the network on the right then corresponds to solution to (COPT) for the network on the left and vice versa

some positive constant to all edge costs. In the single-commodity case the whole network can also be further simplified into a parallel link network.

In situations where we can construct the network by iteratively adding a single node satisfying the assumptions of Lemma 4.32 this now implies existence of an IDE-thin flows comprised of solutions to (COPT).

Lemma 4.34. *Let (f, ξ) be a partial IDE with right-constant flow rates in a feasible network with right-constant network inflow rates and assume that f is a Vickrey edge flow on all edges for all times. If there are no active edges with current travel time zero and the active subgraph of all commodities is acyclic, i.e. we have $E^0(\xi) \cap \bigcup_{i \in I} E_i(\xi) = \emptyset$ and $(V, \bigcup_{i \in I} E_i(\xi))$ is acyclic, then there exists an IDE-thin flow for (f, ξ).*

Proof. Since $(V, \bigcup_{i \in I} E_i(\xi))$ is acyclic, it admits a topological ordering $\succ$ on V such that for every commodity $i \in I$ and edge $e = vw \in E_i$ we have $v \prec w$. Now, enumerate the nodes in V such that we have $v_1 \succ v_2 \succ \cdots \succ v_n$. Then for any $k \in [n]$ the sets $\hat{V} := \{ v_k \}$ and $\check{V} := \{ v_1, \ldots, v_{k-1} \}$ define a partition of the restricted network $\mathcal{N}|_{G[\{ v_1, \ldots, v_k \}]}$ satisfying assumptions (i) and (ii). Thus, by applying Lemma 4.32 n times we can construct an IDE-thin flow for the whole network. □

We can now use this extension-lemma to show existence for the two classes of networks already mentioned as potential use cases for IDE-thin flow augmentations in Example 4.23: Acyclic networks and single-commodity networks:

Theorem 4.35. *Let $\mathcal{N}$ be a feasible acyclic network with strictly positive free flow travel times and right-constant network inflow rates. Then there exists an IDE with right-constant flow rates.*

Proof. As the whole network is acyclic, this is true, in particular, for $(V, \bigcup_{i \in I} E_i(\xi))$. Since all free flow travel times are strictly positive, the set $E^0(\xi)$ is always empty. Thus, we can apply Lemma 4.34 and Proposition 4.22 to extend any partial IDE with right-constant flow rates. Consequently, Theorem 4.6 together with Proposition 4.10/Corollary 4.11 implies the existence of an IDE with right-constant flow rates for all times. □

Theorem 4.36. *Let $\mathcal{N}$ be a feasible single-commodity network with strictly positive free flow travel times and right-constant network inflow rates. Then there exists an IDE with right-constant flow rates.*

Proof. As all free flow travel times are strictly positive, the set $E^0(\xi)$ is always empty. Proposition 2.67n) then guarantees that $(V, E(\xi))$ is always acyclic. Thus, we can apply Lemma 4.34 and Proposition 4.22 to extend any partial IDE with right-constant flow rates. Consequently, Theorem 4.6 together with Proposition 4.10/Corollary 4.11 implies the existence of an IDE with right-constant flow rates for all times. □

We conclude this subsection with two examples showing why the additional assumptions of Theorems 4.35 and 4.36 (the network being acyclic and no zero free flow travel times, respectively) are necessary to allow a node-by-node construction of IDE extensions as in the proofs of those theorems:

Example 4.37. The two-commodity network depicted in Fig. 4.2 demonstrates why in general multi-commodity networks it is not possible to determine an IDE extension on a node per node basis: At time 0 there are two nodes at which a flow distribution has to be determined: The two source nodes s_1 and s_2. Assume that we start at node s_1. Here, we have a network inflow at a rate of 3 for commodity 1 for which there are two possible paths to take: The direct edge towards the sink t_1 or the path s_1v, vs_2, s_2w, wt_1. As we currently do not have any queues in the network, both paths are currently active. Thus, the only possible flow distribution is to send half the flow into each of the two paths, leading to queues growing at a rate of $1/2$ on both edges s_1t_1 and s_1v. Next, we have to determine a flow split at node s_2. Again, two paths (s_2t_2 and s_2w, ws_1, s_1v, vt_2) are currently active. However, for one of those we already know that it will have a queue growing at a rate of $1/2$ on it. In order to compensate for that (and keep the flow split of commodity 2 feasible for some proper interval), we have to send more flow into edge s_2t_2: More precisely, a flow split of $7/4$ into edge s_2t_2 and $5/4$ into s_2w results in a stable flow distribution for commodity 2. However, this still creates an additional queue on edge s_2w which is on one of the paths used by commodity 1. Thus, the previously determined flow distribution for commodity 1 becomes unstable again. Due to the symmetry of the given network, starting with the flow distribution at node s_2 instead of s_1 would not resolve this problem.

Note that this symmetry also makes it fairly easy to determine an actually stable flow distribution—however, only by considering both nodes at the same time: Send flow into edges s_1t_1 and s_2t_2 at a rate of $5/3$ and into edges s_1v and s_2w at a rate of

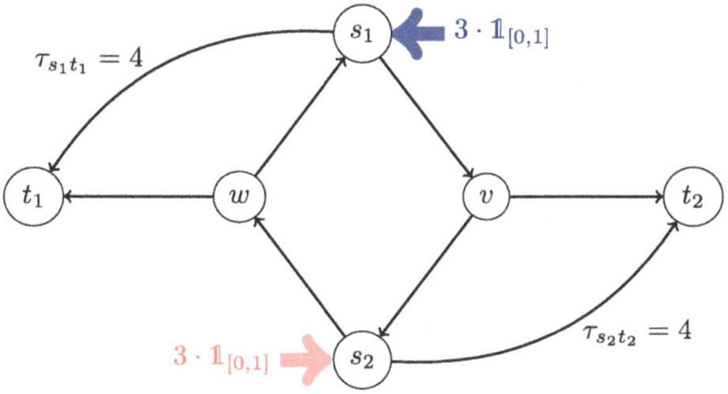

Fig. 4.2 A two-commodity network. All edges have a capacity of 1 and all edges except for s_1t_1 and s_2t_2 have a free flow travel time of 1. For each commodity $i \in \{1, 2\}$ flow enters the network at node s_i at a rate of 3 during the interval $[0, 1]$ with destination t_i

4/3. This means that queues start to grow on the former edges at a rate of 2/3 and on the latter ones at a rate 1/3 leading to a growth rate of the current travel times of 2/3 along all source,sink-paths.

Example 4.38. The single-commodity network depicted in Fig. 4.3 demonstrates why in networks containing edges with zero free flow travel time it can be impossible to find extensions on a node per node basis: At time 0 there are two nodes at which a flow distribution has to be determined: The two source nodes s_1 and s_2. Assume that we start at node s_1. Here, we will have an outflow of 2 for which there are two possible paths to take: The direct edge towards the sink t or the path s_1v, vt. As all edges have a capacity of 1, splitting this flow evenly between the two paths leads to a flow distribution which does not create any queues (and, thus, is the unique feasible flow split from the perspective of node s_1). However, if we now determine a flow split at node s_2 we must send some additional flow along the path s_2, v, t which immediately creates a queue on edge vt. This, in turn, makes the previously determined flow split at node s_1 unstable as it now uses one path with increasing travel time and one with constant travel time. Due to the symmetry of the network, the same effect takes place if we first determine a flow distribution at node s_2 and consider s_1 afterwards.

An actual feasible flow split has the queues on all three edges leading into the sink growing at the same rate, i.e. we send flow into edges s_1t and s_2t at a rate of 4/3 and into edges s_1v and s_2v at rate of 2/3. This leads to queues growing at a rate of 1/3 on all four paths and, thus, a stable situation. Note, that we can only obtain this flow split by considering the nodes s_1 and s_2 at the same time.

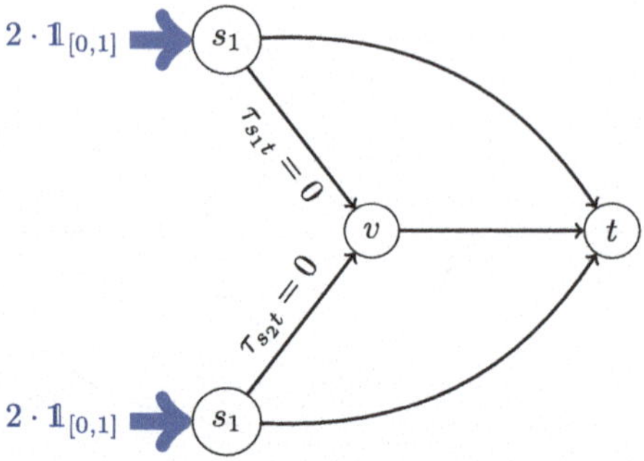

Fig. 4.3 A single-commodity network with edges with zero free flow travel time. All edges have a capacity of 1. The edges s_1v and s_2v both have a free flow travel time of zero while all other edges have a free flow travel time of 1. Flow enters the network at each of the nodes s_1 and s_2 at a rate of 2 during the interval [0, 1] with destination t

4.4 Bibliographic Notes and Open Questions

The first existence result for IDE in single commodity networks with right-constant network inflow rates was published in [GH19, Theorem 1] (which is joint work with Tobias Harks) with a proof similar to the one presented here in Subsection 4.3.3 (more precisely, [GH19, Theorem 1] is exactly Theorem 4.36). For the journal version of this paper ([GHS20]) Leon Sering extended this result to multi-commodity networks with right-constant network inflow rates by transferring the concept of thin flows (which was introduced by Koch and Skutella in [KS11] for the full information model and used by Cominetti, Correa and Larré to show existence of dynamic equilibria in this model in [CCL15]) to the IDE setting. Additionally, he adapted the general existence proof for dynamic equilibria using variational inequalities from [CCL15] to the setting of IDE. The existence result presented here in Subsection 4.3.2 follows the same approach as the former but in a slightly more general setting which also allows for edges of free flow time zero. For the general case considered in Sect. 4.2 we give an alternative proof using the Kakutani–Fan–Glicksberg fixed point theorem instead of an existence theorem for variational inequalities. Note, however, that the requirement for application of either of those two existence results turn out to be very similar (in particular, with regards to the required continuity properties). The connection between IDE thin flow augmentations at single nodes and Wardrop equilibria (Remark 4.33) was pointed out to us by Roberto Cominetti.

The general existence result completely answers the existence question for IDE (except for the technical case of network inflow rates which are locally integrable but not locally p-integrable for any $p > 1$). In fact, as already discussed in Remark 4.16, this result should also hold for other flow models and even other behavioural models and it would be an interesting topic for future research to formulate such a generic existence result. Note, however, that our proof does not completely translate back to equilibria in the full information setting (or, more generally, "non-causal" predictors in the framework introduced in [GHKM23]). This is not so much because of the extension lemma itself but because our version of extending flows does not work well for such equilibria. More precisely, the third property in our meta-existence theorem (Theorem 4.6) is not satisfied any more when the distance labels also depend on the future flow evolution (which is also why in the full information setting partial flows are typically defined differently). However, even in such a setting one might be able to apply the extension-lemma *once* when starting with the zero flow to at least get existence of an equilibrium flow for some *finite* time horizon.

5 Computational Complexity of IDE

In the previous chapter we saw that IDE are guaranteed to exist. However, we do not know yet how to actually compute such an IDE for a given network. In particular, the most general existence result (Theorem 4.15), is highly non-constructive: It uses a fixed point theorem in an infinite dimensional vector space to construct single extension and then Zorn's Lemma to "glue" those together to obtain an IDE for all times. Nevertheless, this still suggests a natural way splitting of up the task of computing IDE: First, we have to find a way of computing a single extension. Second, we have to show that a finite number of such extensions suffices to at least cover a given finite time horizon. Whether such a finite time horizon suffices, i.e. whether IDE terminate in finite time, will be the focus of the next chapter (Chap. 6), while we will focus on the first two poin ts in this chapter.

As we are interested in exact (combinatorial) algorithms here, we will only consider networks with right-constant network inflow rates and flows with right-constant flow rates (as for general measurable inflow rates it would not even be clear how to represent them exactly in finite space). For this case we have seen in Sect. 4.3 that extending a partial IDE is equivalent to finding an IDE-thin flow. This leads to the generic algorithm for computing right-constant IDE described in Algorithm 1.

L. Graf, *Dynamic Network Flows with Adaptive Route Choice based on Current Information*, Mathematische Optimierung und Wirtschaftsmathematik | Mathematical Optimization and Economathematics,
https://doi.org/10.1007/978-3-658-44948-3_5

Algorithm 1: A generic algorithm for computing IDE

Input : A network $\mathcal{N}$ with right-constant network inflow rates and a time horizon $T \geq 0$

Output : A partial IDE (f, T) with right-constant flow rates in $\mathcal{N}$

Set $(f, \xi) \leftarrow (0, 0)$
while $\xi < T$ **do**
  Compute an IDE-thin flow (x^+, x^-, a) for (f, ξ)
  Let g be the flow defined by extending (f, ξ) with (x^+, x^-, a) (cf. (4.3))
  Determine some $\varepsilon > 0$ such that $(g, \xi + \varepsilon)$ is a partial IDE
  $(f, \xi) \leftarrow (g, \xi + \varepsilon)$
end while
return (f, ξ)

This algorithm is clearly correct (by Proposition 4.22), however, it still has two short-comings: First, we do not know yet how to actually compute IDE-thin flows and, second, it is not obvious whether a finite number of extensions suffices to reach the desired time horizon T. We will address these two problems separately in the following two sections. In the third section of this chapter we will then turn to questions on the complexity of computing IDE and, in particular, show the hardness of several natural decision problems involving IDE.

5.1 Computing IDE-Thin Flows

In Sect. 4.3 we saw two different ways of proving existence if IDE-thin flows. One for general networks using Kakutani's fixed point theorem (Subsection 4.3.2) and one for more restricted networks (acyclic or single-commodity) using a convex optimization problem (Subsection 4.3.3). Similarly, we will now present two different ways of computing IDE-thin flows—each of them following one of the two existence proofs.

5.1.1 Multi-Commodity Networks

According to Proposition 4.22 the task of computing an IDE-thin flow for multi-commodity networks is equivalent to finding a solution to eqs. (4.4) to (4.10) and (4.12). We know from Corollary 4.30 that such a solution must exist. However, the proof of Corollary 4.30 and, more importantly, Lemma 4.29 does not tell us anything about how to actually find such a solution as its existence just follows from Kakutani's Fixed Point Theorem.

Hence, we have to take a closer look at the system of equations itself and see if we can somehow solve it in a more constructive way. Here, the main obstacle for doing that seems to be the constraint

$$x^-_{e,i} = \frac{x^+_{e,i} \cdot \nu_e}{\max\{x^+_e, \nu_e\}} \text{ for all } i \in I, e \in E^0(\xi) \qquad (4.4)$$

as it is both non-linear and non-convex. Because of this, we will now restrict ourselves to situations where there are no active edges of current travel time zero. In such a situation we can not only omit (4.4) but also simplify (4.9) using Proposition 2.67o) to obtain the following system of equations:

$$\begin{aligned}
x^-_{e,i} &= 0 && \text{f.a. } i \in I, e \in E^0(\xi) && (4.4')\\
x^-_{e,i} &= f^-_{e,i}(\xi) && \text{f.a. } i \in I, e \in E \setminus E^0(\xi) && (4.5)\\
\sum_{e\in\delta^+(v)} x^+_{e,i} &= u_{v,i}(\xi) + \sum_{e\in\delta^-(v)} x^-_{e,i} && \text{f.a. } i \in I, v \in V \setminus T_i && (4.6)\\
\sum_{e\in\delta^+(t)} x^+_{e,i} &\leq u_{t,i}(\xi) + \sum_{e\in\delta^-(t)} x^-_{e,i} && \text{f.a. } i \in I, t \in T_i && (4.7)\\
x^+_{e,i} &= 0 && \text{f.a. } e \in E \setminus E_i(\xi) && (4.8)\\
a_{v,i} &= \min\left\{ \psi_e(x^+_e) + a_{w,i} \;\middle|\; e = vw \in E_i \right\} && \text{f.a. } i \in I, v \in V \setminus V^\dagger_i && (4.9')\\
a_{v,i} &= 0 && \text{f.a. } i \in I, v \in V^\dagger_i \cup T_i && (4.10')\\
a_{v,i} &= \psi_e(x^+_e) + a_{w,i} \text{ if } x^+_{e,i} > 0 && \text{f.a. } i \in I, v \in V, &&\\
& && e = vw \in \delta^+(v). && (4.12)
\end{aligned}$$

Now, the only remaining difficulties are to decide which edges to send flow into and under which regime the queues on edges with currently empty queue will be (increasing or depleting). Once these decisions are made the remaining problem becomes a linear problem. Thus, we equivalently write the above system of constraints as the following MIP:

$$
\begin{aligned}
x^-_{e,i} &= 0 && \text{for all } i \in I, e \in E^0(\xi)\\
x^-_{e,i} &= f^-_{e,i}(\xi) && \text{for all } e \in E \setminus E^0(\xi)\\
\sum_{e\in\delta^+(v)} x^+_{e,i} &= u_{v,i}(\xi) + \sum_{e\in\delta^-(v)} x^-_{e,i} && \text{for all } i \in I, v \in V \setminus T_i\\
\sum_{e\in\delta^+(t)} x^+_{e,i} &\leq u_{t,i}(\xi) + \sum_{e\in\delta^-(t)} x^-_{e,i} && \text{for all } i \in I, t \in T_i\\
x^+_{e,i} &= 0 && \text{for all } e \in E \setminus E_i(\xi)\\
z_e &\in \{0, 1\} && \text{for all } i \in I, e \in E \text{ with } Q_e(\xi) = 0\\
q_e &= x^+_e - \nu_e && \text{for all } e \in E \text{ with } Q_e(\xi) > 0\\
q_e &= \begin{cases} x^+_e - \nu_e, & \text{if } z_e = 1\\ 0, & \text{else}\end{cases} && \text{for all } e \in E \text{ with } Q_e(\xi) = 0\\
x^+_e &\leq \nu_e \text{ if } z_e = 0 && \text{for all } e \in E \text{ with } Q_e(\xi) = 0\\
x^+_e &\geq \nu_e \text{ if } z_e = 1 && \text{for all } e \in E \text{ with } Q_e(\xi) = 0\\
a_{v,i} &\leq \frac{q_e}{\nu_e} + a_{w,i} && \text{for all } i \in I, v \in V \setminus \{t_i\}, e = vw \in E_i(\xi)\\
a_{i,v} &= 0 && \text{for all } i \in I, v \in V^\dagger_i \cup T_i\\
y_{e,i} &\in \{0, 1\} && \text{for all } i \in I, e \in E_i(\xi)\\
x^+_{e,i} &= 0 \text{ if } y_{e,i} = 0 && \text{for all } i \in I, e \in E_i(\xi)\\
a_{v,i} &= \frac{q_e}{\nu_e} + a_{w,i} \text{ if } y_{e,i} = 1 && \text{for all } i \in I, e = vw \in E_i(\xi)
\end{aligned}
\tag{TF-MIP}
$$

Here, the variable q_e denotes the rate at which the queue length on edge e will change (i.e. the value of $\nu_e \cdot \psi_e(x^+_e)$), the binary variable z_e denotes for queues which are empty at time ξ whether they will remain empty ($z_e = 0$) or whether a queue will start to form after time ξ ($z_e = 1$) and the binary variable $y_{e,i}$ indicates whether a currently active edge will stay active after time ξ and, thus, whether we are allowed to send flow into it ($y_{e,i} = 1$ means that e remains active for commodity i).

Lemma 5.1. *Let (f, ξ) be a partial IDE with right-constant flow rates in a network with right-constant network inflow rates such that $E_i(\xi) \cap E^0(\xi) = \emptyset$ for all commodities $i \in I$. Then a vector (x^+, x^-, a) is an IDE-thin flow for (f, ξ) if and only if there exist binary vectors $(y_{e,i}) \in \{0, 1\}^{\sum_{i\in I}\{i\}\times E_i(\xi)}$ and*

$(z_e) \in \{0,1\}^{\{e \in E \mid Q_e(\xi)=0\}}$ *and a vector* $(q_e) \in \mathbb{R}^E$ *such that* (x^+, x^-, a, q, y, z) *solves (TF-MIP).*

Proof. It is easy to see that (TF-MIP) is equivalent to (4.4′), (4.5), (4.6), (4.7), (4.8), (4.9′), (4.10′) and (4.12) by observing that $y_{e,i} = 1$ is equivalent to e being an edge which remains active for commodity i during the extension phase and $z_e = 1$ is equivalent to $x_e^+ \geq \nu_e$.

Thus, it suffices to show that in the setting of the lemma constraint (4.4′) is equivalent to (4.4) while (4.9′) and (4.10′) are equivalent to (4.9). The former is the case since the lemma ensures that all edges in $E^0(\xi)$ are inactive (for all commodities) and, therefore, have zero inflow by (4.8). For the latter, we observe that for any commodity i the set of active edges $E_i(\xi)$ is acyclic by Proposition 2.67n) (using the assumption that there are no active edges with zero travel time). Thus, we can apply Proposition 2.67o) to the subgraph $(V \setminus V_i^\dagger, E[V \setminus V_i^\dagger] \cap E_i(\xi))$ with edge costs $(\psi_e(x_e^+))$ (see Proposition 4.20) to prove that $(a_{v,i})_{v \in V \setminus V_i^\dagger}$ solves (4.9) if and only if it solves (4.9′) and (4.10′). Finally, for any dead-end node $v \in V_i^\dagger$ constraints (4.10′) and (4.10) are just the same. □

Now, for any given vectors $(y_{e,i})$ and (z_e) the above problem becomes a linear problem (and, thus, efficiently solvable). Since there are only finitely many such binary vectors we can, in principle determine IDE-thin flows in situations where there are no active edges of current travel time zero by solving the remaining linear problem for all possible vectors $(y_{e,i})$ and (z_e) and stop as soon as we found a solution. Corollary 4.30 guarantees that this will happen eventually. This immediately implies the following corollary:

Corollary 5.2. *Let* (f, ξ) *be a partial IDE with right-constant flow rates in a feasible network with right-constant network inflow rates such that* $E^0(\xi) = \emptyset$. *Then we can compute an IDE-thin flow for* (f, ξ) *in finite time.*

In particular, we can always compute IDE-thin flows in feasible networks where all free flow travel times are strictly positive. □

Of course, in general, such an algorithm will have an exponential worst case runtime. However, at least for small instances it might still be feasible. In particular, since we do not really have to try *all* possible vectors $(y_{e,i})$. For example, we know that for any node $v \in V \setminus V_i^\dagger$ and any commodity i we have to set $y_{e,i} = 1$ for at least one edge $e \in E_i(\xi) \cap \delta^+(v)$—and this set, the set of active edges leaving a given node v, might often be quite small, maybe even containing only a single edge (after all, this is the set of starting edges of shortest v, T_i-paths). Moreover, if we only have a single edge with $y_{e,i} = 1$ for each commodity at some node, the value of z_e is also directly determined. Such observations have also been made by Hagenmaier who implemented for his master's thesis [Hag23] an algorithm for computing (approximate) IDE in multi-commodity settings.

Moreover, in certain situations (e.g. acyclic networks) we might be able to break down the problem (TF-MIP) into smaller augmentation-subproblems which can be solved separately using the partition approach from Lemma 4.26. Nevertheless, for general networks the worst case number of potential choices for $y_{e,i}$ and z_e will grow exponentially with the size of the network.

5.1.2 Single-Commodity Networks

In this subsection we will now focus on the case of single-commodity networks (with right-constant network inflow rates). Additionally, we will again assume that there are no active edges of current travel time zero. As seen in Subsection 4.3.3, this situation allows us to determine IDE-thin flows on a node-by-node basis by finding solutions to the optimization problem (COPT) (cf. Lemmas 4.32 and 4.34). As this problem is a convex optimization problem, approximate solutions can be found efficiently using tools from convex optimization (see e.g. [BV04]). Even better, though, we will show in this subsection that we can also find exact solutions using a simple water-filling procedure.

More precisely, we want to find a feasible solution $(x_e^+)_{e \in \delta^+(v)}$ to (COPT) which also satisfies the necessary KKT-conditions. Looking at the proof of Lemma 4.32 we can see that this suffices to obtain an IDE-thin flow augmentation. Thus, our goal is to find a solution $(x_e^+)_{e \in \delta^+(v) \cap E(\xi)} \in \mathbb{R}_{\geq 0}^{\delta^+(v) \cap E(\xi)}$ and $a_v \in \mathbb{R}$ satisfying

$$\sum_{e \in \delta^+(v) \cap E(\xi)} x_e^+ = u_v(\xi) + \sum_{e \in \delta^-(v) \setminus E^0(\xi)} f_e^-(\xi) \tag{5.1}$$

$$\psi_e(x_e^+) + \check{a}_w = a_v \quad \text{for all } e = vw \in \delta^+(v) \cap E(\xi) \text{ with } x_e^+ > 0 \tag{5.2}$$

$$\psi_e(x_e^+) + \check{a}_w \geq a_v \quad \text{for all } e = vw \in \delta^+(v) \cap E(\xi) \tag{5.3}$$

where $\check{a}_w$ are the already determined node labels at the nodes in the inner part of the network. In words: We want to distribute the incoming flow $(u_v(\xi) + \sum_{e \in \delta^-(v) \setminus E^0(\xi)} f_e^-(\xi))$ to the currently active outgoing edges in such a way that all used edges are the start of a v, T-path with the same change of current travel time $(\psi_e(x_e^+) + \check{a}_w)$ while all unused edges only lead to v, T-paths with equal or larger such change. Such a solution can now be determined quite straightforwardly thanks to the simple form of the left hand side of (5.2) and (5.3) as functions in x_e^+: Namely, these functions all consist of a constant part followed by a linearly increasing part (see (4.11) for the definition of ψ_e). We will call such function truncated linear:

Definition 5.3. We call a function $\lambda : \mathbb{R} \to \mathbb{R}$ *truncated linear* if it is of the form

$$\lambda(x) = \begin{cases} \alpha, & \text{if } x \leq \gamma \\ \alpha + \beta(x - \gamma), & \text{if } x \geq c \end{cases}$$

for some constants $\alpha, \beta, \gamma \in \mathbb{R}$ (see Fig. 5.1 for an example of such a function).

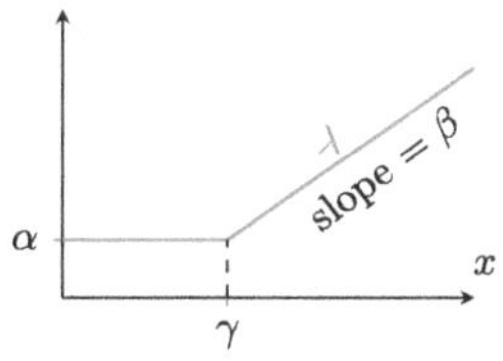

Fig. 5.1 An example of a truncated linear function with parameters α, β, γ

Thus, our task now reduces to finding for any given value $b \geq 0$ and set of truncated linear functions $\lambda_1, \ldots, \lambda_m$ a partition of b into $x_1, \ldots, x_m \geq 0$ such that $\sum_{e=1}^m x_e = b$ and $\lambda_e(x_e) \leq \lambda_{e'}(x_{e'})$ for any $e, e' \in [m]$ with $x_e > 0$. Intuitively, we can find such a partition by first sorting the functions by increasing α_e. Starting from $x_1 = x_2 = \cdots = x_m = 0$ we then increase x_1 until we reach $\lambda_1(x_1) = \lambda_2(0) = \alpha_2$. After that, we simultaneously increase x_1 and x_2 in such a way that we always have $\lambda_1(x_1) = \lambda_2(x_2)$ until we reach $\lambda_3(0)$. From here on, we simultaneously increase x_1, x_2 and x_3 and so on until we have found a partition of b (see Fig. 5.2 for a graphic depiction of this process). We formalize this approach in Algorithm 2.

Algorithm 2: Water filling procedure

Input : A number $b \geq 0$ and a finite set of truncated linear functions $\lambda_1, \dots, \lambda_m$ with parameters $(\alpha_e, \beta_e, \gamma_e)$ satisfying $\beta_e > 0$ and $\gamma_e \geq 0$ for all $e \in [m]$

Output : Values $x_1, \dots, x_m \geq 0$ such that $\sum_{e=1}^{m} x_e = b$ and $\lambda_e(x_e) \leq \lambda_{e'}(x_{e'})$ for any $e, e' \in [m]$ with $x_e > 0$

Renumber the functions such that $\lambda_1(0) \leq \lambda_2(0) \leq \dots \leq \lambda_m(0)$

Determine the maximal $k \in \{0, 1, \dots, m\}$ with $\sum_{e=1}^{k} \max\{x \mid \lambda_e(x) \leq \lambda_k(0)\} \leq b$

if $k < m$ *and* $\sum_{e=1}^{k} \max\{x \mid \lambda_e(x) \leq \lambda_{k+1}(0)\} \leq b$ **then**

$$x_e \leftarrow \begin{cases} \max\{x \mid \lambda_e(x) \leq \lambda_{k+1}(0)\}, & \text{for } e \leq k \\ b - \sum_{e' \leq k} x_{e'}, & \text{for } e = k+1 \\ 0 & \text{for } e > k+1 \end{cases}$$

else

$$x'_e \leftarrow \begin{cases} \max\{x \mid \lambda_e(x) \leq \lambda_k(0)\}, & \text{for } e \leq k \\ 0 & \text{for } e > k \end{cases}$$

$b' \leftarrow b - \sum_{e'=1}^{k} x'_{e'}$

$x_e \leftarrow x'_e + \frac{1/\beta_e}{\sum_{e' \leq k} 1/\beta_{e'}} \cdot b'$ for $e \leq k$

end if

return $x_1, \dots, x_m$

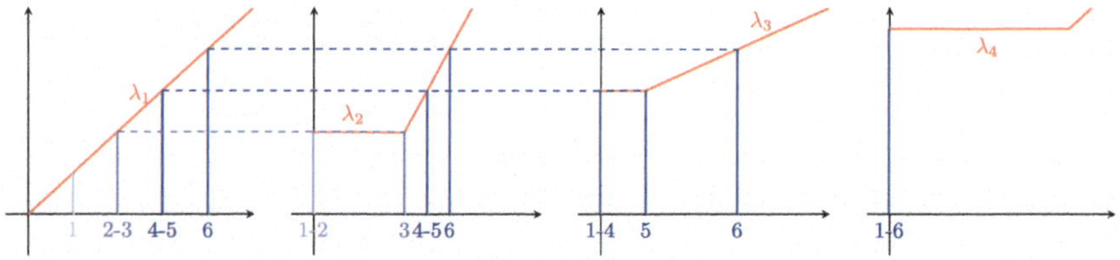

Fig. 5.2 A graphic depiction of the water filling procedure applied to four truncated linear functions. The blue vertical lines denote the values of x_1 to x_4 at six intermediate steps

Remark 5.4. This approach can actually be useful for other convex optimization problem of similar form as well and is often called "water filling" there as well—see e.g. [BV04, Example 5.2]. This name comes from the following interpretation of the procedure: We fill a volume b into a basin which is divided into m parts of different depth. More precisely, part e has a depth of $D - \alpha_e$ (where D is the maximal depth of the basin) and its floor has an area of $1/\beta_e$ (cf. Fig. 5.3). We now start to fill water into this basin at a rate of 1. At first only the part with the largest depth fills up and the water level rises at a speed of β_e. As soon as the water level reaches the

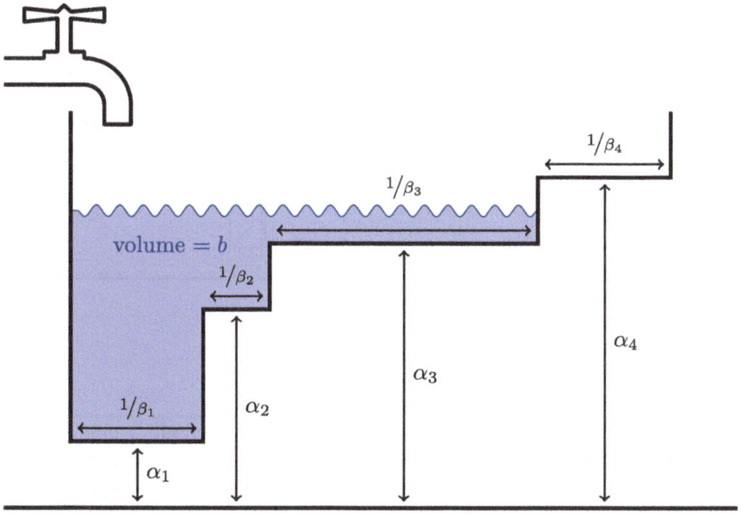

Fig. 5.3 A basin filled with water corresponding to a solution to the problem of finding values $x_1, \ldots, x_4 \geq 0$ such that $x_1 + \cdots + x_4 = b$ and $\lambda_e(x_e) = \alpha_e + \beta_e x_e \leq \lambda_{e'}(x_{e'})$ for all $e, e' \in [4]$ with $x_e > 0$

floor of the part with the next largest depth, this part starts to fill up as well and the water level now rises at a speed of $1/({}^1/_{\beta_e} + {}^1/_{\beta_{e'}})$. This process continues until all water is in the basin.

Note, that this intuition only works for the case where we have $\gamma_e = 0$ for all $e \in [m]$. Otherwise, we can imagine that for any part of the basin an (initially empty) tank of volume γ_e is attached to the bottom of its floor. As soon as the water level reaches such a floor any additional water first fills up this tank and the water level remains constant until the tank is full—at which point the water level starts to rise again as described previously. Figure 5.4 shows this for the situation of Fig. 5.2.

Proposition 5.5. *Algorithm 2 is correct and has a worst case runtime of* $\mathcal{O}(m^2)$.

Proof. We first observe that $\max\{x \mid \lambda_e(x) \leq c\}$ is well defined for any $e \in [m]$ and $c \geq \lambda_e(0)$ (since λ_e is continuous, non-decreasing and unbounded) and can be computed in constant time (due to the simple form of λ_e). Thus, line 2 has a worst case runtime of $\mathcal{O}(m^2)$. Since this also clearly bounds the runtime of any other line, this already determines the worst case runtime for the whole algorithm.

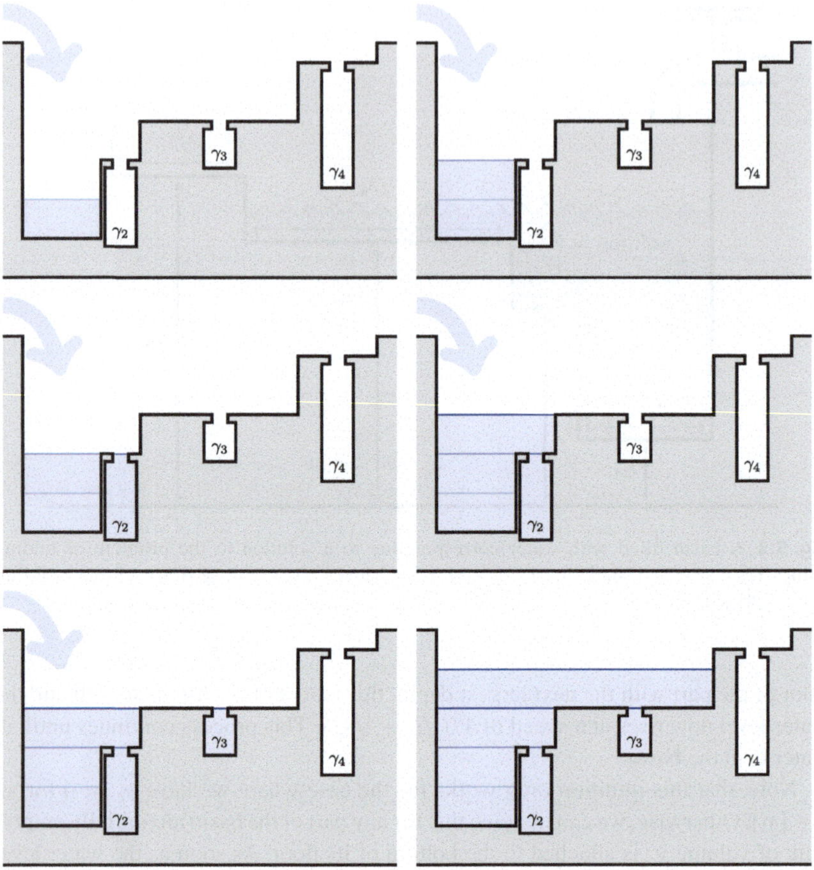

Fig. 5.4 An alternative depiction of the water filling procedure depicted in Fig. 5.2 using the basin-intuition described in Remark 5.4

To show correctness we follow the same case distinction as the algorithm itself:

1. **Case: x_e defined in line 4:** In this case we directly get $x_e \geq 0$ for all $e \in [m]$ as well as

$$\sum_{e=1}^{m} x_e = \sum_{e \leq k} x_e + x_{k+1} = b.$$

Since we have $x_e = \max\{x \mid \lambda_e(x) \leq \lambda_{k+1}(0)\}$ for all $e \leq k$, the maximality of k ensures that $x_{k+1} < \max\{x \mid \lambda_{k+1}(x) \leq \lambda_{k+1}(0)\} = \gamma_{k+1}$ and, therefore, $\lambda_{k+1}(x_{k+1}) = \alpha_{k+1}$. At the same time we have $\lambda_e(x_e) = \lambda_{k+1}(0) = \alpha_{k+1}$ for all $e \leq k$ by the choice of x_e and $\lambda_e(x_e) = \lambda_e(0) \geq \lambda_{k+1}(0) = \alpha_{k+1}$ for all $e > k+1$ due to the order chosen in line 1. Thus, the output is correct in this case.

2. **Case: x_e defined in lines 6 and 8:** In this case we have $k > 0$ and $b' \geq 0$ due to the choice of k in line 2. From this, we directly get $x_e \geq 0$ for all $e \in [m]$ as well as

$$\sum_{e=1}^{m} x_e = \sum_{e=1}^{k} x_e = \sum_{e=1}^{m}\left(x'_e + \frac{1/\beta_e}{\sum_{e' \leq k} 1/\beta_{e'}} \cdot b'\right) = b' + \sum_{e=1}^{m} x'_e = b.$$

Furthermore, we have $x_e \geq x'_e \geq \gamma_e$ and, hence,

$$\lambda_e(x_e) = \lambda_e(x'_e) + \beta_e \cdot (x_e - x'_e) = \lambda_k(0) + \frac{b'}{\sum_{e' \leq k} 1/\beta_{e'}} =: d$$

for all $e \leq k$. At the same time the fact that we are in the else-branch of Algorithm 2 guarantees that if $k < m$ holds, then there exists at least one $e' \leq k$ with $d = \lambda_{e'}(x_{e'}) \leq \lambda_{k+1}(0)$. The order chosen in line 1 then ensures that we have $\lambda_e(x_e) = \lambda_e(0) \geq \lambda_{k+1}(0) \geq d$ for all $e \geq k+1$. Thus, the output in this case is correct as well. □

As described before, this algorithm now allows us to compute IDE-thin flow augmentations in polynomial time:

Lemma 5.6. *Let (f, ξ) be a partial IDE with right-constant flow rates in a feasible single-commodity network with right-constant network inflow rates, $v \in V$ a node such that $\delta^-(v) \cap E(\xi) = \emptyset$ and $\delta^+(v) \cap E(\xi) \cap E^0(\xi) = \emptyset$. Furthermore, let $(\check{x}^+, \check{x}^-, \check{a})$ be an IDE-thin flow for $(f|_{G[V \setminus \{v\}]}, \xi)$. Then we can compute an IDE-thin flow augmentation for $(\check{x}^+, \check{x}^-, \check{a})$ in $\mathcal{O}(1 + |\delta^+(v) \cup \delta^-(v)| + |\delta^+(v) \cap E(\xi)|^2)$.*

Moreover, the value $\hat{a}_v$ of the IDE-thin flow augmentation is uniquely determined by $\check{a}$, $u_v(\xi)$, $f_e^-(\xi)$ for $e \in \delta^-(v) \setminus E^0(\xi)$ and the sets $\delta^+(v) \cap E(\xi)$ and $\{e \in \delta^+(v) \mid Q_e(\theta) > 0\}$.

The first part of this lemma can be shown using the proof of Lemma 4.32 together with the intuition given at the start of this subsection. Here, however, we will provide a more direct proof which is independent of the proof of Lemma 4.32.

Proof. We start by observing that for the given situation (a single-commodity and $\hat{V} = \{ v \}$) equations (4.4*) to (4.12b*) reduce to

$$\begin{aligned}
\hat{x}_e^- &= 0 && \text{for all } e \in (\delta^+(v) \cup \delta^-(v)) \cap E^0(\xi) \\
\hat{x}_e^- &= f_{e,i}^-(\xi) && \text{for all } e \in (\delta^+(v) \cup \delta^-(v)) \setminus E^0(\xi) \\
\sum_{e \in \delta^+(v)} \hat{x}_e^+ &= u_v(\xi) + \sum_{e \in \delta^-(v)} \hat{x}_e^- && \text{if } v \notin T \\
\sum_{e \in \delta^+(v)} \hat{x}_e^+ &\leq u_v(\xi) + \sum_{e \in \delta^-(v)} \hat{x}_e^- && \text{if } v \in T \\
\hat{x}_e^+ &= 0 && \text{for all } e \in (\delta^+(v) \cup \delta^-(v)) \setminus E(\xi) \\
\hat{a}_v &= \min \left\{ \check{a}_w + \psi_e(\hat{x}_e^+) \,\middle|\, \begin{array}{l} e \in \delta^+(v) \cap E(\xi), \\ e = vw,\, w \notin V^\dagger \end{array} \right\} && \text{if } v \notin T \cup V^\dagger \\
\hat{a}_v &= 0 && \text{if } v \in T \cup V^\dagger \\
\hat{a}_v &= \check{a}_w + \psi_e(\hat{x}_e^+) \text{ if } x_e^+ > 0 && \text{for all } vw \in \delta^+(v)
\end{aligned}$$

Thus, computing an IDE-thin flow augmentation in the given situation is equivalent to finding a solution to the above system of equations.

1. **Case: $v \in T$:** Here, we can directly set $\hat{x}_e^+ := 0$ for all $e \in \hat{E} = \delta^+(v) \cup \delta^-(v)$, $\hat{x}_e^- := f_e^-(\xi)$ for all $e \in \hat{E} \setminus E^0(\xi)$, $\hat{x}_e^- := 0$ for all $e \in \hat{E} \cap E^0(\xi)$ and $\hat{a}_v := 0$ to obtain such a solution.
2. **Case: $v \in V^\dagger$:** In this case we have $u_v(\xi) = 0$ since the whole network is feasible and we have $\sum_{e \in \delta^-(v)} f_e^-(\xi) = 0$ due to Proposition 3.66. Thus, we can choose $(\hat{x}^+, \hat{x}^-, \hat{a})$ in the same way as in the previous case.
3. **Case: $v \notin T \cup V^\dagger$ and $u_v(\xi) + \sum_{e \in \delta^-(v)} f_e^-(\xi) = 0$:** Again, we can directly set $\hat{x}_e^+ := 0$ for all $e \in \hat{E}$, $\hat{x}_e^- := 0 = f_e^-(\xi)$ for all $e \in \hat{E} \setminus E^0(\xi)$ and $x_e^- := 0$ for all $e \in \hat{E} \cap E^0(\xi)$. Additionally, we define

$$\hat{a}_v := \min \{ \check{a}_w + \psi_e(\hat{x}_e^+) \mid e = vw \in \delta^+(v) \cap E(\xi),\, w \notin V^\dagger \},$$

where we note that the minimum is well defined as, according to Proposition 2.67e) there exists at least one active edge leaving v. This then gives us a solution to the given set of equations.

4. **Case: $v \notin T \cup V^\dagger$ and $u_v(\xi) + \sum_{e\in\delta^-(v)} f_e^-(\xi) > 0$:** In this case we define functions

$$\lambda_e : \mathbb{R} \to \mathbb{R}, x \mapsto \psi_e(x) + \check{a}_w \text{ for } e \in \delta^+(v) \cap E(\xi)$$

and observe that those are truncated linear functions with parameters

$$\alpha_e := \begin{cases} \check{a}_w - 1, & \text{if } Q_e(\xi) > 0 \\ \check{a}_w, & \text{if } Q_e(\xi) = 0 \end{cases}, \quad \beta_e := \frac{1}{\nu_e} \text{ and } \gamma_e := \begin{cases} 0, & \text{if } Q_e(\xi) > 0 \\ \nu_e, & \text{if } Q_e(\xi) = 0 \end{cases}.$$

As in the previous case we have $\delta^+(v) \cap E(\xi) \neq \emptyset$ since v is neither a sink nor a dead-end node. Thus, we can use Algorithm 2 (with worst case runtime $\mathcal{O}(|\delta^+(v) \cap E(\xi)|^2)$ according to Proposition 5.5) to obtain a solution $(x_e)_{e\in\delta^+(v)\cap E(\xi)}$ with $\sum_{e\in\delta^+(v)\cap E(\xi)} x_e = u_v(\xi) + \sum_{e\in\delta^-(v)} f_e^-(\xi)$. We now define

$$\hat{x}_e^+ := \begin{cases} x_e, & \text{for } e \in \delta^+(v) \cap E(\xi) \\ 0, & \text{for } e \in \hat{E} \setminus E(\xi) \end{cases} \quad \text{and} \quad \hat{x}_e^- := \begin{cases} f_e^-(\xi), & \text{for } e \in \hat{E} \setminus E^0(\xi) \\ 0, & \text{for } e \in \hat{E} \cap E^0(\xi) \end{cases}.$$

Finally, we pick any edge $e = vw \in \delta^+(v) \cap E(\xi)$ with $x_e > 0$ and set $\hat{a}_v := \lambda_e(x_e) = \check{a}_w + \psi_e(\hat{x}_e^+)$. This then defines a solution to the system of equations stated at the beginning of the proof.

Now, for the second path of the lemma, i.e. the uniqueness of $\hat{a}_v$, let $(\hat{x}^+, \hat{x}^-, \hat{a})$ and $(\hat{x}'^+, \hat{x}'^-, \hat{a}')$ be two IDE-thin flow augmentations (or, equivalently, solutions to the system of equations given at the beginning of the proof) and assume for contradiction that $\hat{a}_v < \hat{a}'_v$. We note that this is certainly impossible if $v \in T \cup V^\dagger$ or if there is no outflow to be distributed from node v at time ξ. So, we can assume that there is at least one edge $e = vw \in \delta^+(v)$ with $\hat{x}_e^+ > 0$. For all those edges we then have

$$\check{a}_w + \psi_e(\hat{x}_e^+) = \hat{a}_v < \hat{a}'_v \leq \check{a}_w + \psi_e(\hat{x}_e'^+)$$

and, therefore, $\psi_e(\hat{x}_e^+) < \psi_e(\hat{x}_e'^+)$. Since all ψ_e are non-decreasing functions, this implies $\hat{x}_e^+ < \hat{x}_e'^+$ for all such edges and, thus,

$$\sum_{e\in\delta^+(v)} \hat{x}_e^+ < \sum_{e\in\delta^+(v)} x_e'^+,$$

which is a contradiction to both $(\hat{x}^+, \hat{x}^-, \hat{a})$ and $(\hat{x}'^+, \hat{x}'^-, \hat{a}')$ being IDE-thin flow augmentations for the same situation. □

Corollary 5.7. *Let (f, ξ) be a partial IDE with right-constant flow rates in a feasible single-commodity network with right-constant network inflow rates such that $E^0(\xi) = \emptyset$. Then we can compute an IDE-thin flow for (f, ξ) in $\mathcal{O}(|V|+|E|+|E(\xi)|^2)$ time.*

Proof. Since there are no edges of current travel time zero Proposition 2.67n) guarantees that the set of active edges is acyclic. Thus, there exists a topological order $v_1 \succ v_2 \succ \dots \succ v_n$ on $(V, E(\xi))$ and we have $\delta^-_{G_{k-1}}(v_k) \cap E(\xi) = \emptyset$ where $G_{k'} := G[\{v_1, \dots, v_{k'}\}]$ is the subgraph of the first k nodes. Therefore, starting with the empty vector (which is an IDE-thin flow for G_0) we can iteratively apply Lemma 5.6 to obtain IDE-thin flows for all G_k until, after $|V|$ steps, we finally have an IDE-thin flow for the whole network. □

Remark 5.8. While Lemma 5.6 guarantees that IDE-thin flows in single-commodity networks are unique with respect to the node labels a_v, the same need not be true with respect to the flow rates x_e^+ and x_e^-. This is because of the constant part of the functions ψ_e which allow situations in which different flow distributions lead to the same value for the node label a_v.

It is also interesting to note that for the full information setting the analogous partial uniqueness statement for thin flows holds as well ([CCL15, Theorem 4]). There this result then directly implies that for any given single-source single-sink network with constant network inflow rate all equilibrium flows with right-continuous flow rates induce the same travel times in the network ([CCL15, Theorem 6]).[1] This, however, is not true for IDE as one can already see in quite simple instances like the one from Example 3.65.

The reason for this discrepancy lies, roughly speaking, in the different way extensions work in the two models: In the full informations an extension completely determines the whole source-sink path for any particle handled in this extension. Thus, the only way different choices of the flow distribution within such an extension could

[1] Actually, the statement is also true without the restriction to right-continuous flow rates—however, this follows not as directly from the uniqueness of the thin flows (see [OSK22]).

affect future extension would be via the induced travel times in the network—but those are unique since the node labels are unique. An IDE-extension, on the other hand, only determines the next edge for every involved particle. Thus, different flow distributions can easily lead to different travel times further down the line without any effect on the node labels during the extension period.

On the flip side, for IDE we will see in the next section that we can use the uniqueness of the node labels in IDE-thin flows to show that a finite number of extensions suffices to construct an IDE for any finite time horizon—a question that is still open in the full information setting.

5.2 Bounding the Number of Extensions

In the previous section we saw that in many cases we can compute a single extension in finite time (and even efficiently for single-commodity networks). In order to deduce from this that we can also compute IDE over a finite time horizon in finite time using Algorithm 1, we still have to show that a finite number of such extensions suffices. This, as well as providing some upper and lower bounds on the number of required extensions, will be the focus of this section.

5.2.1 Upper Bound for Single-Commodity Networks

Given some partial IDE (f, ξ) with right-constant flow rates there are two natural ways to estimate how many extensions are necessary in order to compute this flow with an extension algorithm of the form of Algorithm 1:

- Count the number of (maximal) intervals with constant flow rates: As each such interval clearly requires its own extension, this gives us a lower bound for the number of required extensions.
- Count the number of times anything happens which could make a current extension invalid, i.e. any change in network inflow or edge outflow rates, a new edge becoming active or a queue depleting (note that these are exactly the four types of events we used in the proof of Proposition 4.22 to determine the length of the extension period). Since we can then always use a single extension to get from one such time to the next, this gives us an upper bound on the number of required extensions.

We formalize these two concepts in the following definition:

Definition 5.9. Let (f, ξ) be a partial flow with right-constant flow rates. A *phase* of (f, ξ) is a maximal proper subinterval $[a, b) \subseteq [0, \xi)$ such that all flow rates of f are constant during $[a, b)$.

We say that that a *(simple) event* occurs at time $\theta < \xi$ at some node $v \in V$ if there exist two phases $[a, \theta)$ and $[\theta, b)$ or a single phase $[a, b)$ with $\theta \in (a, b)$ and one of the followings things happen at time θ:

- The outflow rate of one of the edges leading towards v changes, i.e. there exists an edge $e \in \delta^-(v)$ and a commodity $i \in I$ with $f^-_{e,i}(a) \neq f^-_{e,i}(\theta)$.
- The network inflow rate $u_{v,i}$ at node v changes for one of the commodities.
- A queue on one of the edges leaving v depletes, i.e. we have $Q_e(\theta) = 0$ and $\partial_- Q_e(\theta) < 0$ for some edge $e \in \delta^+(v)$.
- An edge leaving v becomes newly active, i.e. there exists a commodity i, an edge $e \in \delta^+(v)$ and some time $\vartheta < \theta$ such that $e \in E_i(\theta)$ while e was inactive for i during $[\vartheta, \theta)$.

If an event occurs at a node v, we also say that v is responsible for this event. Moreover, when counting events we will always count all events happening at the same time as one event.

We say that a *Zeno-event* occurs at time $\theta < \xi$ if there exists no phase $[a, \theta)$.

Finally, we call a partial IDE (f, ξ) a *simple IDE* if each of its phases begins with a simple event.

A Zeno-event happens at a time θ if the phases leading up to this time show a Zeno-type behaviour, i.e. there is an infinite sequence of phases $[a_0, a_1)$, $[a_1, a_2)$, $[a_2, a_3)$, ... with $a_k \to \theta$. Such a flow then has an infinite number of phases and clearly cannot be computed by an extension based algorithm in finite time. Note that it is easy to construct such non-simple IDE even in very simple networks: Just consider a network consisting of a source node, a sink node and two parallel edges with equal free flow travel time connecting them. If flow then enters the network at a rate lower than the capacity of each of the two edges, then any flow split between the two is an IDE. In particular, we may partition the interval $[0, 1]$ into an infinite number of shorter and shorter intervals (e.g. $[0, \frac{1}{2}), [\frac{1}{2}, \frac{3}{4}), [\frac{3}{4}, \frac{7}{8}), \dots$) and then alternating between sending all flow into one edge and sending all flow into the other edge results in an IDE which has a Zeno-event at time $\theta = 1$.

Thus, it is clearly not possible to construct *all* IDE with Algorithm 1. Instead, we want to determine in which networks simple IDE are guaranteed to exist and whether we can then compute such an IDE with our extension based algorithm. Towards this goal, we will specialize Algorithm 1 a bit further in order to give it the best possible chance in succeeding at this task. Namely, we always want to choose the length of the extension phase as long as it is safely possible, that is until the next event happens. This leads to Algorithm 3.

Algorithm 3: An extension based algorithm for computing IDE

Input : A network $\mathcal{N}$ with right-constant network inflow rates and a time horizon $T \geq 0$

Output : A partial IDE (f, T) with right-constant flow rates in $\mathcal{N}$

1 Set $(f, \xi) \leftarrow (0, 0)$
2 **while** $\xi < T$ **do**
3 Compute an IDE-thin flow (x^+, x^-, a) for (f, ξ)
4 Let g be the flow defined by extending (f, ξ) with (x^+, x^-, a) (cf. (4.3))
5 Choose $\varepsilon \leftarrow \min\{\, \varepsilon > 0 \mid \text{ an event occurs in } g \text{ at time } \xi + \varepsilon \,\}$
6 $(f, \xi) \leftarrow (g, \xi + \varepsilon)$
7 **end while**
8 **return** (f, ξ)

Lemma 5.10. *Let $\mathcal{N}$ be a feasible network with right-constant network inflow rates. Let $(f^{(k)}, \xi_k)_k$ be the sequence of partial IDE computed by Algorithm 3. Then* $\lim_k (f^{(k)}, \xi_k)$ *is a simple IDE.*

Proof. First of all, every $(f^{(k)}, \xi_k)$ is a partial IDE since we start with a partial IDE and every extension is correct by the same proof as for the sufficiency part of Proposition 4.22 (the choice of ε there exactly matches that in line 5 of Algorithm 3). They are also simple by construction.

Then since all $(f^{(k)}, \xi_k)$ are partial IDE with right-constant flow rates, so is $(f, \xi) := \lim_k (f^{(k)}, \xi_k)$ by Proposition 4.10. To show that it is also simple let $[a, b) \subseteq [0, \xi)$ be any phase of (f, ξ). Then there exists some $k \in \mathbb{N}_0$ such that $a < \xi_k$. Hence, $[a, \min\{\, b, \xi_k \,\})$ is a phase of the simple IDE $(f^{(k)}, \xi_k)$ and, therefore, starts with a non-degenerate event. Since we have $(f^{(k)}, \xi_k) \preceq (f, \xi)$, this is then also true for (f, ξ). □

Note that the lemma neither states that Algorithm 3 terminates nor that the computed sequence converges to a partial IDE until T. It does, however, suggest a way of showing termination. Namely, if we can show that in a given network and for a given time horizon there is a finite bound on the number of events any simple IDE in this network can have before T, then this immediately shows that Algorithm 3 is guaranteed to terminate in this network. Moreover, this bound then also immediately gives us a bound on the number of extensions performed by Algorithm 3.

In the following we will do exactly that for the case of single-commodity networks with right-constant network inflow rates with finitely many discontinuities and strictly positive free flow travel times. Note, that in this case Corollary 5.7 ensures that we can compute each individual extension in polynomial time.

Lemma 5.11. *Let (f, ξ) be a simple IDE in a single-commodity network, $v \in V \setminus T$ a non-sink node with at least one outgoing edge, $\tilde{E}_v \subseteq \delta^+(v)$ some subset of outgoing edges from v and $[a, b) \subseteq [0, \xi)$ an interval satisfying the following properties:*

- *u_v is constant on $[a, b)$,*
- *f_e^- is constant on $[a, b)$ for all $e \in \delta^-(v)$,*
- *L_w are affine functions on $[a, b)$ for all $vw \in \tilde{E}_v$ and*
- *$\delta^+(v) \cap E(\theta) \subseteq \tilde{E}_v$ for all $\theta \in [a, b)$.*

Then v is responsible for at most $2|\tilde{E}_v|^{4^{|\tilde{E}_v|}+1}$ events during $[a, b)$.

Proof. We enumerate the edges in $\tilde{E}_v$ by $vw_1, \dots, vw_K$ and define for each of those edges vw_k a function

$$\Lambda_k : [a, b) \to \mathbb{R}, \theta \mapsto C_{vw_k}(\theta) + L_{w_k}(\theta).$$

Since f is a flow with right-constant flow rates without any Zeno-events, these functions are all continuous and piecewise linear. In particular, their left- and right-derivatives exist at all times $\theta \in (a, b)$. Furthermore, we observe the following properties of these functions for all edges $vw_k \in \delta^+(v)$ and all times $\theta \in (a, b)$:

(i) vw_k is active at time θ if and only if $\Lambda_k(\theta) = \min\left\{ \Lambda_{k'}(\theta) \mid k' \in [K] \right\}$.
(ii) vw_k becomes newly active at time θ if and only if $vw_k \in E(\theta)$ and $\partial_-\Lambda_k(\theta) < \partial_- L_v(\theta)$.
(iii) vw_k becomes inactive at time θ if and only if $vw_k \in E(\theta)$ and $\partial_+\Lambda_k(\theta) > \partial_+ L_v(\theta)$.

We will now show two further key properties of these functions (cf. Figs. 5.6 and 5.5 for a visualization of all these properties):

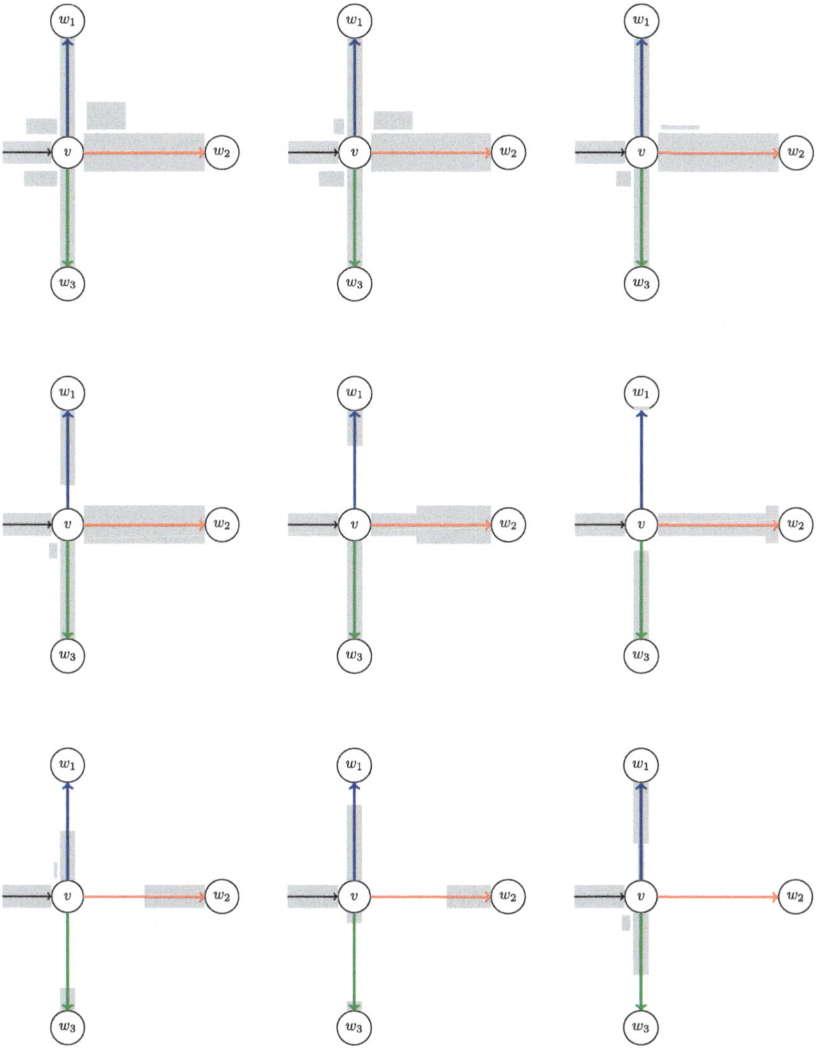

Fig. 5.5 An example for the setting of Lemma 5.11: A single node v with a constant inflow rate (here only over a single edge) and with three outgoing edges vw_1, vw_2 and vw_3. The sequence of pictures shows snapshot of all the times where an event happens at node v. The functions L_{w_k} and Λ_k corresponding to this sequence are depicted in Fig. 5.6

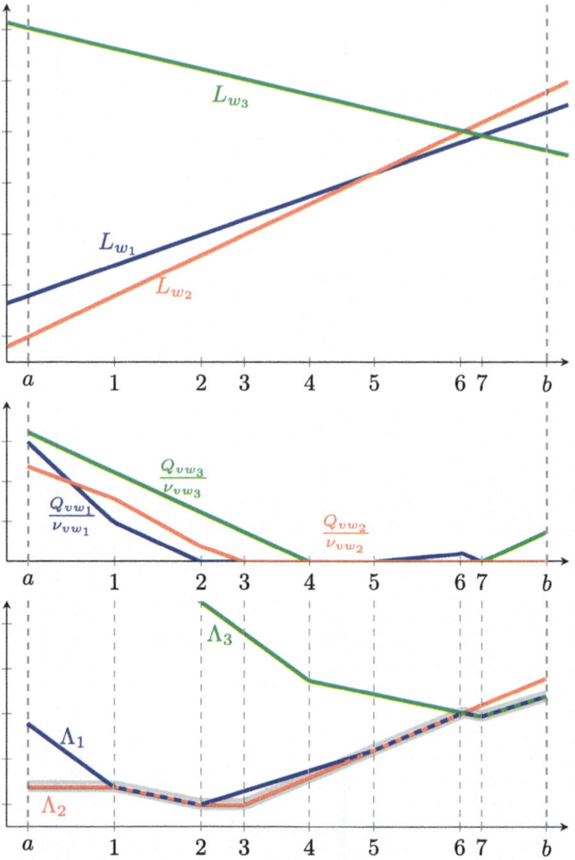

Fig. 5.6 An example for the functions Λ_k defined in the proof of Lemma 5.11 for the situation depicted in Fig. 5.5. The topmost diagram depicts the distance label functions L_{w_k} (which are linear for the whole interval $[a, b]$). The middle diagram displays the waiting times on the edges vw_i resulting from the flow distribution at node v shown in Fig. 5.5. The bottommost diagram then shows the resulting functions Λ_k. The numbers 1 to 7 indicate the events occurring at node v during (a, b). E.g. at event 1 edge vw_1 becomes newly active and we can see that property (ii) holds here. We can also get an intuitive idea from this diagram for why Claims 5.1 to 5.4 hold, e.g. Λ_3 is convex between before event 6 (as stated in Claim 5.1) and the two phases where ∂L_v is minimal (between events 1 and 2 and events 6 and 7) are started by different edges becoming newly active (which is the important consequence of Claim 5.3 we will use in the proof of Lemma 5.11)

Claim 5.1. *If an edge vw_k is inactive during some subinterval $(a', b') \subseteq [a, b)$, then Λ_k is convex on (a', b').*

Claim 5.2. *We have* $\min \{ \partial_- \Lambda_k(\theta) \mid vw_k \in E(\theta) \} \leq \partial_+ L_v(\theta)$ *for all* $\theta \in (a, b)$.

Proof of Claim 5.1. Since edge vw_k is inactive during (a', b') no flow enters this edge during this time. Thus, the queue length function of this edge consists of at most two linear segments during this interval: One where the queue depletes (at a rate of $-\nu_{vw_k}$) and one where it stays empty. The combination of these to segments is clearly a convex function. ■

Proof of Claim 5.2. We define the set $E^+(\theta) := \{vw_k \in E(\theta) | \partial_+ \Lambda_k(\theta) = \partial_+ L_v(\theta)\}$ and observe that this set contains exactly those edges which are active right after time θ. Thus, flow may only enter those edges immediately after time θ. Since the total inflow into v is constant on (a, b), this means that there must be at least one edge $vw_{k'} \in E^+(\theta)$ such that there is at least as much inflow into this edge after θ as before. This directly implies $\partial_- \Lambda_{k'}(\theta) \leq \partial_+ \Lambda_{k'}(\theta)$ and, therefore,

$$\begin{aligned}\min \{ \partial_- \Lambda_k(\theta) \mid vw_k \in E(\theta) \} &\leq \min \{ \partial_- \Lambda_k(\theta) \mid vw_k \in E^+(\theta) \} \\ &\leq \partial_- \Lambda_{k'}(\theta) \leq \partial_+ \Lambda_{k'}(\theta) = \partial_+ L_v(\theta).\end{aligned}$$

■

Now, since the total inflow into node v is constant during $[a, b)$ we get from Lemma 5.6 that there are only finitely many possible values for $\partial_+ L_v$ during this interval. The following claim then implies that the lowest of those values can only appear a finite number of times. Iteratively applying this claim then shows that this is true for all of those finitely many values. This then reduces the lemmas statement to showing that only finitely many events can occur during any interval with constant ∂L_v.

Claim 5.3. *Let $(a', b') \subseteq [a, b)$ be some subinterval with $b' < b$ and $\alpha \in \mathbb{R}$ some value such that we have $\partial_+ L_v(\theta) \geq \alpha$ for all $\theta \in (a', b')$ as well as $\partial_- L_v(b') > \alpha = \partial_+ L_v(b')$. Then there exists an edge vw_k which is inactive during (a', b') and becomes newly active at time b'.*

Proof. According to Claim 5.2 there exists some edge $vw_k \in E(\theta)$ with $\partial_- \Lambda_k(b') \leq \partial_+ L_v(b') = \alpha < \partial_- L_v(b')$. Thus, vw_k becomes newly active at time b' by property (ii). Now, assume that this edge was active at some time during (a', b') and let

$c \in (a', b')$ be the last such time. Then Λ_k is convex on (c, b') by Claim 5.1 and, hence, we have $\partial_-\Lambda_k(\theta) \leq \partial_-\Lambda_k(b') \leq \alpha$ for all $\theta \in (c, b')$. At the same time we have $\partial_- L_v(\theta) \geq \alpha$ for those θ and $\partial_- L_v(\theta) > \alpha$ for some proper subinterval. Together, this is now a contradiction to $\Lambda_k(c) = L_v(c)$ and $\Lambda_k(b') = L_v(b')$ (which holds due to property (i) since vw_k is active at both times). Thus, vw_k cannot been active at any time during (a', b') and, therefore, proves our claim. ■

Claim 5.4. *Let* $(a', b') \subseteq [a, b)$ *be a subinterval such that* ∂L_v *is constant on* (a', b')*. Then* v *is responsible for at most* $2K$ *events during this interval.*

Proof. We first observe that if an edge vw_k changes from being active to being inactive at some point θ during such an interval, then it will remain inactive after θ until at least b'. This follows from the assumption that ∂L_v is constant, property (iii) and Claim 5.1.

Thus, there can be at most K events at node v because an edge in $\delta^+(v)$ becomes active. Furthermore, if an event occurs because the queue of such an edge depletes, then this edge becomes inactive immediately after. Hence, such events can occur at most K times as well. ■

Combining Claims 5.3 and 5.4 now allows us to prove the lemma: Since the total inflow into node v is constant during $[a, b)$, Lemma 5.6 implies that $\partial_+ L_v(\theta)$ is uniquely determined by which edges in $\delta^+(v)$ are active at time θ and which have a non-empty queue. Thus, $\partial_+ L_v(\theta)$ can take at most $(2^K - 1) \cdot 2^K \leq 4^K - 1$ different values during this interval (note that at least one edge is always active). Let α be the smallest such value. Then we can partition (a, b) into at most K subintervals during which $\partial_+ L_v$ is constantly α and at most K subintervals where $\partial_+ L_v$ is strictly larger than α. This is because if we have K maximal intervals with $\partial_+ L_v \equiv \alpha$, then, according to Claim 5.3, there must exist at least K different edges in $\delta^+(v)$ which are always inactive between time a and the start of the first of these intervals. But since there must always be at least one active edge, this implies that a is the start of the first of these intervals.

Now, using the same argument for the second smallest value $\partial_+ L_v$ can attain and each of the at most K intervals where $\partial_+ L_v$ is strictly larger than α, gives us at most K^2 more intervals with constant $\partial_+ L_v$ as well as at most K^2 remaining intervals with even larger value of $\partial_+ L_v$. Continuing this way we find that (a, b) can be partitioned into at most

$$K + K^2 + \cdots + K^{4^K - 1} \leq K^{4^K}$$

subintervals of constant $\partial_{-}L_v$. Finally, Claim 5.4 tells us that each of those intervals contains at most $2K$ events for which v is responsible. This gives us the desired bound. □

As our next step we lift this bound on the number of events occurring at a single node to the whole network using induction over a suitable chosen order of the nodes:

Lemma 5.12. *Let $\mathcal{N}$ be a single-commodity network with strictly positive free flow travel times, (f, ξ) a simple IDE, $\tilde{E} \subseteq E$ some subset of edges and $[a, b) \subseteq [0, \xi)$ a time interval satisfying the following properties:*

- *All network inflow rates u_v are constant during $[a, b)$,*
- *all edge outflow rates f_e^- are constant during $[a, b)$,*
- *$E(\theta) \subseteq \tilde{E}$ for all $\theta \in [a, b)$ and*
- *$\tilde{E}$ is acyclic.*

Then at most $(2\Delta^{4^\Delta+1})^{|V|-1}$ events occur during $[a, b)$, where we denote by $\Delta := \max\{\,|\delta^+(v)| \mid v \in V\,\}$ the maximal out-degree in the network.

Proof. Since $\tilde{E}$ is acyclic, we can choose a topological order $v_1 \succ v_2 \succ \cdots \succ v_n$ in $(V, \tilde{E})$. We will now show the claimed bound on the number of event via induction on this order. More precisely, we show that for any $\ell \in [n]$ at most $(2\Delta^{4^\Delta+1})^{\ell-1}$ events occur during $[a, b)$ at the nodes $v_1, \ldots, v_\ell$:

Base Case ($\ell = 1$): As v_1 is the last node in the topological order, it has no outgoing edges and, therefore, cannot be responsible for any events during an interval with fixed inflow into this node (except for maybe one event at the start of the interval).

Induction Step ($\ell \to \ell + 1$): We first note that, if $v_{\ell+1}$ is a sink node, then no flow will ever enter any edges leaving $v_{\ell+1}$ since we do not have edges of zero travel time here. Thus, such a node cannot be responsible for any events during (b, c) and the induction step becomes trivial. As the same is true for nodes without outgoing edges, we can now restrict ourselves to the case where $v_{\ell+1}$ is a non-sink node with at least one outgoing edge.

For such a node we define $\tilde{E}_{v_{\ell+1}} := \tilde{E} \cap \delta^+(v_{\ell+1})$ and observe that this set clearly contains all outgoing edges from $v_{\ell+1}$ which are active at any point during $[a, b)$. Furthermore, for any such edge $vw_k \in \tilde{E}_{v_{\ell+1}}$ the distance label L_{w_k} can only have a break point at some time $\theta \in (a, b)$ if an event occurs at a node on some active w, T-path at that time. As all those node are higher in the topological order than $v_{\ell+1}$ we know by induction that there are at most $(2\Delta^{4^\Delta+1})^{\ell-1}$ such times. Thus, we can subdivide $[a, b)$ into at most $(2\Delta^{4^\Delta+1})^{\ell-1}$ subinterval such that for each of those we can apply Lemma 5.11 in order to bound the number of events occurring at $v_{\ell+1}$ by $2\Delta^{4^\Delta+1}$. In total, this gives us at most $(2\Delta^{4^\Delta+1}) \cdot (2\Delta^{4^\Delta+1})^{\ell-1} = (2\Delta^{4^\Delta+1})^{\ell}$ events occurring at nodes $v_1, \ldots, v_{\ell+1}$ during $[a, b)$.

Choosing $\ell = |V|$ now directly implies the lemma. □

If we now want to apply this lemma to bound the number of events in a simple IDE, we still need to find some set $\tilde{E}$ which contains all active edges for as long a time as possible while also being acyclic. Intuitively, a good choice for such a set should include all edges which are active or "close" to being active at time a. Moreover, we want to define "closeness" here as wide as possible without risking to include any cycles. The following lemma makes this idea into a formal definition and also provides us with a lower bound for the length of the interval for which we can then use such a set:

Lemma 5.13. *Let (f, ξ) be any partial Vickrey flow in a single-commodity network without cycles of free flow travel time zero and without any dead-end nodes. Furthermore, let $\varepsilon > 0$ be some constant such that we have $\sum_{e\in c} \tau_e \geq |c|\varepsilon$ for all cycles c. Then for any time $\theta < \xi$ the set*

$$\tilde{E} := \{\, e = vw \in E \mid L_v(\theta) > C_e(\theta) + L_w(\theta) - \varepsilon \,\}$$

is acyclic. If, additionally, there exists some constant $M \geq 0$ such that we have $\sum_{v\in V} u_v(\zeta) \leq M$ for all times $\zeta < \xi$, then $\tilde{E}$ contains all edges which are active at any point in time during the interval $\left[\theta, \min\{\, \theta + \frac{\varepsilon}{\sum_{e\in E} \nu_e + |E| + M}, \xi \,\}\right)$.

Proof. First, note that we have $L_v(\zeta) < \infty$ for all $v \in V$ and $\zeta \in \mathbb{R}_{\geq 0}$ since there are no dead-end nodes. Now, to show that $\tilde{E}$ is acyclic take any cycle c in E. Then we have

$$\sum_{e=vw\in c} (L_w(\theta) - L_v(\theta) + C_e(\theta)) = \sum_{e\in c} C_e(\theta) \geq \sum_{e\in c} \tau_e \geq |c|\varepsilon = \sum_{e\in c} \varepsilon.$$

Thus, c must contain at least one edge $e = vw$ with $L_w(\theta) - L_v(\theta) + C_e(\theta) \geq \varepsilon$ and, therefore, $e \notin \tilde{E}$. This shows that $\tilde{E}$ is acyclic.

For the second part of the lemma take any time $\zeta \in [\theta, \min\{\theta + \frac{\varepsilon}{\sum_{e'\in E} \nu_{e'}+|E|+M}, \xi\})$ and any active edge $e = vw \in E(\zeta)$. We need to show that this edge is then also contained in $\tilde{E}$. For this let p be an active v, T-path at time θ and p' an active w, T-path at time ζ. Using Proposition 3.56 we now get the following bounds:

$$\begin{aligned} L_v(\theta) = C_p(\theta) = \sum_{e'\in p} C_{e'}(\theta) &\overset{\text{Prop. 3.56}}{\geq} \sum_{e'\in p} C_{e'}(\zeta) - (\zeta - \theta)\cdot\Big(\sum_{e'\in E}\nu_{e'} + M\Big) \\ &= C_p(\zeta) - (\zeta-\theta)\cdot\Big(\sum_{e'\in E}\nu_{e'} + M\Big) \geq L_v(\zeta) - (\zeta-\theta)\cdot\Big(\sum_{e'\in E}\nu_{e'} + M\Big) \end{aligned} \tag{5.4}$$

and

$$\begin{aligned} L_w(\zeta) = C_{p'}(\zeta) = \sum_{e'\in p'} C_{e'}(\zeta) &\overset{\text{Prop. 3.56}}{\geq} \sum_{e'\in p'} C_{e'}(\theta) - (\zeta-\theta)\cdot|p'| \\ &\overset{(*)}{\geq} L_w(\theta) - (\zeta-\theta)\cdot(|E|-1), \end{aligned} \tag{5.5}$$

where use at $(*)$ that w, T-path (like p') cannot contain any cycles and, in particular, not any edge leading towards w (like e). Together with $C_e(\zeta) \geq C_e(\theta) - (\zeta - \theta)$ (again from Proposition 3.56) this now implies

$$\begin{aligned} L_v(\theta) &\overset{(5.4)}{\geq} L_v(\zeta) - (\zeta-\theta)\cdot\Big(\sum_{e'\in E}\nu_{e'} + M\Big) \\ &= L_w(\zeta) + C_e(\zeta) - (\zeta-\theta)\cdot\Big(\sum_{e'\in E}\nu_{e'} + M\Big) \\ &\overset{(5.5)}{\geq} L_w(\theta) - (\zeta-\theta)\cdot(|E|-1) + C_e(\theta) - (\zeta-\theta) - (\zeta-\theta)\cdot\Big(\sum_{e'\in E}\nu_{e'} + M\Big) \\ &= L_w(\theta) + C_e(\theta) - (\zeta-\theta)\cdot\Big(\sum_{e'\in E}(\nu_{e'}+1) + M\Big) \\ &> L_w(\theta) + C_e(\theta) - \frac{\varepsilon}{\sum_{e'\in E}\nu_{e'} + |E| + M}\cdot\Big(\sum_{e'\in E}(\nu_{e'}+1) + M\Big) \\ &= L_w(\theta) + C_e(\theta) - \varepsilon. \end{aligned}$$

Hence, we have $e \in \tilde{E}$. □

By combining Lemmas 5.12 and 5.13 we can now finally show that in single-commodity networks without edges with free flow time zero only finitely many events occur during any finite time interval of a simple IDE.

Theorem 5.14. *Let $\mathcal{N}$ be a feasible single-commodity network with right-constant network inflow rates with R discontinuities and strictly positive free flow travel times. Then the number of events in any simple IDE (f, ξ) with $\xi < \infty$ is bounded by*

$$\mathcal{O}\left(\left(2\cdot\left(2\Delta^{4^\Delta+1}\right)^{|V|}\right)^{\left\lceil \xi\cdot\frac{\sum_{e\in E}\nu_e+|E|+M}{\tau_{\min}}\right\rceil+R}\right),$$

where

- $\tau_{\min} := \min\{\,\tau_e \mid e \in E\,\} > 0$ *is the shortest free flow travel time,*
- $\Delta := \max\{\,|\delta^+(v)| \mid v \in V\,\}$ *denotes the maximal out-degree and*
- $M := \max\{\,\sum_{v\in V} u_v(\theta) \mid \theta < \xi\,\} \geq 0$ *is a constant bounding the sum of all network inflow rates at all times before ξ.*

Proof. We first note that we can assume without loss of generality that $\mathcal{N}$ does not contain any dead-end nodes: Since no IDE flow could reach such a node anyway (see Proposition 3.66), we can just remove them without changing the IDE in this network.

We now partition $[0, \xi)$ into $K := \left\lceil \xi \cdot \frac{\sum_{e\in E}(\nu_e+1)+M}{\tau_{\min}}\right\rceil + R$ subintervals $[a_0, a_1)$, $[a_1, a_2), \dots, [a_{K-1}, a_K)$ of length at most $\frac{\tau_{\min}}{\sum_{e'\in E}(\nu_{e'}+1)+M}$ such that during each such subinterval all network inflow rates are constant. We will now inductively show that for any $k \in \{0, \dots, K\}$ the number of events before a_k is at most $(2\cdot(2\Delta^{4^\Delta+1})^{|V|})^k$.

Base Case (*k*=0): Since no event can occur before $a_0 = 0$, the bound trivially holds here.

Induction Step (*k* → *k*+1): Since $[a_k, a_{k+1})$ has length at most $\tau_{\min}$ the edge outflow rates during this interval are completely determined by the edge inflow rates before time a_k (cf. Corollary 3.21). By induction and the assumption that (f, ξ) is simple, those have at most $(2\cdot(2\Delta^{4^\Delta+1})^{|V|})^k$ discontinuities. Thus, the edge outflow rates have at most twice this number of discontinuities by Proposition 3.22 and we can further subdivide $[a_k, a_{k+1})$ into at most $2\cdot(2\cdot(2\Delta^{4^\Delta+1})^{|V|})^k$ intervals with constant edge outflow rates. Additionally, for each such interval $[b, c)$ we know from Lemma 5.13 that the set

$\tilde{E} := \{e = vw \in E \mid L_v(b) > C_e(b) + L_w - \tau_{\min}\}$ is acyclic and contains all active edges during this interval.

Thus, we can apply Lemma 5.12 to each of those subintervals $[b, c)$ and show that for each of those at most $2 \cdot (2 \cdot (2\Delta^{4^\Delta+1}))^{|V|})^k$ subintervals in our partition of $[a_k, a_{k+1})$ at most $(2\Delta^{4^\Delta+1})^{|V|-1}$ events occur anywhere in the whole network. This, now, gives us the desired bound of at most

$$(2 \cdot (2\Delta^{4^\Delta+1}))^{|V|})^k + 2 \cdot (2 \cdot (2\Delta^{4^\Delta+1}))^{|V|})^k \cdot (2\Delta^{4^\Delta+1})^{|V|-1}$$
$$\leq 2 \cdot (2 \cdot (2\Delta^{4^\Delta+1}))^{|V|})^k \cdot (2\Delta^{4^\Delta+1})^{|V|} = (2 \cdot (2\Delta^{4^\Delta+1}))^{|V|})^{k+1}$$

events before a_{k+1}.

Finally, choosing $k = \left\lceil \xi \cdot \frac{\sum_{e \in E}(\nu_e+1)+M}{\tau_{\min}} \right\rceil + R$ gives us the bound stated in the theorem. □

Remark 5.15. If the network $\mathcal{N}$ itself is already acyclic, we do not need Lemma 5.13 in the above proof and, thus, can start with a partition of $[0, \xi)$ into "only" $K := \left\lceil \frac{\xi}{\tau_{\min}} \right\rceil + R$ subintervals of length at most $\tau_{\min}$. This then leads to a slightly better bound of

$$\mathcal{O}\Big((2 \cdot (2\Delta^{4^\Delta+1}))^{|V|}\Big)^{\left\lceil \frac{\xi}{\tau_{\min}} \right\rceil + R}\Big).$$

Corollary 5.16. *Let $\mathcal{N}$ be a feasible single-commodity network with right-constant network inflow rates with R discontinuities and strictly positive free flow travel times. Then, for any given time horizon $T \geq 0$ Algorithm 3 computes an IDE until T and has a worst case runtime of*

$$\mathcal{O}\Big((2 \cdot (2\Delta^{4^\Delta+1}))^{|V|}\Big)^{\left\lceil T \cdot \frac{\sum_{e \in E} \nu_e + |E| + M}{\tau_{\min}} \right\rceil + R} \cdot |E|^2\Big),$$

where Δ, $\tau_{\min}$ and M are defined as in Theorem 5.14.

Proof. According to Lemma 5.10 Algorithm 3 computes a simple IDE. Thus, we can apply Theorem 5.14 to bound the number of events and, consequentially, the number of extension steps performed by the algorithm to obtain this partial IDE. Furthermore, according to Proposition 5.5 we can use Algorithm 2 to compute each individual extension in $\mathcal{O}(|E|^2)$. □

5.2.2 Lower Bounds

In the previous subsection we saw that in single-commodity networks with non-zero free flow travel times, there always exists an IDE with a finite number of phases (for any finite time horizon). The following example shows that at least one of these two assumptions is necessary, i.e. the existence of simple IDE is not guaranteed any more for multi-commodity networks with zero free flow travel times:

Example 5.17. Consider the network depicted in Fig. 5.7. It has three commodities which share a common source node s and each have a single sink node. The source node is directly connected to the sink nodes of commodities 1 (blue) and 3 (green) via edges of free flow travel time zero and capacities 6 and 3, respectively. The sink node of commodity 2 (red) is reachable from both these sink nodes via an additional edge with free flow time 1 and capacities 2 and 3, respectively. This means that particles of the blue and the green commodity each only have one single path while particles of the red commodity may choose between two different paths. When the first of these particles start to enter the network at time $\theta = 1$ the particles of the other two commodities already build up queues on the two edges st_1 and st_2 (see Fig. 5.8 for a visual representation). Both queues have the same length (6), but the capacity of edge st_1 is large and so the current travel time along the upper s, t_2-path is strictly shorter than along the bottom one. Thus, at first all red particles enter the upper path leading to the waiting time on edge st_1 to stay constant while the waiting time on edge st_3 starts to decrease. At time $\theta = 2$ then, both paths

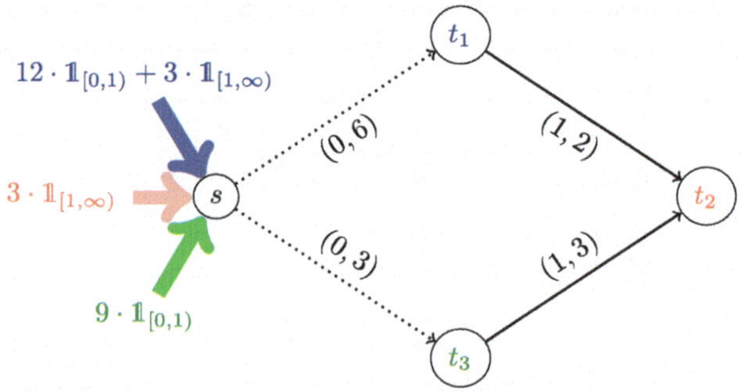

Fig. 5.7 A multi-commodity network in which there exists no simple IDE until any time > 5. The edge labels on the edges denote free flow travel time and capacity in the form (τ_e, ν_e). In particular, the dotted edges are zero free flow travel time edges

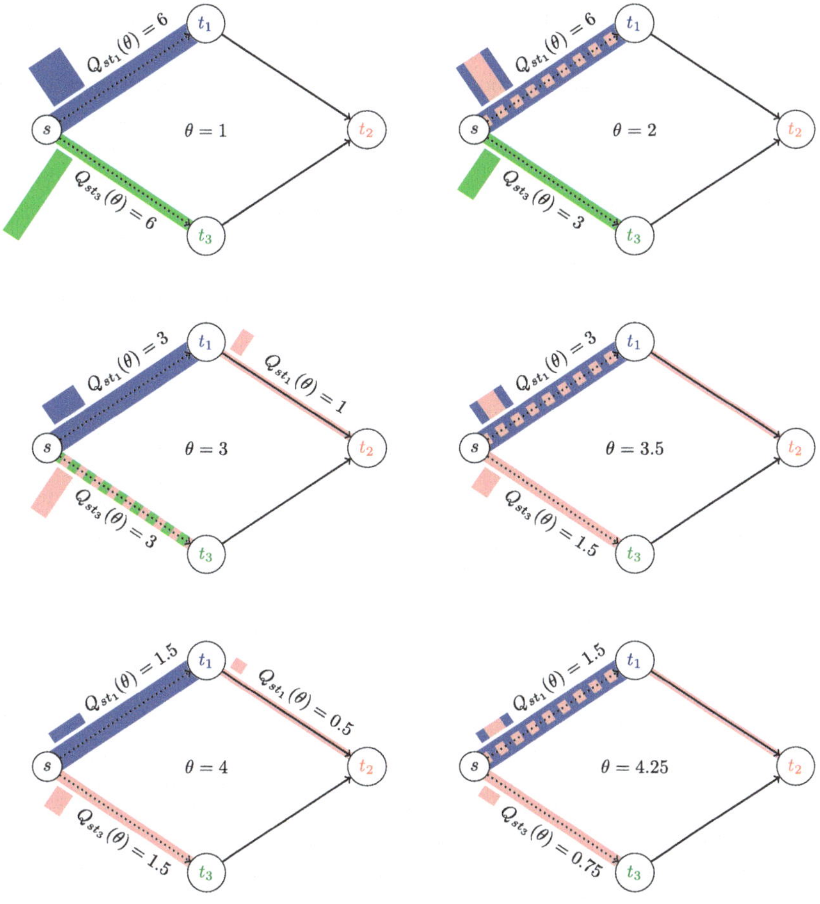

Fig. 5.8 The first six events in the network from Fig. 5.7. Note, that the picture is slightly misleading insofar as the dashed edges should have a physical length of zero. In particular, there is never any proper volume of flow on such an edge (only flow waiting in its queue). The flow shown in the pictures on these edges is, therefore, only meant to indicate at which rate flow will pass through this edge during the next phase but should be seen has having measure zero right now

have the same current travel time. At the same time the first red particles arrive at node t_1 and start to form a queue on edge t_1t_2. This has the effect that for the next phase all red particles have to choose the lower path. This lasts until time $\theta = 3$ at which point the last red particles arrive at node t_1 and, hence, the queue on edge t_1t_2 starts to decrease again. Because of this, the red particles may now only use the upper path again. This pattern of switching between the two paths now continues, however, with ever shorter length of the two types of phases: The next switches happen at times 3.5 and 4, the ones after that at times 4.25 and 4.5 and so on. Thus, the unique IDE in this network exhibits the flow pattern depicted in Fig. 5.9.

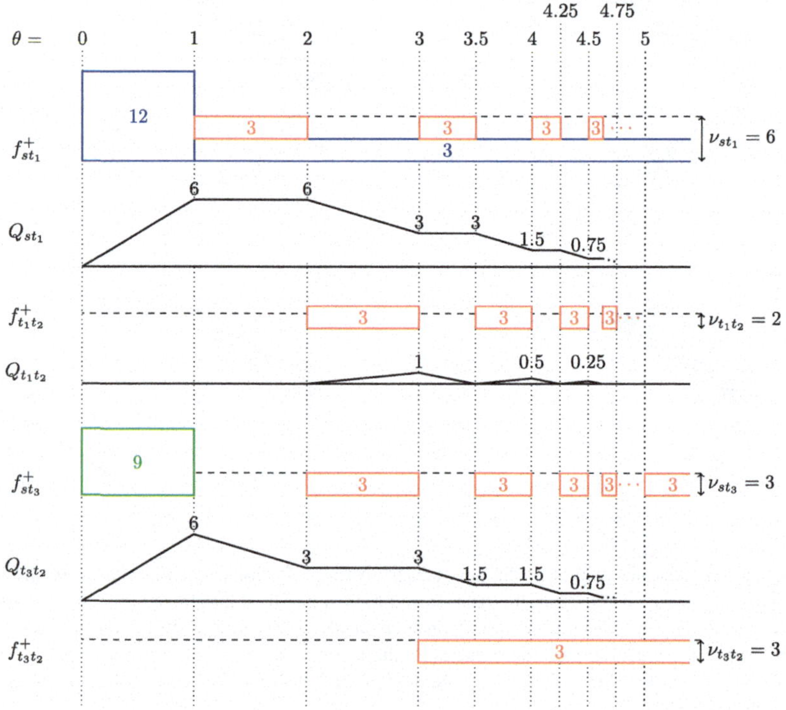

Fig. 5.9 The flow pattern exhibited by the unique IDE in the network depicted in Fig. 5.7 as described in Example 5.17. The height of the rectangles (as well as the numbers inside of them) indicate the inflow rates into the respective edges (with the colours corresponding to the respective commodities) and the height of the graphs for the queues denote their length at any time. The horizontal dashed lines show the capacity of each edge, i.e. whenever the combined inflow rates exceed such a line, the queue on the respective edge grows

We can from this that this IDE requires an infinite number of phases to reach time $\theta = 5$, i.e. it has a Zeno-event at time $\theta = 5$.

This example immediately implies the following theorem:

Theorem 5.18. *There exists an acyclic multi-commodity network with edges of free flow travel time zero in which every IDE until a certain time $\xi < \infty$ requires an infinite number of phases. In particular, we cannot use an extension based approach like Algorithm 1 to compute IDE until any time horizon $T \geq \xi$ in this network.* □

As shown in the previous section, this kind of Zeno-type behaviour is not possible in single-commodity networks without zero free flow travel time edges. However, individual phases might still become arbitrarily small over time even in very simple networks:

Example 5.19. Consider the network depicted in Fig. 5.10 with a single source node s with a constant, infinitely lasting inflow rate of 2. There are two paths from the source s to the sink t: The upper one via node v with a free flow travel time 2 and the lower one via nodes w and x with a free flow travel time of 3. Both start with an edge with capacity 2 which is followed by an edge of capacity 1. Hence, in the empty network at time $\theta = 0$ only the upper path is active and all flow has to take this path (see Fig. 5.11 for a visual depiction). Beginning with time $\theta = 1$ this flow enters the second edge vt of the upper path resulting in a queue forming on this edge. By time $\theta = 2$ this queue reaches length 1 and, thus, the lower path

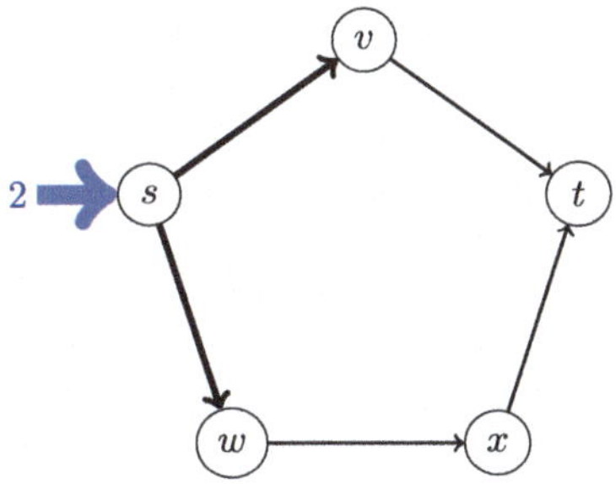

Fig. 5.10 A single-commodity network which can create arbitrarily short extension phases. All free flow travel times are 1. The capacities of the thick edges sv and sw are 2 while all other capacities are 1

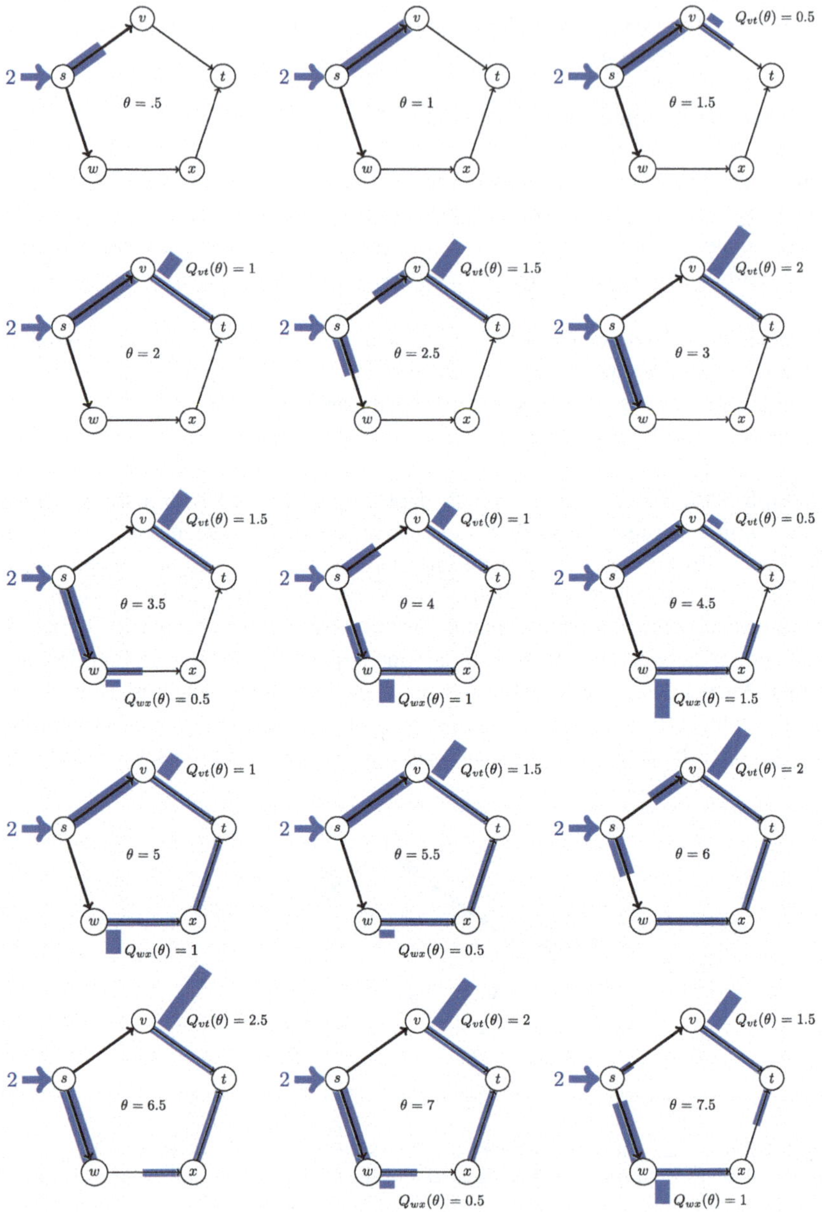

Fig. 5.11 Snapshots of the unique IDE flow in the network depicted in Fig. 5.10

becomes active while the upper one becomes inactive immediately after as the queue on edge vt continues to grow because of the particles already on the upper path's first edge sv. Consequently, after time $\theta = 2$ all particles have to use the lower path. At time $\theta = 3$ a new queue begins to grow on edge wx while at the same time the last particles from edge sv arrive at node v and, thus, the queue on edge vt starts to deplete after that. By time $\theta = 3.5$ those queues reach a length of 0.5 and 1.5, respectively, making the upper path active again while the lower path becomes inactive immediately after. This pattern of switching active paths continues forever, i.e. the unique IDE during the initial interval [0, 3.5] follows the pattern depicted in Fig. 5.12 and for every interval $[4k + 2^{-k} - 1, 4(k+1) + 2^{-(k+1)} - 1]$ with $k \in \mathbb{N}^*$ after that by the pattern shown in Fig. 5.13. Figure 5.14 shows the resulting queue length functions on the edges vt and wx as well as the current travel times of the two paths for the times between 0 and 15. Figure 5.11 depicts snapshots of this IDE between times 0.5 and 7.5.

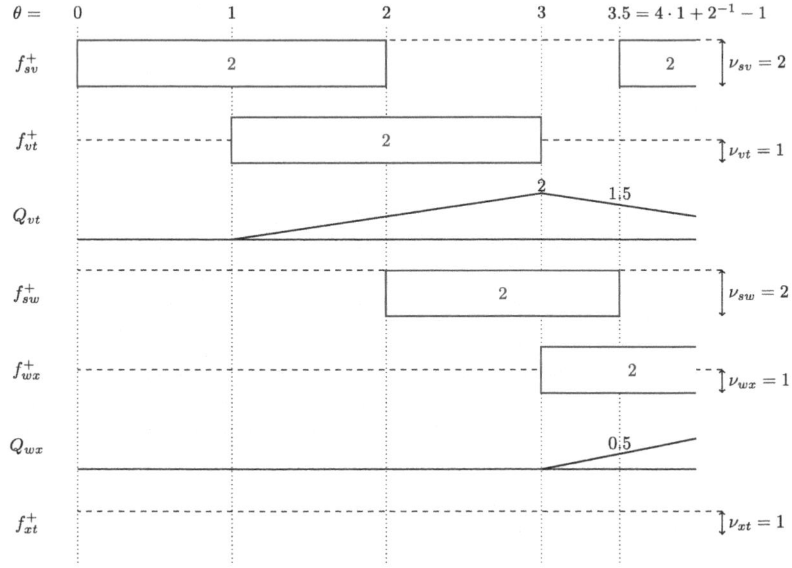

Fig. 5.12 The initial flow pattern exhibited by the unique IDE in the network depicted in Fig. 5.10 as described in Example 5.19

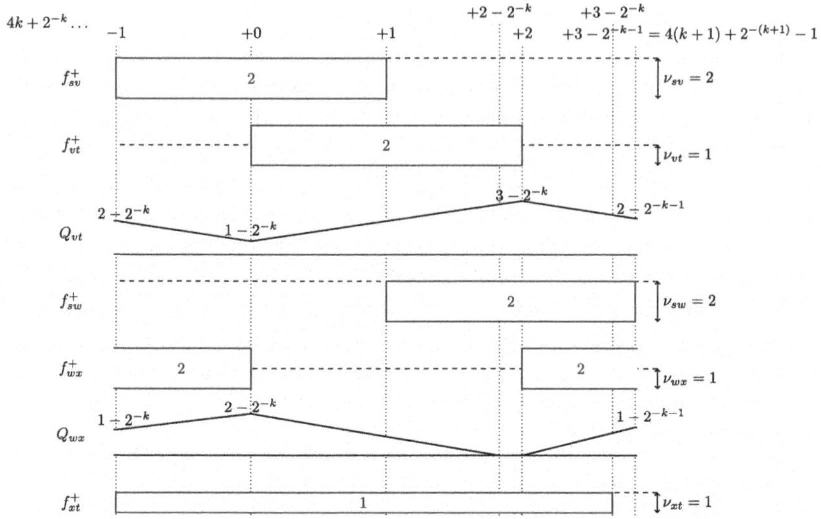

Fig. 5.13 The flow pattern exhibited by the unique IDE in the network depicted in Fig. 5.10 for interval $[4k + 2^{-k} - 1, 4(k+1) + 2^{-(k+1)} - 1]$ with $k \in \mathbb{N}^*$

This example directly implies several negative results on IDE: Over time, extension phases can become arbitrarily small, the number of necessary extension phases can be large compared to the size of the instance and it is not guaranteed that an IDE ever reaches a steady state where, following [CCO22a] we say that a dynamic flow reaches a steady state if all queue lengths (and, consequently, all current travel times) remain constant forever after a certain point in time. Additionally, we also define the weaker notion of a periodic state wherein queues may still change but only in a periodic way. Both kinds of states are of interest for computing flows as, if such a state is reached (and detected), one can stop the computation and still know the complete flow, i.e. one can compute an IDE for all times by only computing it for some finite time.

Definition 5.20. Let f be a dynamic flow. We say that f reaches

- a *steady state* if there exists some time ξ such that we have $Q_e(\theta) = Q_e(\xi)$ for times $\theta \geq \xi$ and all edges $e \in E$.
- a *periodic state* if there exists some time ξ and a periodicity $\pi \in \mathbb{R}_{\geq 0}$ such that we have $Q_e(\theta + k\pi) = Q_e(\theta)$ for times $\theta \geq \xi$, all $k \in \mathbb{N}_0$ and all edges $e \in E$.

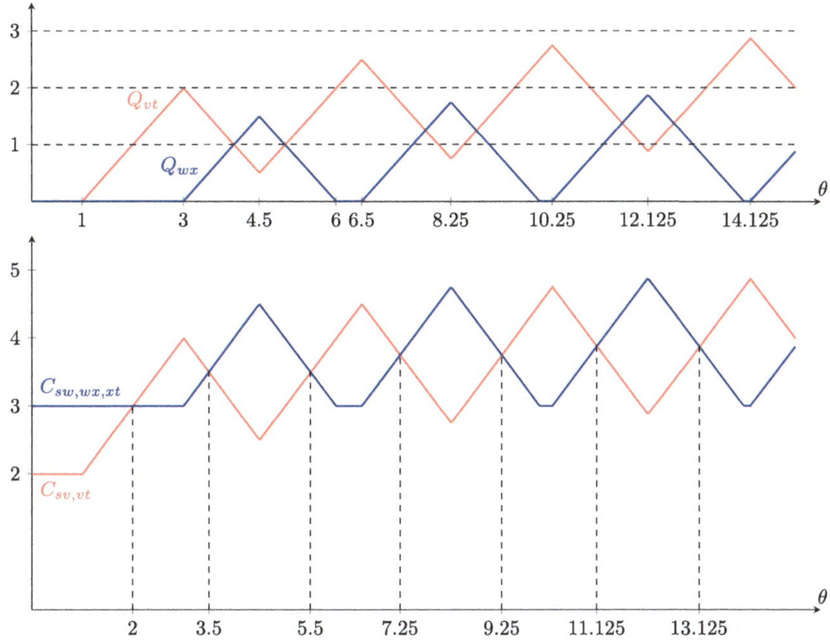

Fig. 5.14 Queue lengths on edges vt and wx (top) and current travel times on the paths sv, vt and sw, wx, xt, respectively (bottom)

Note that a steady state is, in particular, also a periodic state (with arbitrary periodicity).

Theorem 5.21. *Considering only single-commodity networks with constant network inflow rates and strictly positive, integer capacities and free flow travel times. Then the following statements are true:*

a) There exists no general non-trivial lower bound for the length of the required extension phases in this network depending only on capacities and free flow travel times.

b) There exists a network where the number of extensions required to reach an IDE up to any given time horizon T is in $\Omega(T)$.

c) There exists a network wherein no IDE ever reaches a periodic steady state even though the minimal cut has higher capacity than the network inflow rate.

d) *The worst case output complexity of any algorithm computing IDE in such network is not polynomially bounded in the encoding size of the instances—even if we allow the algorithm to use simple periodicity to reduce the size of the output.*

Proof. The network from Example 5.19 clearly exhibits the first two properties. For the third property, we add an edge st with free flow travel time ≥ 5 and any capacity > 0 to this network. Since the current travel time along both path sv, vt and sw, wx, xt never reaches 5 this does not change the unique IDE in the instance (which clearly never reaches a periodic state). The fourth property follows directly from the second one as the phases in the unique IDE in this network are all different and, thus, we the output has to explicitly describe each of them individually. □

Remark 5.22. The third property is not true for dynamic equilibria in the full information setting since such equilibrium flows are guaranteed to reach a steady state (under the same assumptions on the network) by [CCO22a, Theorem 3]. The other three properties are, to our knowledge, still open questions in this model.

5.3 NP-Hardness of IDE-Decision Problems

While we have shown in Sects. 5.1 and 5.2 that in single-commodity networks we can, in principle, always compute IDE with Algorithm 1, the runtime bound provided in Corollary 5.16 is very large (and, in particular, superpolynomial). Partly, the fault for this large bound certainly lies in the rather rough estimates used in the proof of this result and it seems unlikely that there exists any instance where such a runtime is actually achieved. On the other hand. Theorem 5.21 already suggests that at least a polynomial time algorithm for computing IDE is unlikely to exist as even just writing down a complete IDE (in a non-compressed way) requires superpolynomial space (and, therefore, time).

An even stronger result in this direction would be to show that computing IDE is (in some sense) NP-hard. To do that, we first have to formalize a corresponding decision problem. The natural problem would be

IDE EXISTENCE

- **Input:** A feasible single-commodity network $\mathcal{N}$ with right-constant network inflow rates
- **Question:** Does there exist an IDE in $\mathcal{N}$?

However, this is actually a trivial problem: The answer is always "yes" since we have shown in Chap. 4 that IDE are always guaranteed to exist (cf. Theorem 4.36). This situation is quite reminiscent of that for (mixed) Nash equilibria in finite game: Existence of mixed Nash equilibria is guaranteed by a famous theorem by Nash ([Nas51, Theorem 1]) and proven using a fixed point theorem. However, all known algorithms for computing such equilibria have superpolynomial worst case runtime. Thus, to show that this problem is hard (and, hereby, justifying the large worst case runtimes) one has to go different ways: One way is to introduce a new complexity class and show that a) it contains other problems assumed to be hard and b) that the given problem is a hard problem within this class. For the problem of finding mixed Nash equilibria this has been done by [DGP10] with the class PPAD (cf. [DGP10, Theorem 6.1]). Another way is to create other, non-trivial decision problems by not just asking for the existence of any equilibrium, but equilibria with certain additional properties (e.g. a Nash equilibrium where a certain strategy is used or which achieves a certain social welfare—cf. [CS03]). We refer to [NRTV07, Chapter 2] for a nice introduction to the topic of the complexity of computing Nash equilibria.

For thin flows in the full information setting Cominetti, Correa and Larré argue in [CCL15, p. 27] that because their existence can be shown using Kakutani's fixed point theorem (which itself is PPAD-complete according to [Pap94, p. 526], [PVZ23, Theorem 3.17, Lemma 3.18]), the problem of finding those thin flows is also contained in PPAD. Thus, the same should be true for IDE-thin flows by the proof of Lemma 4.29. Note, however, that giving a formal proof for either ot these two claims is by no means straightforward—see [Mar20, Abschnitt 5.2.2] for a discussion of the problems arising thereby for the case of the full information setting.

We will now, instead, follow the second approach and adapt it to our setting. Namely, we will show that all of the following problems are NP-hard:

SC-IDE WITHOUT GIVEN EDGE:

- **Input:** A single-commodity network $\mathcal{N}$ with right-constant network inflow rates and an edge e in the network
- **Question:** Does there exist an IDE in $\mathcal{N}$ not using edge e?

SC-IDE WITH GIVEN EDGE:

- **Input:** A single-commodity network $\mathcal{N}$ with right-constant network inflow rates, an edge e in the network and a constant $L > 0$
- **Question:** Does there exist an IDE in $\mathcal{N}$ such that at some time θ we have $F_e^\Delta(\theta) \geq L$?

SC-IDE TERMINATING:

- **Input:** A single-commodity network $\mathcal{N}$ with right-constant network inflow rates with bounded support and some time horizon $T > 0$
- **Question:** Does there exist an IDE in $\mathcal{N}$ terminating before time T?

SC-IDE WITH FEW PHASES:

- **Input:** A single-commodity network $\mathcal{N}$ with right-constant network inflow rates and some number $L \in \mathbb{N}_0$
- **Question:** Does there exist an IDE in $\mathcal{N}$ with at most L phases?

We will show NP-hardness of these problems by reducing the NP-complete problem 3SAT to them.

Proof idea: Since the main part of the reduction will be the same for all these problems, we start by explaining this part: The goal here is to construct for any given 3SAT-formula Φ a network $\mathcal{N}_\Phi$ which has the following property: There exists some special indicator edge $\hat{e}$ in $\mathcal{N}_\Phi$ such that every satisfying interpretation of Φ can be translated into an IDE flow not using edge $\hat{e}$ and every IDE flow not using this edge can be translated into a satisfying interpretation of Φ. This construction directly proves that IDE WITHOUT GIVEN EDGE is NP-hard. For the rest we can use the indicator edge as an inflow switch for a source node $\tilde{s}$ in another network $\tilde{\mathcal{N}}$: If the indicator edge is not used by $\mathcal{N}_\Phi$ it can be used for flow normally entering $\tilde{\mathcal{N}}$ at $\tilde{s}$ to bypass the whole network. If, however, the indicator edge is used by $\mathcal{N}_\Phi$, a queue builds up on this edge forcing the flow to use the network as usual (see Fig. 5.15 for a schematic overview).

We now start the formal proof by constructing the network $\mathcal{N}_\Phi$:

Lemma 5.23. *There exists a mapping $\Phi \mapsto (\mathcal{N}_\Phi, \hat{e})$ between 3SAT-formulas and feasible networks with a special edge $\hat{e}$ satisfying the following properties:*

a) $(\mathcal{N}_\Phi, \hat{e})$ can be constructed from Φ in polynomial time (in the length of Φ).
b) For every formula Φ the network $\mathcal{N}_\Phi$ is an acyclic single-commodity network with constant, finitely lasting network inflow rates and non-zero integer free flow times and capacities.

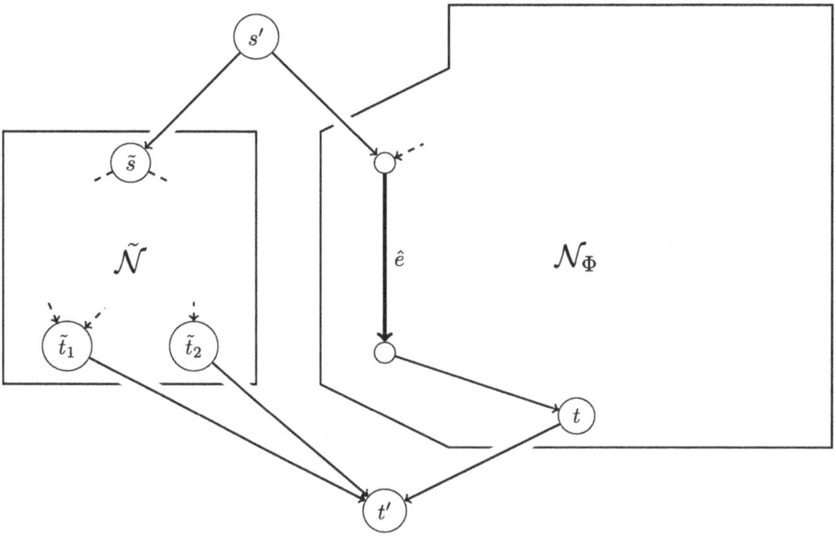

Fig. 5.15 A schematic overview of the type of network we will construct to show NP-hardness of several decision problems involving IDE. The right part ($\mathcal{N}_\Phi$) will be the same for each of those problems and we describe its construction in the proof Lemma 5.23. The left side ($\tilde{\mathcal{N}}$) will be chosen differently for each of the different decision problems

c) *If Φ is satisfiable, then there exists an IDE in $\mathcal{N}_\Phi$ not using edge $\hat{e}$. Additionally, this IDE can be chosen such that it has at most $9K + 6$ phases and a termination time of at most $9K + 6$ where K is the number of clauses in Φ.*
d) *If Φ is not satisfiable, then every IDE in $\mathcal{N}_\Phi$ sends flow of volume at least 1 into edge $\hat{e}$ between times 7 and 8.*
e) *Changing the capacity of edge $\hat{e}$ or any edge between $\hat{e}$ and the sink node t of $\mathcal{N}_\Phi$ and/or putting additional flow on any of those edges after time 6 does not change properties c) and d).*

Proof. Let $\Phi = c_1 \wedge c_2 \wedge \cdots \wedge c_K$ be a 3SAT-formula using variables $x_1, \ldots, x_N$. We then construct $\mathcal{N}_\Phi$ using two types of gadgets: K clause gadgets (one for each clause in Φ) and N variable gadgets (one for every variable in Φ). We will first describe them separately and then explain how to connect them:

A **clause gadget** C (cf. Fig. 5.16 (left)) consists of a source node s with a network inflow rate $u_s \equiv 9 \cdot \mathbb{1}_{[0,1]}$, three more nodes v_1, v_2, v_3 and three edges sv_1, sv_2, sv_3 each with a free flow travel time of 1 and a capacity of 9. We will embed this gadget

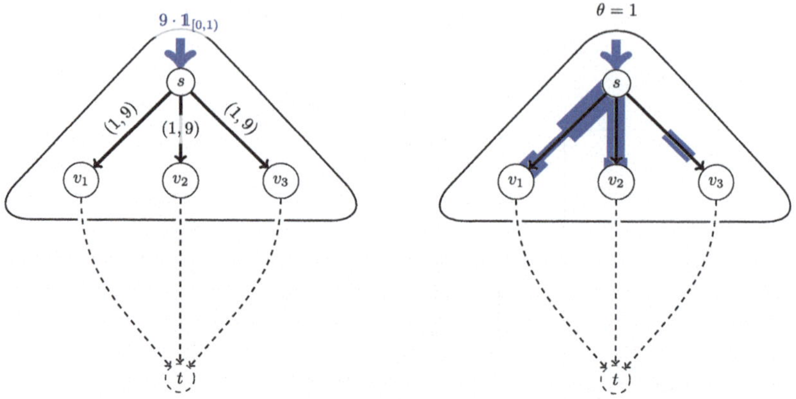

Fig. 5.16 Left side: The clause gadget C. The dashed edges as well as node t are not part of the gadget and only indicate how to embed C into a larger network. Right side: A possible flow in C. When translating this flow to an interpretation of the given 3SAT-formula we would interpret the first two literals of the corresponding clause to be true

into a larger feasible network in such a way that there are no incoming edges into C and exactly one outgoing edge from each of the three nodes v_ℓ. Furthermore, we will ensure that there is a shortest v_ℓ, T-path from each of those nodes with equal free flow travel time and which in every IDE has no flow on it until at least time $\theta = 1$.

Claim 5.5. *Let C be a clause gadget correctly embedded into a larger feasible network $\mathcal{N}$ with locally p-integrable network inflow rates for some $p > 1$. Then the following two properties hold:*

i) In every IDE there is at least one edge sv_ℓ with $F^-_{sv_\ell}(2) \geq 3$.
ii) For every set of non-negative measurable functions $f^+_{sv_1}, f^+_{sv_2}, f^+_{sv_3}$ satisfying $\sum_\ell f^+_{sv_\ell} \equiv u_s$, there exists an IDE with those functions as the edge inflow rates in C.

Proof. Property (i) follows immediately from the observation that due to strong flow conservation we must have $F^+_{sv_\ell}(1) \geq \frac{9}{3}$ for at least one of the three edges. As the capacities on those edges are large enough such that no queue can ever form there, this directly implies $F^-_{sv_\ell}(2) = F^+_{sv_\ell}(1) \geq 3$.

For property (ii) let $(f, 1)$ be some partial IDE in $\mathcal{N}$ (which exists by Theorem 4.15). Since there are no edges leading into C, the edge inflow rates on the edges sv_ℓ inside gadget C does not affect whether or not $(f, 1)$ is an IDE in the remaining network. Furthermore, our assumption on the correct embedding of C guarantees that all three edges are active during $[0, 1]$. Thus, we can change the edge inflow rates on these edges in any way we want and obtain a new partial IDE $(f', 1)$. Extending this to an IDE for all times (which is always possible according to Theorem 4.15) gives us an IDE in $\mathcal{N}$ with the desired flow rates on the edges in C. ■

A **variable gadget** X (cf. Fig. 5.17 (left)) consists of two input nodes x and $\bar{x}$, a mixing node y and two output nodes z and $\hat{z}$ which are connected as follows: there are three edges xy, $\bar{x}y$ and yz with free flow travel times 1 and capacities 1 as well as an edge $y\hat{z}$ with free flow travel time 1 and capacity 2. We will embed this gadget in a larger network in such a way that the only incoming edges connect to the input nodes x and $\bar{x}$ and the only outgoing edges start at the output nodes z and $\hat{z}$. Moreover, there is a shortest z, T-path p_z and a shortest $\hat{z}$, T-path $p_{\hat{z}}$ such that the free flow travel time along $p_{\hat{z}}$ is exactly one more than that along p_z and

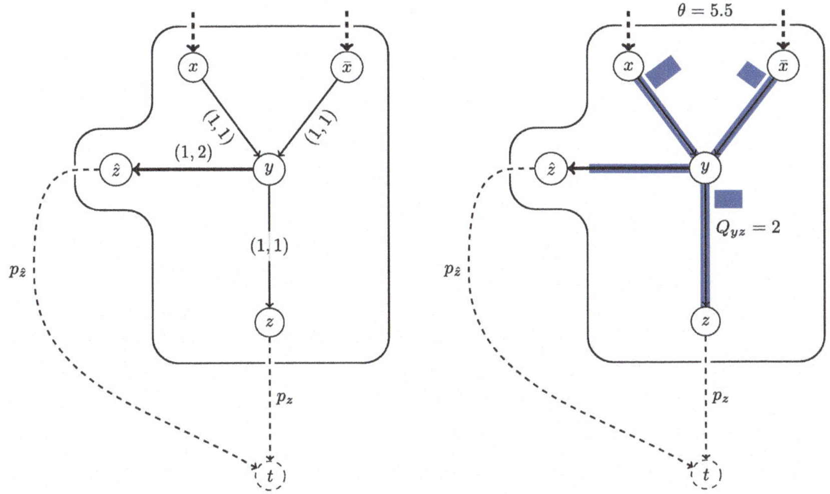

Fig. 5.17 Left side: The variable gadget X. The dashed edges as well as node t are not part of X and only indicate how to embed this gadget into a larger network. Right side: A prototypical example of an IDE where flow of volume at least 3 entered the variable gadget through each of the two input nodes during the time interval [2, 3]

in every IDE there is never any queue on path p_z and on path $p_{\hat{z}}$ there is never any queue before time 6.

Claim 5.6. *Let X be a clause gadget correctly embedded into a larger network $\mathcal{N}$ and f any IDE in this network.*

iii) If no flow ever arrives at either node x or $\bar{x}$, then no flow will ever use edge $y\hat{z}$.
iv) If a flow volume of at least 3 arrives at each of the two input nodes during $[2, 3]$, then the edge inflow rate into edge $y\hat{z}$ is at least 1 almost everywhere during $[5, 6]$.
v) If no flow enters the gadget before time 2, then no flow leaves edge $y\hat{z}$ before time 5. This is true even without the embedding-assumption that there are no queues on path $p_{\hat{z}}$ before time 6.

Proof. If flow only ever enters the gadget over node x, this flow arrives at node y at a rate of at most 1 (since $\nu_{xy} = 1$). Thus, no queue can ever form on edge yz. As there is also never a queue on path p_z while path $p_{\hat{z}}$ is strictly longer, yz will always be the only active edge starting at y. Therefore, no flow ever enters edge $y\hat{z}$. Exactly the same is true if flow only ever enters the gadget via node $\bar{x}$. Thus, property (iii) is satisfied.

If, on the other hand, at each node a flow volume of at least 3 arrives during $[2, 3]$, both edges xy and $\bar{x}y$ have a queue of length at least 2 by time $\theta = 3$. Since the queues operate at capacity, this implies that each of the edges has an outflow rate of 1 during $[4, 6]$. As long as the queue on edge yz is smaller than 1 all this flow enters this edge (as it is the only active one then). Once the queue length reaches 1 (which happens by time 5 at the latest), edge $y\hat{z}$ becomes active as well. Whenever this is the case, flow can enter edge yz at a rate of at most 1 (to keep the queue from growing any further) and the rest enters edge $y\hat{z}$. In particular, during $[5, 6]$ the queue on edge yz will always have a length of 1 and the flow arriving at a rate of 2 at node y must split equally between the two edges (note that no queue can ever forms on edge $y\hat{z}$ and no queue exists on path $p_{\hat{z}}$ until at least time $\theta = 6$ by our assumptions). Thus, property (iv) holds as well. An example for such a flow is depicted in Fig. 5.17 (right).

Finally, for property (v) we once more use the observation that flow may only enter edge $y\hat{z}$ if there is a queue of length at least 1 on edge yz. Under the assumption that no flow enters the network before time 2 this cannot be the case before time 4 as the inflow into node y can then only start at time 3 and be at a rate of at most 2. Thus, flow cannot enter edge $y\hat{z}$ before time 4 and, hence, not leave it before time 5.

We now construct the whole network Φ as follows (cf. Fig. 5.18): We take one copy of the clause gadget C for every clause of Φ and one copy of the variable gadget X for every variable in Φ and connect them as follows: If the ℓ-th literal in some clause c of Φ is a positive variable x we connect node v_ℓ in the clause gadget corresponding to c to node x in the variable gadget corresponding to the variable x. If the ℓ-th literal is a negated variable $\neg x$ we connect to the node $\bar{x}$ in the corresponding variable gadget instead. In both cases the connecting edge has a free flow travel time of 1 and a capacity of 9. Next, we add three more nodes a, b and t and declare t the only sink node in this network. Furthermore, we add an edge zt with free flow time 1 and capacity 1 as well as an edge $\hat{z}a$ of free flow time 1 and capacity 2 from every variable gadget. Finally we add an edge $\hat{e} := ab$ and an edge bt, both with free flow time 1 and capacity 1.

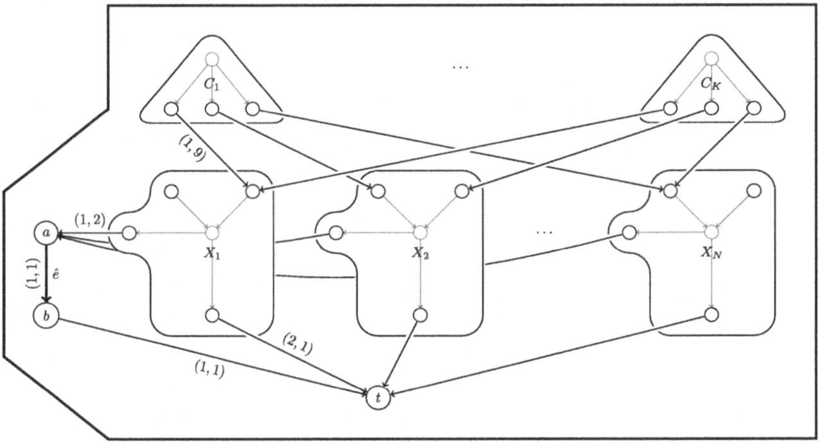

Fig. 5.18 The complete network $\mathcal{N}_\Phi$ for a 3SAT-formula consisting of K clauses in N variables

This construction is clearly possible in polynomial time, i.e. the mapping $\Phi \mapsto (\mathcal{N}_\Phi, \hat{e})$ satisfies a). It is also obvious that the network $\mathcal{N}_\Phi$ satisfies all the properties in b). So, it remains for us to show that properties c) and d) are satisfied as well. For this, we first observe that both types of gadgets are embedded into $\mathcal{N}_\Phi$ in such a way as to satisfy the assumptions of Claims 5.5 and 5.6: For every node v_ℓ in any of the clause gadgets the unique physically shortest v_ℓ, t-path is either a path of the form $v_\ell x, xy, yz, zt$ or $v_\ell \bar{x}, \bar{x}y, yz, zt$ (depending on whether the ℓ-th literal in the corresponding clause is positive or negative). All these paths have the same free flow travel time (namely, 5). Since the only nodes with positive network inflow are the nodes s in the clause gadgets, there can be no flow outside these gadgets before time $\theta = 1$. So, the clause gadgets are correctly embedded in the network. The shortest z, t-path for any variable gadget is just the edge zt (with free flow time 2) while the shortest $\hat{z}, t$-path is $\hat{z}a$, ab, bt (with free flow time 3). Furthermore, the capacities of edges zt, $y\hat{z}$ and $\hat{z}a$ guarantee that no queue ever forms on them. Finally, no flow can reach edge ab or bt before time $\theta = 6$ by property (v) and, thus, no queue can form on those edges before time 6. Thus, the gadgets X are correctly embedded as well. Note, that changing the capacity of edge $\hat{e} = ab$ or edge bt or putting additional flow on them after time 6 does not change any of this. Hence, property e) holds as well.

Using Claims 5.5 and 5.6 we can now show that $\mathcal{N}_\Phi$ satisfies properties c) and d):

1. **Case: Φ satisfiable:** Let $\beta : \{x_1, \ldots, x_N\} \to \{\text{TRUE}, \text{FALSE}\}$ be a satisfying interpretation of Φ. Then, we can choose for every clause of Φ one literal which is true under β. According to Claim 5.5(ii) there exists then an IDE f in $\mathcal{N}_\Phi$ wherein in every clause gadget we only send flow towards the node v_ℓ corresponding to the chosen literal in this clause. This then implies that in every variable gadget flow only arrives either at node x or at node $\bar{x}$. Thus, by Claim 5.6(iii) no flow will ever use any of the edges $y\hat{z}$ and, therefore, no flow will ever arrive at node a. Hence, edge $\hat{e}$ will always be empty in f.
   For the bound on the termination time we note that the total flow volume ever in the network is $9K$. If, in the worst case, all this flow goes through a single input node x or $\bar{x}$, then the last particle arriving there at time $\theta = 3$ would have a waiting time of $9K - 1$. Together with the free flow travel time of 4 for the remaining path xy, yz, zt, this means the last particle will arrive at t at time $3 + 9K - 1 + 4 = 9K + 6$. The bound on the number of phases now follows by observing that in the flow constructed above events only happen at integer times.
2. **Case: Φ unsatisfiable:** Let f be any IDE in $\mathcal{N}$. If there is a variable gadget with flow of volume at least 3 arriving at each of its two input nodes during $[2, 3]$, then a flow of volume at least 1 will leave this gadget at node $\hat{z}$ during $[6, 7]$ by Claim 5.6(iv). This flow will then enter edge $\hat{e}$ during $[7, 8]$. If, on the other hand, there were no such variable gadget, then we could construct a satisfying interpretation of Φ as follows (which is then a contradiction to Φ being unsatisfiable): For any variable x set $\beta(x) := \text{TRUE}$ if if the total inflow into node x is at least 3 and $\beta(x) := \text{FALSE}$ if the total inflow into $\bar{x}$ is at least 3. If neither of those is the case, we may choose any interpretation for the variable x. This interpretation now satisfies Φ since, according to Claim 5.5(i),

for every clause the corresponding gadget has at least one node v_ℓ which receives a flow of at least 3 which then enters the corresponding variable gadget. Thus, the literal corresponding to this node v_ℓ is true under β. □

With this lemma we now immediately get NP-hardness of SC- IDE WITHOUT GIVEN EDGE. For the reductions to the other problems we need one additional construction: Namely, we want to use $\mathcal{N}_\Phi$ as a switch for the network inflow in another network $\tilde{\mathcal{N}}$, i.e. construct a new network $\tilde{\mathcal{N}}_\Phi$ in which there exists an IDE where a certain part of the flow completely bypasses $\tilde{\mathcal{N}}$ if and only if Φ is satisfiable.

Lemma 5.24. *Let $\tilde{\mathcal{N}}$ be a single-commodity network where all network inflow rates are zero before time $\theta = 8 + M$ for some $M \in \mathbb{N}_0$, there is no network inflow at any sink node and $\tilde{s}$ is a source node in $\tilde{\mathcal{N}}$ such that the network inflow rate $\tilde{u}_{\tilde{s}}$ at $\tilde{s}$ is bounded and has its essential support in $[8 + M, 8 + 2M]$. Then there is a mapping $\Phi \mapsto \tilde{\mathcal{N}}_\Phi$ between 3SAT-formulas and feasible networks containing $\tilde{\mathcal{N}}$ as subnetwork satisfying the following properties:*

a) $\tilde{\mathcal{N}}_\Phi$ can be constructed from Φ in polynomial time (in the length of Φ).
b) If $\tilde{\mathcal{N}}$ is acyclic, has non-zero integer free flow travel times and/or capacities, then the same holds for $\tilde{\mathcal{N}}_\Phi$.
c) If Φ is satisfiable, then there exists an IDE in $\tilde{\mathcal{N}}_\Phi$ such that its restriction to $\tilde{\mathcal{N}}$ is an IDE where $\tilde{u}_{\tilde{s}}$ has been replaced by the zero function. If $\tilde{s}$ was the only node with positive network inflow in $\tilde{\mathcal{N}}$ then the IDE in $\tilde{\mathcal{N}}_\Phi$ can be chosen such that

- *it has a termination time of at most* $\max\{9K + 6, 9 + 2M + \tilde{\tau}_{p_{\min}}\}$, *where $\tilde{\tau}_{p_{\min}}$ is the free flow travel time along a physically shortest $\tilde{s}, T$-path in $\tilde{\mathcal{N}}$ and K the number of clauses in Φ.*
- *it has at most $9K + 16$ phases if, additionally, $\tilde{u}_{\tilde{s}}$ is constant during $[8 + M, 8 + 2M]$.*

d) If Φ is not satisfiable, then every IDE in $\tilde{\mathcal{N}}_\Phi$ restricted to $\tilde{\mathcal{N}}$ is an IDE in this subnetwork.

Proof. Given a 3SAT-formula Φ we construct $\tilde{\mathcal{N}}_\Phi$ as follows (cf. Fig. 5.19): Let $B \in \mathbb{N}^*$ be some bound on the network inflow rate $\tilde{u}_{\tilde{s}}$ at $\tilde{s}$ in $\tilde{\mathcal{N}}$, $\tilde{\tau}_{p_{\min}} := \min\{\sum_{e\in p} \mid p \text{ a } \tilde{s}, T\text{-path in } \tilde{\mathcal{N}}\}$ the free flow travel time along a physically shortest $\tilde{s}, T$-path in $\tilde{\mathcal{N}}$ and assume without loss of generality that $\tilde{\tau}_{p_{\min}} \geq 2$, which is also the free flow travel time along a physically shortest a, t-path in $\mathcal{N}_\Phi$.

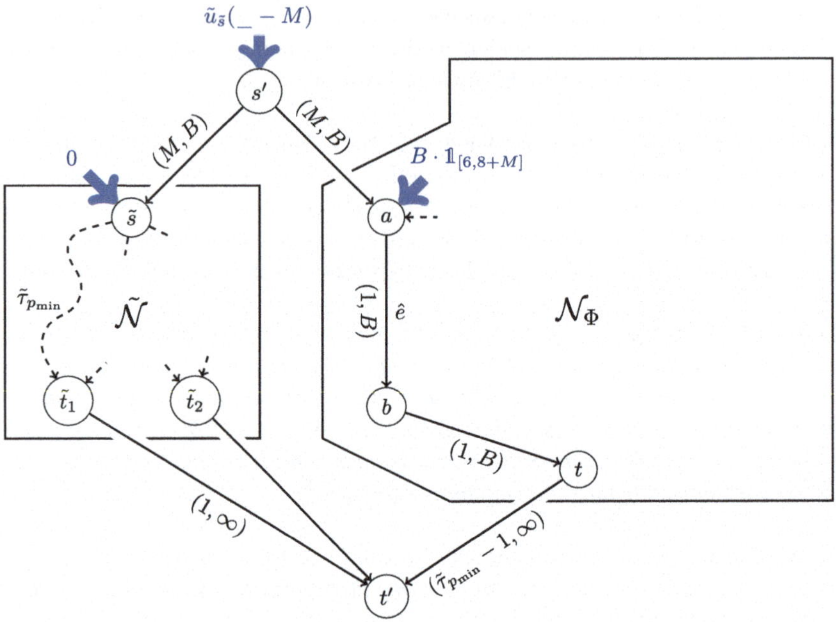

Fig. 5.19 The network $\tilde{\mathcal{N}}_\Phi$ constructed in Lemma 5.24 for a given network $\tilde{\mathcal{N}}$ and a 3SAT-formula Φ. Capacities of ∞ denote capacities chosen in such a way as to guarantee no queue ever forms on these edges, i.e. at least the sum of the capacities of all edges leading into the respective edge

We take the (disjoint) union of the networks $\tilde{\mathcal{N}}$ and $\mathcal{N}_\Phi$ and add two extra nodes s' and t'. We connect s' to $\tilde{s}$ in $\tilde{\mathcal{N}}$ and a in $\mathcal{N}_\Phi$ with edges of travel time M and capacity B. Furthermore we add an edge $\tilde{t}t'$ with free flow travel time 1 and capacity $\sum_{e \in \delta^-(\tilde{t})} \nu_e$ for every sink node $\tilde{t}$ in $\tilde{\mathcal{N}}$ and an edge tt' with free flow travel time $\tilde{\tau}_{p_{\min}} - 1$ and capacity $\sum_{e \in \delta^-(t)} \nu_e$ for the sink node t in $\mathcal{N}_\Phi$. Finally, we change the capacities of $\hat{e} = ab$ and bt in $\mathcal{N}_\Phi$ to B, the network inflow rates at $\tilde{s}$ in $\tilde{\mathcal{N}}$ to zero, at a in $\mathcal{N}_\Phi$ to $B \cdot \mathbb{1}_{[6, M+8]}$ and at s' to $\tilde{u}_{\tilde{s}}(\_ - M)$.

Now, we observe that since the only changes made to $\mathcal{N}_\Phi$ are moving the sink further away, changing the capacity of edges $\hat{e}$ and bt and adding additional flow to these edges after time 6, the IDE flows in $\mathcal{N}_\Phi$ still exhibit the same properties as in Lemma 5.23. Furthermore, the capacities of the new edges are chosen such that no queue can ever form on them while the free flow travel times are chosen such that without any queues the edges $s'\tilde{s}$ and $s'a$ are both active.

Therefore, if Φ is satisfiable, then according to Lemma 5.23c) there exists an IDE such that the only flow from $\mathcal{N}_\Phi$ entering $\hat{e}$ is the one entering the network at node a. Thus, the queue on $\hat{e}$ is empty until at least time M and edge $s'a$ is active throughout this period. Hence we can send all network inflow at node s' along this edge, hereby bypassing network $\tilde{\mathcal{N}}$. As this flow will not encounter any queues on its way, its last particle has reached the sink node t' by time $\theta = 8 + M + M + 1 + 1 + \tilde{\tau}_{p_{\min}} - 1 = 9 + 2M + \tilde{\tau}_{p_{\min}}$, which, together with the bound in Lemma 5.23c), gives us the desired bound on the termination time for the case where there are no other nodes with positive network inflow in $\tilde{\mathcal{N}}$. If, additionally, $\tilde{u}_{\tilde{s}}$ is constant, then the bypassing flow is responsible for at most 2 events on each of the 5 nodes on its path. Together with the bound in Lemma 5.23c) this again gives us the desired bound.

If, on the other hand, Φ is unsatisfiable, we know from Lemma 5.23d) that a queue of length at least 1 will have formed on edge $\hat{e}$ by time 8 and remain there until at least time $8 + M$. Since no flow enters network $\tilde{\mathcal{N}}$ before time $8 + M$, this means edge $s'a$ will be inactive throughout $[8, 8 + M]$ and all network inflow at node s' has to travel along edge $s'\tilde{s}$. Thus, this flow enters $\tilde{\mathcal{N}}$ in exactly the same way as described by $\tilde{u}_{\tilde{s}}$.

Hence, $\tilde{\mathcal{N}}_\Phi$ satisfies both properties c) and d). Properties a) and b) are directly clear from the construction. □

Theorem 5.25. *The problems* SC-IDE WITHOUT GIVEN EDGE, SC-IDE WITH GIVEN EDGE, SC-IDE TERMINATING *and* SC-IDE WITH FEW PHASES *are all NP-hard.*

SC-IDE WITH FEW PHASES *is NP-complete when restricted to instances where K is polynomially bounded in the size of $\mathcal{N}$.*

All this remains true even when restricted to only acyclic networks with strictly positive, integer free flow travel times and capacities.

Proof. NP-hardness of IDE WITHOUT GIVEN EDGE follows directly from Lemma 5.23 and the NP-hardness of 3SAT (Proposition 2.73) as $\mathcal{N}_\Phi$ has an IDE not using edge $\hat{e}$ if and only if Φ is satisfiable.

For IDE WITH GIVEN EDGE we use Lemma 5.24 with a network $\tilde{\mathcal{N}}$ consisting of a single edge $\tilde{s}\tilde{t}$ and a network inflow rate of $\tilde{u}_{\tilde{s}} := \mathbb{1}_{[8+M,8+2M]}$. Then, we can send all network inflow at node s' in $\tilde{\mathcal{N}}_\Phi$ into edge $s'a$ (hereby having $F^{\Delta}_{s'a}(8+M) = M$) if and only if Φ is satisfiable.

For IDE TERMINATING we use Lemma 5.24 with a network $\tilde{\mathcal{N}}_K$ consisting of a single edge $\tilde{s}\tilde{t}$ with free flow travel time 1 and capacity 1 and a network inflow rate of $\tilde{u}_{\tilde{s}} := (9K + 6) \cdot \mathbb{1}_{[8+1,8+2]}$ and choose $T := 9K + 7$. Then, we can bypass $\tilde{\mathcal{N}}$ and achieve a termination time of $\max\{9K + 6, 12\} \leq T$ if Φ is satisfiable, while having to send all network inflow at node s' through $\tilde{\mathcal{N}}$ and get a termination time of $9 + 9K + 5 + 2 > T$, otherwise.

Finally, for IDE WITH FEW PHASES we use Lemma 5.24 and the network from Example 5.19 with inflow rate $2 \cdot \mathbb{1}_{[3K+11,6K+15)}$ as $\tilde{\mathcal{N}}$ and choose $L = 9K + 16$. Then, we get an IDE flow with at most $9K + 16 = L$ phases by bypassing $\tilde{\mathcal{N}}$ if Φ is satisfiable. If, on the other hand, Φ is unsatisfiable, then any IDE in $\tilde{\mathcal{N}}$ has at least $4 \cdot (3K + 4) = 12K + 16 > L$ phases.

If K is polynomially bounded in the length of Φ, then IDE WITH FEW PHASES becomes NP-complete as the IDE flow corresponding to a satisfying interpretation can be described (and checked) in polynomial size (and time) in the number of its phases. □

Clearly, many more decision problems involving IDE can be shown to be NP-hard using the same approach as in the above proof. Essentially, every property for which the following two assumption hold, can be used to create such an NP-hard decision problem:

- The IDE in $\mathcal{N}_\Phi$ corresponding to a satisfying interpretation of Φ has this property.
- There exists a feasible network wherein no IDE has this property.

As a final remark we want to mention that the switch-construction in the proof of Lemma 5.24 clearly also works for multi-commodity networks. Thus, the following decision problems can easily be shown to be NP-hard as well:

MC-IDE WITH FINITELY MANY PHASES
- **Input:** A multi-commodity network $\mathcal{N}$ with right-constant network inflow rates with bounded support
- **Qestion:** Does there exist an IDE in $\mathcal{N}$ with finitely many phases?

MC-IDE EVENTUALLY TERMINATING
- **Input:** A multi-commodity network $\mathcal{N}$ with right-constant network inflow rates with bounded support
- **Question:** Does there exist an IDE in $\mathcal{N}$ which eventually terminates?

For MC- IDE WITH FINITELY MANY PHASES we can use the network from Example 5.17 as $\tilde{\mathcal{N}}$ while for MC- IDE EVENTUALLY TERMINATING we can use the network we will use in the following chapter to prove that multi-commodity networks are not guaranteed to terminate (Theorem 6.18).

5.4 Bibliographic Notes and Open Questions

The results of this chapter are mostly based on joint work with Tobias Harks published in [GH23]. The possibility of computing IDE-thin flows for multi-commodity networks without edges of free flow travel time zero using a MIP was already observed by our coauthor Leon Sering in [GHS20, Section 5.1]. A first result for bounding the number of phases in an IDE was obtained by Kraus in his master's thesis for a certain subclass of series parallel graphs ([Kra20, Corollary 2.12]). His proof already highlighted how one can show properties of IDE by starting with a local analysis of these properties (e.g. at a single node) and then lifting them by some inductive argument to the whole network.

This "localness" of IDE is also a crucial ingredient to most of the proofs in this chapter: E.g. we compute IDE-thin flows by computing them on a node-by-node bases, we bound the number of events in an IDE by bounding it separately at every node and we construct (and show correctness of) networks for our NP-hardness results out of small gadgets. It is also because of this reliance on local analysis that it seems unlikely that one can transfer these results directly to dynamic equilibria in the full information setting where most of the questions answered in this chapter are still open. In particular, a polynomial algorithm for computing thin flows in the full information setting is only known for the case of series-parallel networks ([Kai22a]) while the question of whether a finite number of extensions suffices to construct a full equilibrium flow is still completely open (see e.g. [Kai22b, Section 4.6] or [OSK22, Section 7]).

There are, of course, also still open question when it comes to the computational complexity of IDE: Our bounds for the worst case runtime of an algorithm constructing IDE in a single-commodity network given in Corollary 5.16 is extremely large and seems unlikely to be tight. Improvements on this bound, therefore, would certainly seem plausible. Note that our lower bound from Example 5.19 would even allow for a weakly polynomial algorithm. Another potential direction for future research could be studying whether IDE in networks with zero free flow travel time edges or multiple commodities can still be computed in finitely many extensions (for networks with both of these we know this not to be the case by Example 5.17). In particular, for acyclic multi-commodity networks it seems quite possible that one could adapt our proof for the single-commodity case as the only place we use a specific property of single-commodity IDE is for Claim 5.3 in the proof of Lemma 5.11 where we used the uniqueness of the node-labels in single-commodity IDE-thin flows. Finally, it would also be interesting to obtain stronger hardness results for IDE like, for example, PPAD-hardness.

Quality of IDE

6

In this chapter we study the quality of IDE in terms of the two quality measures defined in Subsection 3.2.2: Makespan and total travel time. More specifically, we want to answer the following questions:

- Are IDE guaranteed to terminate (and, therefore, have finite makespan and total travel time)?
- If so, what lower and upper bounds can we find for makespan/total travel time of IDE?
- What is the worst ratio of makespan/total travel time of an IDE compared to an optimal flow (i.e. the Price of Anarchy)?

Of course, all these question are only relevant for networks with finitely lasting network inflow rates. Hence, with the exception of Proposition 6.3 and Theorem 6.5 we will only consider those throughout this chapter. We will start by considering the first two questions first for acyclic networks, then for single-commodity networks and, finally, for multi-commodity networks. We will then summarize all our results in terms of the Price of Anarchy for these different classes of instances.

6.1 Upper Bounds

As our first results on the quality of IDE we will show upper bounds for both makespan and total travel time in acyclic networks and single-commodity networks.

L. Graf, *Dynamic Network Flows with Adaptive Route Choice based on Current Information*, Mathematische Optimierung und Wirtschaftsmathematik | Mathematical Optimization and Economathematics,
https://doi.org/10.1007/978-3-658-44948-3_6

6.1.1 Acyclic Networks

For the case of acyclic networks we will show our upper bounds not just for IDE but actually for all Vickrey flows. Given an acyclic network we denote by

$$\tau_{p_{\max}} := \max \{ \sum_{e \in p} \tau_e \mid p \text{ a } v, T_i\text{-path for any } v \in V, i \in I \}$$

the free flow travel time along a physically longest path and by

$$\nu_{\min} := \min \{ \nu_e \mid e \in E \}$$

the smallest capacity in the network. Then, a natural guess of an upper bound for the makespan is $\hat{\theta} + \tau_{p_{\max}} + U$ since this is the arrival time of the last particle that enters the network (at time $\hat{\theta}$) if it chooses the longest possible path (with free flow travel time $\tau_{p_{\max}}$) and, at some point, has to wait behind every other particle of the network (leading to a total waiting time of $\frac{U}{\nu_{\min}}$). Intuitively speaking, this should be the worst possible flow (with respect to the makespan). However, the following example shows that at least proving this upper bound is not as straightforward:

Example 6.1. Consider the network in Fig. 6.1 consisting of a single path consisting of three source nodes s_1, s_2 and s_3 followed by a single sink node t at the end. In the unique Vickrey flow (also depicted in Fig. 6.1) the blue particles entering the network at the leftmost source s_1 first have to wait on edge s_2s_3 behind the red particles entering the network at s_2 and then again on edge s_3t behind the *same* particles as those in turn had to wait behind the green particles which entered the network at the rightmost source s_3.

Nevertheless, the conjectured bound itself still holds in this network and one could even argue that it is still true that every particle is "responsible" for making the blue particles wait at most once as the second time those particles get blocked by red particles, this is only because they, in turn, got blocked, by green particles. For the discrete version of this flow model (with unit sized packets instead of infinitesimal particles) this idea can in fact be formalized by introducing an additional virtual packet (token) which enters and and leaves the network at the same time as the last particle while also travelling through the network in such a way as to only be delayed at most once by any other packet. This idea was used by Cao, Chen, Chen and Wang in [CCCW22] to show an upper bound on the makespan for the discrete flow model. As Sering, Vargas Koch and Ziemke have shown in [SVZ22]

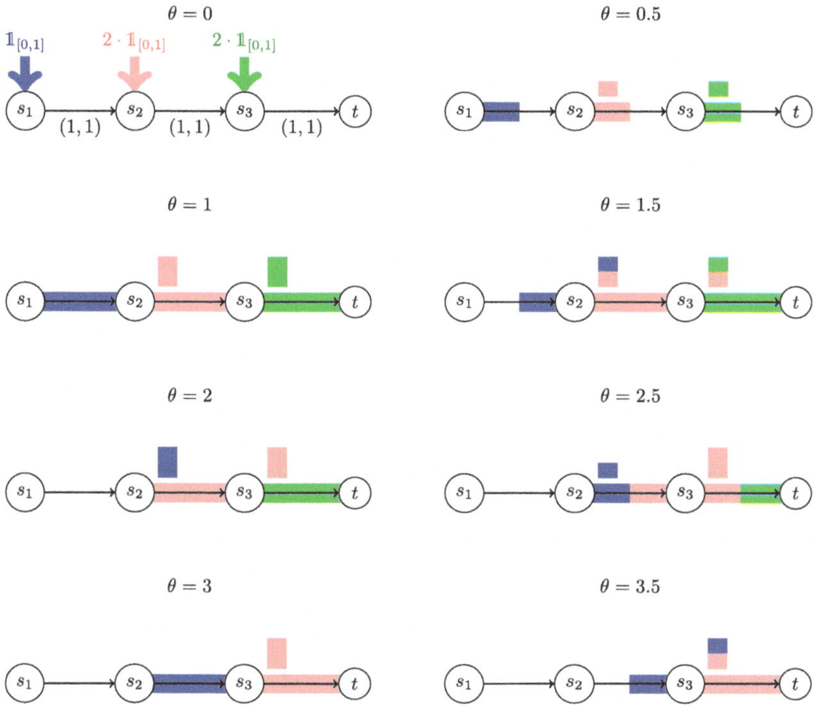

Fig. 6.1 An instance wherein particles of the blue flow are blocked twice by the same particles (of red flow)

that (continuous) Vickrey flows can be approximated by discrete packet routings, this bound can be transferred to our setting as well:

Proposition 6.2. *Let $\mathcal{N}$ be a feasible acyclic network with strictly positive, integer free flow travel times and capacities, a single source,sink-pair for each commodity and finitely lasting network inflow. Then we have*

$$\Psi(f) \leq \hat{\theta} + U + \tau_{p_{\max}}$$

for any Vickrey flow f in $\mathcal{N}$ which has a path decomposition into path flows along simple paths.

Since we have not formally defined the discrete flow model here, we will not provide a full proof for this proposition using the approach explained above. Instead, we will only provide a rough proof sketch and refer to Theorem 6.5 (which generalizes Proposition 6.2) for a formal proof. We refer to [Var20, Section 4] for a good introduction to the discrete packet routing model and to [SVZ22] for a formal description of the discretization of (continuous) Vickrey flows.

Proof sketch. We start with the observation that we can assume without loss of generality that all free flow times and all capacities are exactly 1. If this is not the case, this can be accomplished by subdividing each edge e into first ν_e parallel edges and then each of those into a sequence of τ_e edges. We can also transform f into a Vickrey flow in the new network in the natural way. We observe that this transformation changes neither the free flow travel time of the physically longest path nor the makespan of f.

Now, according to [SVZ22, Theorem 2] there exists a sequence $(r^n)_{n\in\mathbb{N}^*}$ of discrete packet routings such that in r^n we have $\lfloor n^2U \rfloor$ packets of size $\frac{1}{n^2}$ and a time step size of $\frac{1}{n}$. Furthermore, the arrival times $\Psi(r^n)$ of the last packet under r^n converge to the arrival time of the last particle in f, i.e. $\Psi(r^n) \xrightarrow{n} \Psi(f)$. Hence, it suffices to bound the the arrival times $\Psi(r^n)$. To do that for any fixed $n \in \mathbb{N}^*$, we once more transform our network by scaling the time step size and free flow travel times by n (which, in particular, scales both $\Psi(r^n)$ and $\tau_{p_{\max}}$ by n), edge capacities by n (i.e. setting $\nu_e = n$ for all edges) and every packet by n^2 (making them unit sized). Then we subdivide every edge in n consecutive edges (with free flow travel time 1 each) and add, for every packet, a new source node connected to the original one with a path consisting of na edges (with free flow travel time 1 and capacity n) where $a \in \mathbb{N}_0$ is the starting time of the respective packet in the original network. Hence, we can now let all particles start at time 0 and achieve the same routing r^n as before (using the appropriate transformations). Furthermore, the free flow travel time of the longest source-sink path in the new network is clearly bounded by $\lceil n\hat{\theta} \rceil + n\tau_{p_{\max}}$.

For this new routing we can now apply [CCCW22, Theorem 2] to show that the last packet arrives at its sink before time $\lceil n\hat{\theta} \rceil + n\tau_{p_{\max}} + \frac{\lfloor n^2 U \rfloor}{n}$ which immediately implies $\Psi(r^n) \leq \hat{\theta} + \frac{1}{n} + \tau_{p_{\max}} + U$ in the untransformed network. As this holds for any $n \in \mathbb{N}_0$, this shows $\Psi(f) \leq \hat{\theta} + \tau_{p_{\max}} + U$. ○

Using a cut-based approach set completely in the world of continuous flows we can find a more general lower bound that not only bounds the arrival time of the *last* particle but also the amount of flow having reached the sink at *any* given time. This then not only implies the above bound on the makespan but also allows us to derive a bound on the total travel time. We first show this bound only for single-commodity networks and afterwards deduce the bounds for multi-commodity networks from it.

Proposition 6.3. *Let $\mathcal{N}$ be an acyclic single-commodity network with a single sink reachable from every node in $\mathcal{N}$ and $\nu_{\min} \geq 1$. Then we have*

$$Z(\theta) \geq Z(\vartheta) + \min\{\, F^{\Delta}(\zeta), \theta - \vartheta - \tau_{p_{\max}} \,\}$$

for any Vickrey flow f in $\mathcal{N}$ and all times $\theta \geq \vartheta \geq 0$.

Proof idea: Since we want to upper bound the worst case flows, we will consider pessimistic distances, i.e. distance measured in terms of the longest possible path towards the sink. Now, consider the node which is the furthest away from the sink (with respect to this pessimistic distance). Then particles starting at this node have no other choice but to travel towards a node which is strictly closer to the sink. Since all edge capacities are lower bounded by $\nu_{\min}$, this happens at least at a rate of $\nu_{\min}$. Together this gives us a lower bound for how much flow must be closer to the sink than $\tau_{p_{\max}}$. We can now repeat this argument with the node which is second furthest away from the sink and so on until we get our desired bound for flow already at the sink.

In other words, we will actually prove the following stronger claim: For any time and any suitable cut of the network splitting it in nodes closer to the sink and nodes further away, there is some lower bound (given explicitly in Claim 6.2) on how much flow must have crossed this cut by then.

Proof. We start by defining for every node $v \in V$ the *pessimistic remaining distance*

$$\tilde{d}_v := \max\{\, \sum_{e \in p} \tau_e \mid p \text{ a } v,t\text{-path} \,\},$$

where $t \in V$ is the unique sink node in $\mathcal{N}$. This is well-defined as $\mathcal{N}$ is acyclic and t is reachable from every node. Next, we claim that there exists some topological order which is compatible with the pessimistic remaining distance:

Claim 6.1. *There exists a topological order* $v_n \prec \cdots \prec v_1$ *such that we have* $v_1 = t$ *and* $v \prec w \implies \tilde{d}_w \leq \tilde{d}_v$ *for all nodes* v, w *as well as* $\tilde{d}_v \geq \tau_{vw} + \tilde{d}_w$ *for all edges* $e = vw$.

Proof. We can construct an order $\prec$ with the desired properties as follows: Start with any topological order $\prec'$ (which exists by Proposition 2.65 because $\mathcal{N}$ is acyclic) and then sort the nodes first with respect to their (non-increasing) pessimistic remaining distance and second with respect to the initial topological order (i.e. use $\prec'$ to sort nodes with the same pessimistic remaining distance).

Since the network is acyclic, we have $\tilde{d}_t = 0$. Furthermore, as we can reach t from every other node, t is the only node without any outgoing edges. Thus, t is the largest node with respect to the initial topological order. Together these two properties ensure that t is also the largest node with respect to our new ordering $\prec$.

To show that $\prec$ is a topological order, take any edge vw. Then we have $\tilde{d}_v \geq \tau_{vw} + \tilde{d}_w \geq \tilde{d}_w$ and, therefore, $v \prec w$. Finally, $\prec$ is compatible with the pessimistic remaining distance by construction. ■

So, from now on we fix one such topological order and enumerate the nodes such that we have $v_n \prec \cdots \prec v_1 = t$. Let $V_k := \{ v_1, \ldots, v_k \}$ be the set of the k largest nodes with respect to this order. We then clearly have $\tilde{d}_v \leq \tilde{d}_{v_k}$ for all $v \in V_k$ and $\tilde{d}_v \geq \tilde{d}_{v_k}$ for all $v \in V \setminus V_k$. We denote by

$$F_{\leq k}(\theta) := \sum_{v \in V_k} U_v(\theta) + \sum_{e=vw \in \delta^-(V_k)} F_e^-(\theta + \min\{ \tilde{d}_{v_k} - \tilde{d}_w, \tau_e \})$$

the flow volume closer to the sink than v_k (see Fig. 6.2). Note, that we count flow which has already reached the sink as still being in the network here. We now observe the following natural properties of $F_{\leq k}$:

(i) $F_{\leq k} : \mathbb{R}_{\geq 0} \to \mathbb{R}_{\geq 0}$ is non-decreasing for any $k \in [n]$,
(ii) $F_{\leq k}(\theta) \geq F_{\leq k-1}(\theta)$ for all $k \in \{ 2, \ldots, n \}$ and $\theta \in \mathbb{R}_{\geq 0}$,
(iii) $F_{\leq n}(\theta) = U(\theta)$ and
(iv) $F_{\leq 1}(\theta) = Z(\theta)$ for all $\theta \in \mathbb{R}_{\geq 0}$.

Monotonicity of $F_{\leq k}$ follows immediately from the monotonicity of U_v and F_e^-. Furthermore, we have $F_{\leq n}(\theta) = \sum_{v \in V} U_v(\theta) = U(\theta)$ since $V_n = V$ and $\delta^+(V) = \emptyset$ while $F_{\leq 1}(\theta) = Z(\theta)$ holds by strong flow conservation at $v_1 = t$ and the fact that there are no outgoing edges from t. Finally, we have

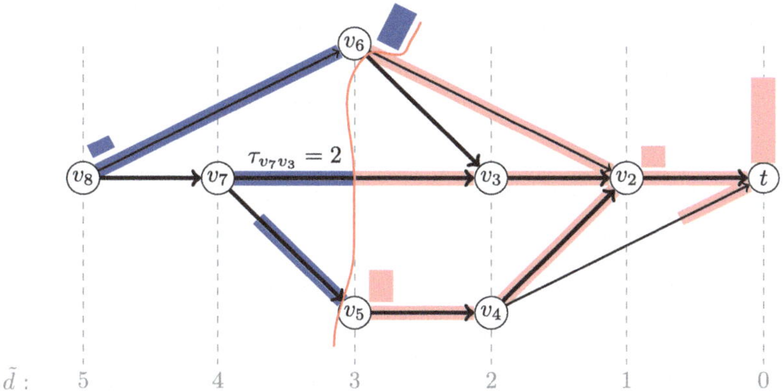

Fig. 6.2 An acyclic single-sink network with nodes placed according to their pessimistic distance $\tilde{d}_v$. All edges except for edge v_7v_3 have free flow travel time 1. Thick edges are tight with respect to pessimistic distance. The red line indicates the cut corresponding to $k = 5$. The colour of the drawn flow shows whether this flow is counted in $F_{\leq 5}(\theta)$ (red) or not (blue)

$$\begin{aligned}
F_{\leq k-1}(\theta) \overset{(\bigcirc)}{=} & \sum_{v \in V_{k-1}} U_v(\theta) + \sum_{e=vw \in \delta^-(V_{k-1}) \cap \delta^-(V_k)} F_e^-(\theta + \min\{\tilde{d}_{v_{k-1}} - \tilde{d}_w, \tau_e\}) \\
& + \sum_{e=v_k w \in \delta^+(v_k)} F_e^-(\theta + \min\{\tilde{d}_{v_{k-1}} - \tilde{d}_w, \tau_e\}) \\
\overset{(\#)}{\leq} & \sum_{v \in V_{k-1}} U_v(\theta) + \sum_{e=vw \in \delta^-(V_{k-1}) \cap \delta^-(V_k)} F_e^-(\theta + \min\{\tilde{d}_{v_k} - \tilde{d}_w, \tau_e\}) \\
& + \sum_{e=v_k w \in \delta^+(v_k)} F_e^-(\theta + \tau_e) \\
\overset{(\triangle)}{\leq} & \sum_{v \in V_{k-1}} U_v(\theta) + \sum_{e=vw \in \delta^-(V_{k-1}) \cap \delta^-(V_k)} F_e^-(\theta + \min\{\tilde{d}_{v_k} - \tilde{d}_w, \tau_e\}) \\
& + \sum_{e=v_k w \in \delta^+(v_k)} F_e^+(\theta) \\
\overset{(*)}{=} & \sum_{v \in V_{k-1}} U_v(\theta) + \sum_{e=vw \in \delta^-(V_{k-1}) \cap \delta^-(V_k)} F_e^-(\theta + \min\{\tilde{d}_{v_k} - \tilde{d}_w, \tau_e\}) \\
& + U_{v_k}(\theta) + \sum_{e \in \delta^-(v_k)} F_e^-(\theta)
\end{aligned}$$

$$= \sum_{v \in V_{k-1}} U_v(\theta) + \sum_{e=vw \in \delta^-(V_{k-1}) \cap \delta^-(V_k)} F_e^-(\theta + \min\{\tilde{d}_{v_k} - \tilde{d}_w, \tau_e\})$$

$$+ U_{v_k}(\theta) + \sum_{e=vv_k \in \delta^-(v_k)} F_e^-(\theta + \min\{\tilde{d}_{v_k} - \tilde{d}_{v_k}, \tau_e\})$$

$$\overset{(\bigcirc)}{=} \sum_{v \in V_k} U_v(\theta) + \sum_{e=vw \in \delta^-(V_k) \cap \delta^-(V_k)} F_e^-(\theta + \min\{\tilde{d}_{v_k} - \tilde{d}_w, \tau_e\})$$

$$= F_{\leq k}(\theta)$$

which shows that (ii) holds as well. Here we used the fact that all edges $v_k w \in \delta^+(v_k)$ satisfy $v_k \prec w$ and, therefore, $w \in V_{k-1}$ at ($\bigcirc$), strong flow conservation at the non-sink node v_k for ($*$), the fact that F_e^- and F_e^+ are non-decreasing at (#) and flow conservation on edges at ($\triangle$). Additionally, we used $\tilde{d}_{v_{k-1}} \leq \tilde{d}_{v_k}$.

Proving the proposition can now be accomplished by providing a suitable lower bound for $F_{\leq 1}$. We will now do that by showing the following lower bound for all $F_{\leq k}$ via downwards induction on k:

Claim 6.2. *For all $k \in [n]$ and times $\theta \geq \vartheta \geq 0$ we have*

$$F_{\leq k}(\theta) \geq M_k(\theta, \vartheta) := \min\{U(\vartheta), F_{\leq k}(\vartheta) + \theta - \vartheta - \tau_{p_{\max}} + \tilde{d}_{v_k}\}. \quad (6.1)$$

Proof. We will show this claim by downwards induction over k:

Base Case ($k = n$): This case is trivial as we have:

$$F_{\leq n}(\theta) \overset{\text{(iii)}}{=} U(\theta) \geq U(\vartheta) \geq \min\{U(\vartheta), F_{\leq n}(\vartheta) + \theta - \vartheta - \tau_{p_{\max}} + \tilde{d}_{v_n}\}$$
$$= M_n(\theta, \vartheta).$$

Induction Step ($k \to k-1$): Fix any $\vartheta \geq 0$. We then have to show

$$F_{\leq k-1}(\theta) - M_{k-1}(\theta, \vartheta) \geq 0$$

for all $\theta \geq \vartheta$. We do this separately for the two intervals $[\vartheta, \vartheta + \tau_{p_{\max}} - \tilde{d}_{v_{k-1}}]$ and $(\vartheta + \tau_{p_{\max}} - \tilde{d}_{v_{k-1}}, \infty)$.

The case $\theta \in [\vartheta, \vartheta + \tau_{p_{\max}} - \tilde{d}_{v_{k-1}}]$ is trivial as we have

$$\begin{aligned} F_{\leq k-1}(\theta) &\overset{\text{(i)}}{\geq} F_{\leq k-1}(\vartheta) \geq F_{\leq k-1}(\vartheta) + \theta - (\vartheta + \tau_{p_{\max}} - \tilde{d}_{v_{k-1}}) \\ &\geq \min\{\, U(\vartheta),\, F_{\leq k-1}(\vartheta) + \theta - \vartheta - \tau_{p_{\max}} + \tilde{d}_{v_{k-1}} \,\} = M_{k-1}(\theta, \vartheta). \end{aligned}$$

For the case $\theta \in (\vartheta + \tau_{p_{\max}} - \tilde{d}_{v_{k-1}}, \infty)$ we want to apply Proposition 2.48. This is possible since $F_{\leq k-1}(\theta) - M_{k-1}(\theta, \vartheta)$ is (as function in θ) absolutely continuous as sum of absolutely continuous functions and we have already shown that $F_{\leq k-1}(\vartheta + \tau_{p_{\max}} - \tilde{d}_{v_{k-1}}) - M_{k-1}(\vartheta + \tau_{p_{\max}} - \tilde{d}_{v_{k-1}}, \vartheta) \geq 0$. Thus, take any time $\theta \in (\vartheta + \tau_{p_{\max}} - \tilde{d}_{v_{k-1}}, \infty)$ where $F_{\leq k-1}(\theta) - M_{k-1}(\theta, \vartheta)$ is differentiable and we have $F_{\leq k-1}(\theta) - M_{k-1}(\theta, \vartheta) < 0$. Then, we have

$$\begin{aligned} M_{k-1}(\theta, \vartheta) &= \min\{\, U(\vartheta),\, F_{\leq k-1}(\vartheta) + \theta - \vartheta - \tau_{p_{\max}} + \tilde{d}_{v_{k-1}} \,\} \\ &\overset{\text{(ii)}}{\leq} \min\{\, U(\vartheta),\, F_{\leq k}(\vartheta) + (\theta + \tilde{d}_{v_{k-1}} - \tilde{d}_{v_k}) - \vartheta - \tau_{p_{\max}} + \tilde{d}_{v_k} \,\} \\ &= M_k(\theta + \tilde{d}_{v_{k-1}} - \tilde{d}_{v_k}, \vartheta) \end{aligned} \tag{6.2}$$

which gives us

$$\begin{aligned} 0 > F_{\leq k-1}(\theta) - M_{k-1}(\theta, \vartheta) &\overset{(6.2)}{\geq} F_{\leq k-1}(\theta) - M_k(\theta + \tilde{d}_{v_{k-1}} - \tilde{d}_{v_k}, \vartheta) \\ &\overset{(\diamond)}{\geq} F_{\leq k-1}(\theta) - F_{\leq k}(\theta + \tilde{d}_{v_{k-1}} - \tilde{d}_{v_k}) \\ &= \sum_{v \in V_{k-1}} \int_{\theta + \tilde{d}_{v_{k-1}} - \tilde{d}_{v_k}}^{\theta} u_v(\zeta) \mathrm{d}\zeta - U_{v_k}(\theta + \tilde{d}_{v_{k-1}} - \tilde{d}_{v_k}) \\ &\quad + \sum_{e = vw \in \delta^-(V_k) \cap \delta^-(V_{k-1})} \Big(F_e^-(\theta + \min\{\, \tilde{d}_{v_{k-1}} - \tilde{d}_w, \tau_e \,\}) \\ &\qquad\qquad - F_e^-(\theta + \tilde{d}_{v_{k-1}} - \tilde{d}_{v_k} + \min\{\, \tilde{d}_{v_k} - \tilde{d}_w, \tau_e \,\}) \Big) \\ &\quad + \sum_{e = vw \in \delta^-(V_{k-1}) \setminus \delta^-(V_k)} F_e^-(\theta + \min\{\, \tilde{d}_{v_{k-1}} - \tilde{d}_w, \tau_e \,\}) \\ &\quad - \sum_{e = vw \in \delta^-(V_k) \setminus \delta^-(V_{k-1})} F_e^-(\theta + \tilde{d}_{v_{k-1}} - \tilde{d}_{v_k} + \min\{\, \tilde{d}_{v_k} - \tilde{d}_w, \tau_e \,\}) \\ &\overset{(\#)}{\geq} -U_{v_k}(\theta + \tilde{d}_{v_{k-1}} - \tilde{d}_{v_k}) \end{aligned}$$

$$
\begin{aligned}
&+\sum_{e=vw\in\delta^-(V_k)\cap\delta^-(V_{k-1})}\Big(F_e^-(\theta+\min\{\tilde d_{v_{k-1}}-\tilde d_w,\tau_e\})\\
&\qquad\qquad -F_e^-(\theta+\min\{\tilde d_{v_{k-1}}-\tilde d_w,\tau_e\})\Big)\\
&+\sum_{e=v_kw\in\delta^+(v_k)}F_e^-(\theta+\min\{\tilde d_{v_{k-1}}-\tilde d_{v_k},\tau_e\})\\
&-\sum_{e=vv_k\in\delta^-(v_k)}F_e^-(\theta+\tilde d_{v_{k-1}}-\tilde d_{v_k}+\min\{\tilde d_{v_k}-\tilde d_{v_k},\tau_e\})\\
&=-U_{v_k}(\theta+\tilde d_{v_{k-1}}-\tilde d_{v_k})\\
&\quad+\sum_{e=v_kw\in\delta^+(v_k)}F_e^-(\theta+\min\{\tilde d_{v_{k-1}}-\tilde d_w,\tau_e\})-\sum_{e\in\delta^-(v_k)}F_e^-(\theta+\tilde d_{v_{k-1}}-\tilde d_{v_k})\\
&\overset{(*)}{=}\sum_{e=v_kw\in\delta^+(v_k)}F_e^-(\theta+\min\{\tilde d_{v_{k-1}}-\tilde d_w,\tau_e\})-\sum_{e=v_kw\in\delta^+(v_k)}F_e^+(\theta+\tilde d_{v_{k-1}}-\tilde d_{v_k})\\
&\overset{(\#),(\bigcirc)}{\geq}\sum_{e=v_kw\in\delta^+(v_k)}\Big(F_e^-(\theta+\min\{\tilde d_{v_{k-1}}-\tilde d_w,\tau_e\})\\
&\qquad\qquad -F_e^+(\theta+\min\{\tilde d_{v_{k-1}}-\tilde d_w-\tau_e,0\})\Big)\\
&=\sum_{e=v_kw\in\delta^+(v_k)}-Q_e(\theta+\min\{\tilde d_{v_{k-1}}-\tilde d_w,\tau_e\}-\tau_e).
\end{aligned}
$$

Here we used induction together with $\theta+\tilde d_{v_{k-1}}-\tilde d_{v_k}\geq\vartheta+\tau_{p_{\max}}-\tilde d_{v_k}\geq\vartheta$ (using our case assumption) at $(\diamond)$. At (#) we used the fact that F_e^+, F_e^- and U_v are non-decreasing and at $(*)$ we used strong flow conservation at the non-sink node v_k. Finally, for $(\bigcirc)$ we observe that $\tilde d_w+\tau_e\leq\tilde d_{v_k}$ for any edge $e=v_kw\in\delta^+(v_k)$ implies $\tilde d_{v_{k-1}}-\tilde d_{v_k}\leq\tilde d_{v_{k-1}}-\tilde d_w-\tau_e$ for such edges. Thus, there is at least one edge $e=v_kw\in\delta^+(v_k)\subseteq\delta^-(V_{k-1})$ which satisfies $Q_e(\theta+\min\{\tilde d_{v_{k-1}}-\tilde d_w,\tau_e\}-\tau_e)>0$. As the queue on this edge operates at capacity, this implies

$$\partial F_e^-(\theta+\min\{\tilde d_{v_{k-1}}-\tilde d_w,\tau_e\})=f_e^-(\theta+\min\{\tilde d_{v_{k-1}}-\tilde d_w,\tau_e\})=\nu_e\geq 1$$

and, therefore,

$$\partial_\theta\left(F_{\leq k-1}(\theta)-M_{k-1}(\theta,\vartheta)\right)\geq f_e^-(\theta+\min\{\tilde d_{v_{k-1}}-\tilde d_w,\tau_e\})-1\geq 0$$

for almost all such θ. Thus, Proposition 2.48 gives $F_{\leq k-1}(\theta) - M_{k-1}(\theta, \vartheta) \geq 0$ for all $\theta \geq \vartheta + \tau_{p_{\max}} - \tilde{d}_{v_{k-1}}$ which proves the claim. ■

The proposition now follows by choosing $k = 1$:

$$\begin{aligned} Z(\theta) &\overset{\text{(iv)}}{=} F_{\leq 1}(\theta) \overset{\text{(Cl. 6.2)}}{\geq} M_1(\theta, \vartheta) = \min\{\, U(\vartheta), F_{\leq 1}(\vartheta) + \theta - \vartheta - \tau_{p_{\max}} + \tilde{d}_t \,\} \\ &\overset{\text{(iv)}}{=} \min\{\, U(\vartheta), Z(\vartheta) + \theta - \vartheta - \tau_{p_{\max}} + 0 \,\} \\ &\overset{\text{Prop. 3.53}}{=} Z(\vartheta) + \min\{\, F^{\Delta}(\vartheta), \theta - \vartheta - \tau_{p_{\max}} \,\}. \end{aligned}$$

□

This result is now easily extended to the multi-commodity case (in acyclic networks!) by adding a super-sink:

Corollary 6.4. *Let $\mathcal{N}$ be an acyclic multi-commodity network where the network inflow rates at all sink nodes are essentially bounded. Then we have*

$$Z(\theta) \geq Z(\vartheta) + \min\{\, F^{\Delta}(\vartheta), \nu_{\min} \cdot (\theta - \vartheta - \tau_{p_{\max}}) \,\}$$

for any Vickrey flow f in $\mathcal{N}$ and all times $\theta \geq \vartheta \geq 0$.

Proof. We want to reduce this corollary to the case of Proposition 6.3. For this we transform the given network $\mathcal{N}$ and flow f into a new single-commodity network $\mathcal{N}'$ and flow f' as follows: First, we scale all edge capacities in $\mathcal{N}$ by $\frac{1}{\nu_{\min}}$ and then add a new sink node t and an edge vt with free flow travel time 0 and capacity

$$\nu'_{vt} := \frac{1}{\nu_{\min}} \Big(\sum_{e \in \delta^-(v)} \nu_e + \sum_{i \in I : v \in T_i} \operatorname{ess\,sup}\{\, u_{v,i}(\theta) \mid \theta \in \mathbb{R}_{\geq 0} \,\} \Big)$$

from every node v in $\mathcal{N}$ to the new node t. Next, we define the node inflow rates in $\mathcal{N}'$ by setting $u'_v := \frac{1}{\nu_{\min}} \sum_{i \in I} u_{v,i}$ for all nodes from the original network and $u_t := 0$ at the new node t and declare t to be the only sink node in $\mathcal{N}'$. Finally, we define a flow f' in $\mathcal{N}'$ by setting $f'^{+}_e := \frac{1}{\nu_{\min}} f^+_e$ and $f'^{-}_e := \frac{1}{\nu_{\min}} f^-_e$ for all edges e of the original network while for any new edge vt setting $f'^{+}_{vt} := f'^{-}_{vt} := \frac{1}{\nu_{\min}} \sum_{i \in I} \partial B_{v,i}$.

The new network $\mathcal{N}'$ is now clearly an acyclic single-commodity network with edge capacities of at least 1 and a single sink node reachable from every other node. Furthermore, we have $\tau'_{p_{\max}} = \tau_{p_{\max}}$ and f' is a Vickrey flow in $\mathcal{N}'$ with $Z' = \frac{1}{\nu_{\min}} \cdot Z$

(note that the edge capacities on the new edges are chosen large enough such that they are not violated by our choice of f'^{-}_{vt}). Thus, we get

$$Z(\theta) = \nu_{\min} \cdot Z'(\theta) \overset{\text{Prop. 6.3}}{\geq} \nu_{\min} \cdot Z'(\vartheta) + \min\{\, \nu_{\min} \cdot F'^{\Delta}(\vartheta), \nu_{\min} \cdot \left(\theta - \vartheta - \tau'_{p_{\max}}\right) \}$$
$$= Z(\vartheta) + \min\{\, F^{\Delta}(\vartheta), \nu_{\min} \cdot \left(\theta - \vartheta - \tau_{p_{\max}}\right) \}$$

for all times $\theta \geq \vartheta \geq 0$. □

Using this result we can now also bound makespan and total travel time of Vickrey flows in general acyclic networks:

Theorem 6.5. *Let $\mathcal{N}$ be an acyclic multi-commodity network with finitely lasting network inflow rates and essentially bounded network inflow rates at all sink nodes. Then we have*

$$\Psi(f) \leq \hat{\theta} + \frac{U}{\nu_{\min}} + \tau_{p_{\max}}, \quad \textit{and} \quad \Xi(f) \leq \frac{U^2}{\nu_{\min}} + U\tau_{p_{\max}}$$

for any Vickrey flow f in $\mathcal{N}$ where $U := U(\hat{\theta})$ is the total flow volume to ever enter the network.

Proof. As the assumptions here are the same as in Corollary 6.4, we can use the lower bound on $Z(\theta)$ from there. The makespan bound then follows by choosing $\vartheta := \hat{\theta}$ and $\theta := \hat{\theta} + \frac{U}{\nu_{\min}} + \tau_{p_{\max}}$ which gives us

$$Z(\theta) = Z(\hat{\theta} + \frac{U}{\nu_{\min}} + \tau_{p_{\max}}) \geq Z(\hat{\theta}) + \min\{\, F^{\Delta}(\hat{\theta}), U \,\} \geq U = U(\theta)$$

and, therefore, $F^{\Delta}(\theta) = U(\theta) - Z(\theta) \leq 0$ by Proposition 3.53. This, then shows $\Psi(f) \leq \theta = \hat{\theta} + \frac{U}{\nu_{\min}} + \tau_{p_{\max}}$.

For the bound on the total travel time we choose $\vartheta := \zeta$ and $\theta := \zeta + \tau_{p_{\max}} + \frac{U}{\nu_{\min}}$ for any $\zeta \geq 0$ to get

$$Z(\zeta + \tau_{p_{\max}} + \frac{U}{\nu_{\min}}) - Z(\zeta) \geq \min\{\, F^{\Delta}(\zeta), U \,\} \geq F^{\Delta}(\zeta).$$

Together with Proposition 3.75 we then get

$$\begin{aligned}
\Xi(f) &= \int_0^{\Psi(f)} F^{\Delta}(\zeta)\mathrm{d}\zeta \leq \int_0^{\Psi(f)} \Big(Z(\zeta + \tau_{p_{\max}} + \tfrac{U}{\nu_{\min}}) - Z(\zeta)\Big)\,\mathrm{d}\zeta \\
&= \int_0^{\Psi(f)} Z(\zeta + \tau_{p_{\max}} + \tfrac{U}{\nu_{\min}})\mathrm{d}\zeta - \int_0^{\Psi(f)} Z(\zeta)\mathrm{d}\zeta \\
&= \int_{\tau_{p_{\max}} + \frac{U}{\nu_{\min}}}^{\Psi(f)+\tau_{p_{\max}}+\frac{U}{\nu_{\min}}} Z(\zeta)\mathrm{d}\zeta - \int_0^{\Psi(f)} Z(\zeta)\mathrm{d}\zeta \\
&\leq \int_{\Psi(f)}^{\Psi(f)+\tau_{p_{\max}}+\frac{U}{\nu_{\min}}} Z(\zeta)\mathrm{d}\zeta \\
&= \Big(\tau_{p_{\max}} + \tfrac{U}{\nu_{\min}}\Big)\,U.
\end{aligned}$$

□

6.1.2 General Single-Commodity Networks

In acyclic networks it is intuitively clear that all Vickrey flows (with finite $\hat{\theta}$) eventually terminate—which is exactly what we showed in the previous section. This is obviously not true any more as soon as our network contains a cycle since then we can just send flow around this cycle forever. Thus, we will now restrict ourselves to flows which are also IDE.

However, even for such flows cycling behaviour is possible (see e.g. the first IDE described in Example 3.65) and, thus, it is not as obvious as it may seem whether IDE in general networks even terminate (and, if so, in what time). In fact, it turns out that the answer to this question is different for single-commodity IDE than it is for multi-commodity networks. We will consider the first case in this section and will come back to the multi-commodity case in Subsection 6.2.2. Since the proof for the upper bounds on makespan and total travel time will be rather lengthy, we first give a rough idea of its main steps:

Proof idea: Our first step towards bounding the makespan (and total travel time) in general single-commodity networks will be the following observation: Whenever there is almost no flow in the network, then flow particles in an IDE cannot be diverted too far away from the physically shortest paths and, therefore, may only use an acyclic subgraph of the whole network. Hence, we can use our bounds for acyclic networks from the last subsection 6.1.1.

Next, we have to show that such a state of an almost empty network is eventually reached. For this we focus on the sink nodes and distinguish to cases: As long as we have steady inflow into the sink nodes, we are also making progress towards an empty network (as the total flow volume in the network is bounded). If, on the other hand, we have only very little inflow into the sink nodes over some extended period of time, this means that there also cannot be much flow on the edges leading towards the sinks (using the no-idling property of Vickrey flows—cf. Corollary 3.24). This then implies that the subgraph consisting of the sink nodes and their closest neighbourhood is almost empty. Hence, no flow can leave this neighbourhood (by the same argument as in the first step). In some sense this subgraph, then, acts like a sink for the rest of the network and we will, therefore, call it a "sink-like-subgraph". Repeating the argument we just used for the sink nodes now for the sink-like subgraph, allows us to iteratively extend this subgraph until we can conclude that the whole network is sink-like and, therefore, almost empty.

Thus, for any long enough time interval during which an IDE does not terminate, we have some lower bound on the amount of flow that must reach a sink during this time. From this we can then immediately deduce upper bounds on makespan and total travel time.

In order to formalize the first step of the proof we will again make use of the set $\tilde{E} \subseteq E$ introduced in Lemma 5.13, only this time specifically for $\theta = 0$, i.e. for the network without flow:

$$\tilde{E} := \{\, e = vw \in E \mid L_v(0) > \tau_e + L_w(0) - \varepsilon \,\}$$

From Lemma 5.13 we already know that this set is acyclic (for a suitable choice of $\varepsilon > 0$) provided that the network does not contain any cycles of free flow travel time zero. Additionally, we will now show that this set also contains all active edges whenever there is almost no flow on any physically shortest path.

Proposition 6.6. *Let $\mathcal{N}$ be a single-commodity network without dead-end nodes, f a Vickrey flow in $\mathcal{N}$ and $\varepsilon > 0$ some constant. If then for some node $v \in V$ and time $\theta \geq 0$ we have $\sum_{e \in p} F_e^{\Delta}(\theta) < \varepsilon \nu_{\min}$ for all physically shortest v, T-paths, then all active edges $e = vw$ leaving v satisfy*

$$L_v(0) > \tau_e + L_w(0) - \varepsilon.$$

Proof. Let $e = vw \in \delta^+(v) \cap E(\theta)$ be an active edges at time θ, p a physically shortest v, T-path and p' an active w, T-path at time θ. Then we have

$$\begin{aligned} L_v(\theta) \leq C_p(\theta) &= \sum_{e \in p} \left(\tau_e + \frac{Q_e(\theta)}{\nu_e} \right) \leq \sum_{e \in p} \tau_e + \sum_{e \in p} \frac{F_e^{\Delta}(\theta)}{\nu_e} \leq \sum_{e \in p} \tau_e + \sum_{e \in p} \frac{F_e^{\Delta}(\theta)}{\nu_{\min}} \\ &= L_v(0) + \tfrac{1}{\nu_{\min}} \sum_{e \in p} F_e^{\Delta}(\theta) < L_v(0) + \varepsilon \end{aligned}$$

as well as

$$L_w(\theta) = C_{p'}(\theta) \geq \sum_{e \in p'} \tau_e \geq L_w(0).$$

Together with vw being active at time θ this gives us

$$L_v(0) > L_v(\theta) - \varepsilon = L_w(\theta) + C_{vw}(\theta) - \varepsilon \geq L_w(0) + \tau_{vw} - \varepsilon.$$

□

For our second step, we now need to make precise the notion of sink-like subgraphs:

Definition 6.7. Let $\mathcal{N}$ be a single-commodity network, f some IDE in $\mathcal{N}$, $[a, b] \subseteq \mathbb{R}_{\geq 0}$ some time interval, $\varepsilon > 0$ some constant, $W \subseteq V$ some subset of nodes and $H := G[W]$ the subgraph induced by W. We call H an *ε-sink-like subgraph on* $[a, b]$ if it satisfies the following three properties:

- $T \subseteq W$.
- For all $v \in W$ we have $\delta^+(v) \cap \tilde{E} \subseteq E(H)$.
- We have

$$\overline{\mathrm{vol}}_H(a, b) := \sum_{e \in E(H)} F_e^{\Delta}(a) + \sum_{e \in \delta^-(W)} \int_a^b f_e^-(\zeta) \mathrm{d}\zeta + \sum_{v \in W} \int_a^b u_v(\zeta) \mathrm{d}\zeta < \frac{\varepsilon \nu_{\min}}{2}.$$

Observation 6.8. As the notation suggests, $\overline{\mathrm{vol}}_H(a,b)$ is an upper bound on the volume of flow in H at any time θ during $[a,b]$:

$$\begin{aligned}
\sum_{e\in E(H)} F_e^{\Delta}(\theta) &\overset{\text{Prop. 3.53}}{=} \sum_{v\in W} U_v(\theta) + \sum_{e\in\delta^-(W)} F_e^-(\theta) - \sum_{e\in\delta^+(W)} F_e^+(\theta) - \sum_{v\in W} B_v(\theta) \\
&\leq \sum_{v\in W} U_v(a) + \sum_{e\in\delta^-(W)} F_e^-(a) - \sum_{e\in\delta^+(W)} F_e^+(a) - \sum_{v\in W} B_v(a) \\
&\quad + \sum_{v\in W} \int_a^\theta u_v(\zeta)\mathrm{d}\zeta + \sum_{e\in\delta^-(W)} \int_a^\theta f_e^-(\zeta)\mathrm{d}\zeta \\
&\overset{\text{Prop. 3.53}}{=} \sum_{e\in E(H)} F_e^{\Delta}(a) + \sum_{v\in W} \int_a^\theta u_v(\zeta)\mathrm{d}\zeta + \sum_{e\in\delta^-(W)} \int_a^\theta f_e^-(\zeta)\mathrm{d}\zeta \leq \overline{\mathrm{vol}}_H(a,b)
\end{aligned}$$

With this observation an immediate consequence of Proposition 6.6 is now that if the whole network is sink-like, then an IDE may only use an acyclic subgraph of the network. If this happens at a time where no new flow enters the network anymore (i.e. after time $\hat{\theta}$), we can apply our termination bound for acyclic networks from the previous section to show that the flow terminates.

Corollary 6.9. *Let $\mathcal{N}$ be a single-commodity network, f an IDE in $\mathcal{N}$ and $\varepsilon > 0$ some constant such that we have $\sum_{e\in c}\tau_e \geq \varepsilon|c|$ for all cycles c in $\mathcal{N}$. If the whole network is an ε-sink-like subgraph on some time interval $[a,b] \subseteq \mathbb{R}_{\geq 0}$ with $b-a \geq \varepsilon + \tau_{\max} + \tau_{p_{\max}}$, where $\tau_{\max} := \max\{\tau_e \mid e \in E\}$ is the free flow travel time of the longest edge in the network, then we have*

$$Z(b) \geq U(a + \tfrac{\varepsilon}{2} + \tau_{\max}).$$

In particular, if there exists some time $\theta \geq \hat{\theta}$ such that the whole network is an ε-sink-like subgraph at time θ (i.e. on the interval $[\theta,\theta]$), then it is sink-like on $[\theta,\vartheta]$ for any $\vartheta \geq \theta$ and f terminates before time $\theta + \varepsilon + \tau_{\max} + \tau_{p_{\max}}$.

Proof. If the whole network is ε-sink-like on $[a,b]$, then Observation 6.8 and Proposition 6.6 together show that during this interval all active edges are in $\tilde{E}$, i.e. f may only enter edges in $\tilde{E}$ in that time interval. Furthermore, Corollary 3.24 ensures that all flow has left the edge it was on at time θ by time $a + \max\{\frac{\varepsilon\nu_{\min}}{2\nu_e} + \tau_e \mid e \in E\} \leq a + \frac{\varepsilon}{2} + \tau_{\max} =: \tilde{a}$. Thus, we know that all edges carrying flow after time $\tilde{a}$ must be in $\tilde{E}$. Since our choice of ε guarantees that $(V,\tilde{E})$ is acyclic (by Lemma 5.13), we can apply Corollary 6.4 to obtain

$$Z(\tilde{a} + \tfrac{\varepsilon}{2} + \tau_{p_{\max}}) \geq Z(\tilde{a}) + \min\{\, F^{\Delta}(\tilde{a}),\, \tfrac{\varepsilon \nu_{\min}}{2} \,\} \overset{\text{Obs. 6.8}}{\geq} Z(\tilde{a}) + F^{\Delta}(\tilde{a}) = U(\tilde{a}).$$

Now, for the second part take any time $\vartheta \geq \theta \geq \hat{\theta}$. Then we have

$$\overline{\mathrm{vol}}_G(\theta, \vartheta) = \sum_{e \in E} F_e^{\Delta}(\theta) + \sum_{v \in V} \int_{\theta}^{\vartheta} u_v(\zeta)\mathrm{d}\zeta = \sum_{e \in E} F_e^{\Delta}(\theta) = \overline{\mathrm{vol}}_G(\theta, \theta) < \tfrac{\varepsilon \nu_{\min}}{2}.$$

Hence, the whole network is an ε-sink-like subgraph on $[\theta, \vartheta]$. Thus, we can apply the first part with $a = \theta$ and $b = \theta + \varepsilon + \tau_{\max} + \tau_{p_{\max}}$ to get

$$Z(\theta + \varepsilon + \tau_{\max} + \tau_{p_{\max}}) \geq U(\theta + \tfrac{\varepsilon}{2} + \tau_{p_{\max}}) = U(\hat{\theta}).$$

Therefore, f terminates before time $\theta + \varepsilon + \tau_{\max} + \tau_{p_{\max}}$. □

Proposition 6.10. *Let $\mathcal{N}$ be a single-commodity network without any dead-end nodes or cycles of free flow travel time zero and f any IDE in $\mathcal{N}$. Furthermore, let $W \subsetneq V$ be some proper subset of nodes, $\varepsilon > 0$ some constant such that we have $\sum_{e \in c} \tau_e \geq \varepsilon |c|$ for any cycle c in $\mathcal{N}$ and $a \in \mathbb{R}_{\geq 0}$ any time. Finally, define*

$$b := a + \sum_{e \in E \setminus E(H)} \left(\tfrac{\varepsilon \nu_{\min}}{2\nu_e} + \tau_e \right).$$

If $H = G[W]$ is an ε-sink-like subgraph on $[a, b]$, then there exists a node $v \in V \setminus W$ such that $H' := G[W']$ with $W' := W \cup \{ v \}$ is an ε-sink-like subgraph on $[a, b']$ where b' is defined for H' analogous to b for H, i.e.

$$b' := a + \sum_{e \in E \setminus E(H')} \left(\tfrac{\varepsilon \nu_{\min}}{2\nu_e} + \tau_e \right).$$

Proof. According to Lemma 5.13 the set $\tilde{E}$ contains no cycles. Hence, there exists a topological order on $(V, \tilde{E})$. Now choose $v \in V \setminus W$ as the last node (with respect to this topological order) which is not in W. We will show that $W' := W \cup \{ v \}$ then defines an ε-sink-like subgraph $H' := [W]$. First of all we clearly have $T \subseteq W \subseteq W \cup \{ v \}$. Next, for any edge $e = vw \in \delta^+(v) \cap \tilde{E}$ we have $w \in W$ (by the choice of v) and, therefore, $e \in E[H']$.

Thus, it now remains to show that H' satisfies the third condition as well: We will accomplish that by showing that all flow currently on the edges in $E(H') \setminus E(H)$ as well as the additional inflow into node v during $[a, b']$ will enter H over the course

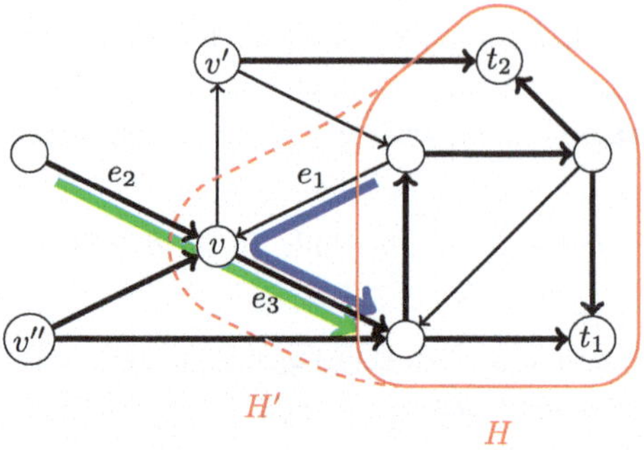

Fig. 6.3 The situation in Proposition 6.10. The thick edges are part of the set $\tilde{E}$. Thus, v' would have been another valid choice for extending H whereas v'' would not be (as v is closer to H than v''). The blue and green arrow exemplifies the statements in Claim 6.3: All flow on edge e_1 at time a reaches v by time $\tilde{b}$ (Claim 6.3(i)), enters edge e_3 (Claim 6.3(iii)) and reaches H by time b (Claim 6.3(iv)). Similarly, all flow reaching v via edge e_2 before time $\tilde{b}$ also enters edge e_3 (Claim 6.3(iii)) and reaches H by time b (Claim 6.3(iv))

of $[a, b]$. Hence, the bound for H gives us the same bound for H'. This will result from the following claim (Fig. 6.3):

Claim 6.3. *Define* $\tilde{b} := b' + \sum_{e \in \delta^+(W) \cap \delta^-(v)} \left(\frac{\varepsilon \nu_{\min}}{2\nu_e} + \tau_e \right)$. *Then the following properties hold:*

(i) All flow on edges from H *to* v *at time* a *reaches* v *before time* $\tilde{b}$*, i.e.*

$$F_e^{\Delta}(a) \leq \int_a^{\tilde{b}} f_e^-(\zeta) \mathrm{d}\zeta \text{ for all } e \in \delta^+(W) \cap \delta^-(v).$$

(ii) Any edges from H *to* v *carries a flow volume of less than* $\frac{\varepsilon \nu_{\min}}{2}$ *at time* a*, i.e.*

$$F_e^{\Delta}(a) < \frac{\varepsilon \nu_{\min}}{2} \text{ for all } e \in \delta^+(W) \cap \delta^-(v).$$

(iii) All flow reaching v *between* a *and* $\tilde{b}$ *enters an edge towards* H*, i.e.*

$$\int_a^{\tilde{b}} u_v(\zeta) \mathrm{d}\zeta + \sum_{e \in \delta^-(v)} \int_a^{\tilde{b}} f_e^-(\zeta) \mathrm{d}\zeta = \sum_{e \in \delta^+(v) \cap \delta^-(W)} \int_a^{\tilde{b}} f_e^+(\zeta) \mathrm{d}\zeta.$$

(iv) *All flow on an edge from v to H during any time in $[a, \tilde{b}]$ reaches H before time b, i.e.*

$$F_e^{\Delta}(a) + \int_a^{\tilde{b}} f_e^+(\zeta)\mathrm{d}\zeta \leq \int_a^b f_e^-(\zeta)\mathrm{d}\zeta \text{ for all } e \in \delta^+(v) \cap \delta^-(W).$$

(v) *Any edges from v to H carries a flow volume of less than $\frac{\varepsilon \nu_{\min}}{2}$ at any time between a and $\tilde{b}$, i.e.*

$$F_e^{\Delta}(\theta) < \tfrac{\varepsilon \nu_{\min}}{2} \text{ for all } e \in \delta^+(v) \cap \delta^-(W), \theta \in [a, \tilde{b}].$$

Proof. First, we observe that due to our choice of b, b' and $\tilde{b}$ we have $a + \frac{\varepsilon \nu_{\min}}{2\nu_e} + \tau_e \leq \tilde{b}$ for all $e \in \delta^+(W) \cap \delta^-(v)$ and $\tilde{b} + \frac{\varepsilon \nu_{\min}}{2\nu_e} + \tau_e \leq b$ for all $e \in \delta^+(v) \cap \delta^-(W)$.

We now show the five statements of the claim in reverse order as we will use (v) to show (iv) and so on.

(v): Take any edge $e \in \delta^+(v) \cap \delta^-(W)$ from v to H and assume for contradiction that at some time $\theta \in [a, \tilde{b}]$ it carries a flow volume of at least $\frac{\varepsilon \nu_{\min}}{2}$. By Corollary 3.24 a flow of at least volume $\frac{\varepsilon \nu_{\min}}{2}$ then leaves edge e between $\theta \geq a$ and $\theta + \frac{\varepsilon \nu_{\min}}{2\nu_e} + \tau_e \leq \tilde{b} + \frac{\varepsilon \nu_{\min}}{2\nu_e} + \tau_e \leq b$. This then implies

$$\overline{\mathrm{vol}}_H(a, b) \geq \int_a^b f_e^-(\zeta)\mathrm{d}\zeta \geq \int_\theta^{\theta + \frac{\varepsilon \nu_{\min}}{2\nu_e} + \tau_e} f_e^-(\zeta)\mathrm{d}\zeta \geq \tfrac{\varepsilon \nu_{\min}}{2},$$

which is a contradiction to H being an ε-sink-like subgraph on $[a, b]$.

(iv): Using (v) we get $\tilde{b} + \frac{F_e^{\Delta}(\tilde{b})}{\nu_e} + \tau_e < \tilde{b} + \frac{\varepsilon \nu_{\min}}{2\nu_e} + \tau_e \leq b$ for any edge $e \in \delta^+(v) \cap \delta^-(W)$. Thus, Corollary 3.24 implies

$$F_e^{\Delta}(\tilde{b}) \leq \int_{\tilde{b}}^{\tilde{b} + \frac{F_e^{\Delta}(\tilde{b})}{\nu_e} + \tau_e} f_e^-(\zeta)\mathrm{d}\zeta \leq \int_{\tilde{b}}^b f_e^-(\zeta)\mathrm{d}\zeta, \tag{6.3}$$

from which we get

$$\begin{aligned}F_e^{\Delta}(a) + \int_a^{\tilde{b}} f_e^+(\zeta)\mathrm{d}\zeta &= F_e^+(a) - F_e^-(a) + F_e^+(\tilde{b}) - F_e^+(a) = F_e^+(\tilde{b}) - F_e^-(a)\\ &= F_e^+(\tilde{b}) - F_e^-(\tilde{b}) + F_e^-(\tilde{b}) - F_e^-(a) = F_e^{\Delta}(\tilde{b}) + \int_a^{\tilde{b}} f_e^-(\zeta)\mathrm{d}\zeta\\ &\overset{(6.3)}{\leq} \int_{\tilde{b}}^b f_e^-(\zeta)\mathrm{d}\zeta + \int_a^{\tilde{b}} f_e^-(\zeta)\mathrm{d}\zeta = \int_a^b f_e^-(\zeta)\mathrm{d}\zeta.\end{aligned}$$

(iii): As H is an ε-sink-like subgraph, all sink nodes are already contained and, in particular, v cannot be a sink node. Hence, strong flow conservation together with f being an IDE imply that it suffices to show that during $[a, \tilde{b}]$ all active edges leaving v lead to H, i.e. we have $\delta^+(v) \cap E(\theta) \subseteq \delta^+(v) \cap \delta^-(W)$ for all $\theta \in [a, \tilde{b}]$. To show this, we want to apply Proposition 6.6. So, let p be a physically shortest v, T-path. Then, we have $L_{v'}(0) = \tau_{e'} + L_{w'}(0)$ for any edge $e' = v'w' \in p$ and, therefore, all edges of p are contained in $\tilde{E}$. Thus, by the choice of v the first edge of p leads into H and, as H is an ε-sink-like subgraph, every subsequent edge is contained in $E(H)$. Thus, writing $p = e', p'$ we get

$$\begin{aligned}\sum_{e\in p} F_e^{\Delta}(\theta) = F_{e'}^{\Delta}(\theta) + \sum_{e\in p'} F_e^{\Delta}(\theta) &\overset{(v)}{<} \frac{\varepsilon\nu_{\min}}{2} + \sum_{e\in p'} F_e^{\Delta}(\theta)\\ &\overset{\text{Obs. 6.8}}{\leq} \frac{\varepsilon\nu_{\min}}{2} + \overline{\mathrm{vol}}_H(a, b) < \varepsilon\nu_{\min}\end{aligned}$$

for any time $\theta \in [a, \tilde{b}]$. Hence, we can apply Proposition 6.6 to get $\delta^+(v) \cap E(\theta) \subseteq \delta^+(v) \cap \tilde{E} \subseteq \delta^+(v) \cap \delta^-(W)$, where the second inclusion holds by the choice of v.

(ii): This is essentially the same proof as for (v): Take any edge $e \in \delta^+(W) \cap \delta^-(v)$ and assume for contradiction that we have $F_e^{\Delta}(a) \geq \frac{\varepsilon\nu_{\min}}{2}$. According to Corollary 3.24 a flow of volume at least $\frac{\varepsilon\nu_{\min}}{2}$ then leaves edge e between a and $a + \frac{\varepsilon\nu_{\min}}{2\nu_e} + \tau_e \leq \tilde{b}$. Using (iii) and (iv) this then leads to a contradiction to H being an ε-sink-like subgraph on $[a, b]$ as follows:

$$\frac{\varepsilon \nu_{\min}}{2} \overset{\text{Cor. 3.24}}{\leq} \int_a^{a+\frac{\varepsilon \nu_{\min}}{2\nu_e}+\tau_e} f_e^-(\zeta)\mathrm{d}\zeta \leq \int_a^{\tilde{b}} f_e^-(\zeta)\mathrm{d}\zeta \leq \sum_{e'\in\delta^-(v)} \int_a^{\tilde{b}} f_{e'}^-(\zeta)\mathrm{d}\zeta$$

$$\overset{\text{(iii)}}{\leq} \sum_{e'\in\delta^+(v)\cap\delta^-(W)} \int_a^{\tilde{b}} f_{e'}^+(\zeta)\mathrm{d}\zeta \overset{\text{(iv)}}{\leq} \sum_{e'\in\delta^+(v)\cap\delta^-(W)} \int_a^{b} f_{e'}^-(\zeta)\mathrm{d}\zeta$$

$$\leq \overline{\mathrm{vol}}_H(a,b).$$

(i): This now follows from (ii) in the same way that (iv) followed from (v): We have $a + \frac{F_e^\Delta(a)}{\nu_e} + \tau_e \overset{\text{(ii)}}{\leq} a + \frac{\varepsilon \nu_{\min}}{2\nu_e} + \tau_e \leq \tilde{b}$ and, thus,

$$F_e^\Delta(a) \overset{\text{(Cor. 3.24)}}{\leq} \int_a^{a+\frac{F_e^\Delta(a)}{\nu_e}+\tau_e} f_e^-(\zeta)\mathrm{d}\zeta \leq \int_a^{\tilde{b}} f_e^-(\zeta)\mathrm{d}\zeta$$

for any edge $e \in \delta^+(W) \cap \delta^-(v)$. ∎

Combining the statements (i), (iii) and (iv) from the above claim, we are now able to show the third property of H' being an ε-sink-like subgraph on $[a, \tilde{b}]$ (and, therefore, on $[a, b']$) by a direct computation:

$$\overline{\mathrm{vol}}_{H'}(a,\tilde{b}) = \sum_{e\in E(H')} F_e^\Delta(a) + \sum_{e\in\delta^-(W')} \int_a^{\tilde{b}} f_e^-(\zeta)\mathrm{d}\zeta + \sum_{w\in W'} \int_a^{\tilde{b}} u_w(\zeta)\mathrm{d}\zeta$$

$$= \sum_{e\in\delta^+(v)\cap\delta^-(W)} F_e^\Delta(a) + \sum_{e\in\delta^+(W)\cap\delta^-(v)} F_e^\Delta(a)$$

$$+ \sum_{e\in\delta^-(v)\setminus\delta^+(W)} \int_a^{\tilde{b}} f_e^-(\zeta)\mathrm{d}\zeta + \int_a^{\tilde{b}} u_v(\zeta)\mathrm{d}\zeta$$

$$+ \sum_{e\in E(H)} F_e^\Delta(a) + \sum_{e\in\delta^-(W)\setminus\delta^+(v)} \int_a^{\tilde{b}} f_e^-(\zeta)\mathrm{d}\zeta + \sum_{w\in W} \int_a^{\tilde{b}} u_w(\zeta)\mathrm{d}\zeta$$

$$\overset{\text{(i)}}{\leq} \sum_{e\in\delta^+(v)\cap\delta^-(W)} F_e^\Delta(a) + \sum_{e\in\delta^+(W)\cap\delta^-(v)} \int_a^{\tilde{b}} f_e^-(\zeta)\mathrm{d}\zeta$$

$$+ \sum_{e\in\delta^-(v)\setminus\delta^+(W)} \int_a^{\tilde{b}} f_e^-(\zeta)\mathrm{d}\zeta + \int_a^{\tilde{b}} u_v(\zeta)\mathrm{d}\zeta$$

$$
\begin{aligned}
&+ \sum_{e \in E(H)} F_e^{\Delta}(a) + \sum_{e \in \delta^-(W) \setminus \delta^+(v)} \int_a^{\tilde{b}} f_e^-(\zeta)\mathrm{d}\zeta + \sum_{w \in W} \int_a^{\tilde{b}} u_w(\zeta)\mathrm{d}\zeta \\
\overset{\text{(iii)}}{=}\ & \sum_{e \in \delta^+(v) \cap \delta^-(W)} F_e^{\Delta}(a) + \sum_{e \in \delta^+(v) \cap \delta^-(W)} \int_a^{\tilde{b}} f_e^+(\zeta)\mathrm{d}\zeta \\
&+ \sum_{e \in E(H)} F_e^{\Delta}(a) + \sum_{e \in \delta^-(W) \setminus \delta^+(v)} \int_a^{\tilde{b}} f_e^-(\zeta)\mathrm{d}\zeta + \sum_{w \in W} \int_a^{\tilde{b}} u_w(\zeta)\mathrm{d}\zeta \\
\overset{\text{(iv)}}{\leq}\ & \sum_{e \in \delta^+(v) \cap \delta^-(W)} \int_a^{b} f_e^-(\zeta)\mathrm{d}\zeta \\
&+ \sum_{e \in E(H)} F_e^{\Delta}(a) + \sum_{e \in \delta^-(W) \setminus \delta^+(v)} \int_a^{\tilde{b}} f_e^-(\zeta)\mathrm{d}\zeta + \sum_{w \in W} \int_a^{\tilde{b}} u_w(\zeta)\mathrm{d}\zeta \\
\leq\ & \sum_{e \in E(H)} F_e^{\Delta}(a) + \sum_{e \in \delta^-(W)} \int_a^{b} f_e^-(\zeta)\mathrm{d}\zeta + \sum_{w \in W} \int_a^{b} u_v(\zeta)\mathrm{d}\zeta \\
=\ & \overline{\mathrm{vol}}_H(a, b) < \tfrac{\varepsilon \nu_{\min}}{2}.
\end{aligned}
$$

Hence, H' is an ε-sink-like subgraph on $[a, b']$. □

This proposition now implies that whenever we have a sufficiently long time interval with very small total inflow into the sinks, then the the whole network is an ε-sink-like subgraph at the beginning of that interval.

Corollary 6.11. *Let $\mathcal{N}$ be a single-commodity network without any dead-end nodes or cycles of free flow travel time zero and no outgoing edges from any sink node. Let f be any IDE in $\mathcal{N}$, $\varepsilon > 0$ some constant such that we have $\sum_{e \in c} \tau_e \geq \varepsilon |c|$ for any cycle c in $\mathcal{N}$ and $[a, b] \subseteq \mathbb{R}_{\geq 0}$ some time interval.*

If we have $b - a \geq \sum_{e \in E} \left(\frac{\varepsilon \nu_{\min}}{2\nu_e} + \tau_e\right)$ and $Z(b) - Z(a) < \frac{\varepsilon \nu_{\min}}{2}$, then the whole network is an ε-sink-like subgraph at time a.

Proof. We start by observing that $H := G[T]$ is an ε-sink-like subgraph: The first two properties are trivially satisfied and for the third we have

$$
\overline{\mathrm{vol}}_H(a, b) = \sum_{e \in \delta^-(T)} \int_a^b f_e^-(\zeta)\mathrm{d}\zeta + \sum_{t \in T} \int_a^b u_t(\zeta)\mathrm{d}\zeta \overset{\delta^+(T) = \emptyset}{=} Z(b) - Z(a) < \tfrac{\varepsilon \nu_{\min}}{2}.
$$

Thus, we can iteratively apply Proposition 6.10 to show that the whole network is an ε-sink-like subgraph on $[a, a]$. □

Now, if the situation from this corollary occurs at any time after $\hat{\theta}$, then Corollary 6.9 implies that the IDE terminates shortly after. This observation now gives us upper bounds on both the makespan and on the total travel time of any single-commodity IDE.

Theorem 6.12. *Let $\mathcal{N}$ be a single-commodity network where all network inflow rates are essentially bounded and finitely lasting, which has no cycles of free flow travel time zero and no edges leaving any sink node. Then for any IDE f in $\mathcal{N}$ the makespan is bounded by*

$$\Psi(f) \leq \hat{\theta} + U \sum_{e \in E} \left(\tfrac{1}{\nu_e} + \tfrac{2\tau_e}{\varepsilon \nu_{\min}} \right) + \varepsilon + \tau_{\max} + \tau_{p_{\max}}$$

while the total travel time is bounded by

$$\Xi(f) \leq U^2 \cdot \left(\sum_{e \in E} \left(\tfrac{1}{\nu_e} + \tfrac{2\tau_e}{\varepsilon \nu_{\min}} \right) + 4 \left(\tfrac{1}{\nu_{\min}} + \tfrac{\tau_{\max}}{\varepsilon \nu_{\min}} + \tfrac{\tau_{p_{\max}}}{\varepsilon \nu_{\min}} \right) \right)$$

where $\varepsilon := \min \{ \frac{1}{|c|} \sum_{e \in c} \tau_e \mid c \text{ a cycle in } \mathcal{N} \}$ is the minimum mean free flow travel time around any cycle in the network.

Proof. We first note that we can assume without loss of generality that $\mathcal{N}$ has no dead-end nodes. Otherwise we could just remove them without changing the flow in the remaining network or the overall makespan (see Proposition 3.66). Now, to show the makespan bound consider the intervals

$$[a_k, a_{k+1}] := \left[\hat{\theta} + k \cdot \sum_{e \in E} \left(\tfrac{\varepsilon \nu_{\min}}{2\nu_e} + \tau_e \right), \hat{\theta} + (k+1) \cdot \sum_{e \in E} \left(\tfrac{\varepsilon \nu_{\min}}{2\nu_e} + \tau_e \right) \right]$$

for $k = 0, 1, \ldots, \lfloor \frac{2U}{\varepsilon \nu_{\min}} \rfloor$. Since Z is non-negative, non-decreasing and upper bounded by U, there must be some $k \leq 2U$ such that we have $Z(a_{k+1}) - Z(a_k) < \frac{\varepsilon \nu_{\min}}{2}$ as otherwise we would have

$$Z\left(\left\lfloor \frac{2U}{\varepsilon\nu_{\min}} \right\rfloor\right) = \sum_{k=0}^{\left\lfloor \frac{2U}{\varepsilon\nu_{\min}} \right\rfloor} (Z(a_{k+1}) - Z(a_k)) \geq \left(\left\lfloor \frac{2U}{\varepsilon\nu_{\min}} \right\rfloor + 1\right) \cdot \frac{\varepsilon\nu_{\min}}{2} > \frac{2U}{\varepsilon\nu_{\min}} \cdot \frac{\varepsilon\nu_{\min}}{2} = U.$$

But then the whole network is an ε-sink-like subgraph at time a_k by Corollary 6.11 which, according to Corollary 6.9, implies that f terminates before

$$a_k + \varepsilon + \tau_{\max} + \tau_{p_{\max}} \leq \hat{\theta} + \left\lfloor \frac{2U}{\varepsilon\nu_{\min}} \right\rfloor \sum_{e\in E} \left(\frac{\varepsilon\nu_{\min}}{2\nu_e} + \tau_e\right) + \varepsilon + \tau_{\max} + \tau_{p_{\max}}.$$

For the bound on the total travel time we fix some (small) $\gamma > 0$ and define a new sequence of intervals $[b_k, b_{k+1}]$ by setting $b_0 := 0$ and then recursively

$$b_{k+1} := \sup\left\{ \theta \in [b_k, \Psi(f)] \;\middle|\; Z(\theta) - Z(b_k) \leq \frac{\varepsilon\nu_{\min}}{2+\gamma} \right\}$$

for $k = 0, 1, \dots, \lfloor \frac{(2+\gamma)U}{\varepsilon\nu_{\min}} \rfloor =: K$. Since Z is continuous, we have $Z(b_{k+1}) - Z(b_k) = \frac{\varepsilon\nu_{\min}}{2+\gamma}$ for all those intervals except for, maybe, the last one. Thus, we have $Z(b_K) = K \cdot \frac{\varepsilon\nu_{\min}}{2+\gamma} \geq U - \frac{\varepsilon\nu_{\min}}{2+\gamma}$ and, therefore, $b_{K+1} = \Psi(f)$.

We now separate these intervals into two groups: We say that an interval $[b_k, b_{k+1}]$ is *short* if

$$b_{k-1} - b_k \leq \alpha := \sum_{e\in E} \left(\frac{\varepsilon\nu_{\min}}{2\nu_e} + \tau_e\right) + \varepsilon + \tau_{\max} + \tau_{p_{\max}}$$

and *long* otherwise. Furthermore, we denote by $S \subseteq \{0, \dots, K\}$ the set of indices of short intervals and by L the set of indices of long intervals. Now, for any long interval $[b_k, b_{k+1}]$ Corollary 6.11 guarantees that the whole network is ε-sink-like on $[b_k, b_{k+1} - \sum_{e\in E}\left(\frac{\varepsilon\nu_{\min}}{2\nu_e} + \tau_e\right)]$. Thus, we can apply Corollary 6.9 for any time in $\theta \in [b_k, b_{k+1} - \sum_{e\in E}\left(\frac{\varepsilon\nu_{\min}}{2\nu_e} + \tau_e\right) - \varepsilon - \tau_{\max} - \tau_{p_{\max}}] = [b_k, b_{k+1} - \alpha]$ to get

$$Z(\theta + \varepsilon + \tau_{\max} + \tau_{p_{\max}}) \geq U(\theta + \tfrac{\varepsilon}{2} + \tau_{\max}) \geq U(\theta) \overset{\text{Prop. 3.53}}{=} Z(\theta) + F^{\Delta}(\theta). \tag{6.4}$$

This, in turn, allows us to upper bound the total travel times incurred during the first part of any long interval by

$$\begin{aligned}
\int_{b_k}^{b_{k+1}-\alpha} F_e^{\Delta}(\zeta)\mathrm{d}\zeta &\overset{(6.4)}{\leq} \int_{b_k}^{b_{k+1}-\alpha} Z(\zeta+\varepsilon+\tau_{\max}+\tau_{p_{\max}}) - Z(\zeta)\mathrm{d}\zeta \\
&= \int_{b_k}^{b_{k+1}-\alpha} Z(\zeta+\varepsilon+\tau_{\max}+\tau_{p_{\max}})\mathrm{d}\zeta - \int_{b_k}^{b_{k+1}-\alpha} Z(\zeta)\mathrm{d}\zeta \\
&= \int_{b_k+\varepsilon+\tau_{\max}+\tau_{p\max}}^{b_{k+1}-\alpha+\varepsilon+\tau_{\max}+\tau_{p\max}} Z(\zeta)\mathrm{d}\zeta - \int_{b_k}^{b_{k+1}-\alpha} Z(\zeta)\mathrm{d}\zeta \\
&\leq \int_{b_{k+1}-\alpha}^{b_{k+1}-\alpha+\varepsilon+\tau_{\max}+\tau_{p\max}} \underbrace{Z(\zeta)}_{\leq U(\zeta)}\mathrm{d}\zeta \\
&\leq U\cdot(\varepsilon+\tau_{\max}+\tau_{p_{\max}}).
\end{aligned}$$

Using this bound together with the trivial upper bound of $F^{\Delta}(\theta) \leq U$ during the first parts of all long intervals as well as for all short intervals now gives us the following upper bound for the total travel times of f:

$$\begin{aligned}
\Xi(f) &\overset{\text{Prop. 3.75}}{=} \int_0^{\Psi(f)} F^{\Delta}(\zeta)\mathrm{d}\zeta = \sum_{k=0}^{K}\int_{b_k}^{b_{k+1}} F^{\Delta}(\zeta)\mathrm{d}\zeta \\
&= \sum_{k\in L}\int_{b_k}^{b_{k+1}-\alpha} F^{\Delta}(\zeta)\mathrm{d}\zeta + \sum_{k\in L}\int_{b_{k+1}-\alpha}^{b_{k+1}} F^{\Delta}(\zeta)\mathrm{d}\zeta + \sum_{k\in S}\int_{b_k}^{b_{k+1}} F^{\Delta}(\zeta)\mathrm{d}\zeta \\
&\leq \sum_{k\in L} U\cdot(\varepsilon+\tau_{\max}+\tau_{p_{\max}}) + \sum_{k\in L} U\cdot\alpha + \sum_{k\in S} U\cdot\alpha \\
&\leq K\cdot U\cdot(\varepsilon+\tau_{\max}+\tau_{p_{\max}}+\alpha) \\
&\leq \tfrac{(2+\gamma)U}{\varepsilon\nu_{\min}}\cdot U\cdot\Big(\sum_{e\in E}\Big(\tfrac{\varepsilon\nu_{\min}}{2\nu_e}+\tau_e\Big)+2\left(\varepsilon+\tau_{\max}+\tau_{p_{\max}}\right)\Big).
\end{aligned}$$

As this bound holds for any $\gamma > 0$, letting γ go to zero finally gets us the desired bound of

$$\Xi(f) \leq U^2\cdot\Big(\sum_{e\in E}\Big(\tfrac{1}{\nu_e}+\tfrac{2\tau_e}{\varepsilon\nu_{\min}}\Big)+4\Big(\tfrac{1}{\nu_{\min}}+\tfrac{\tau_{\max}}{\varepsilon\nu_{\min}}+\tfrac{\tau_{p\max}}{\varepsilon\nu_{\min}}\Big)\Big).$$

□

Remark 6.13. If all free flow travel times and capacities are lower bounded by 1, then we can simplify the bounds from Theorem 6.12 to

$$\Psi(f) \leq \hat{\theta} + U \cdot \Big(|E| + 2\sum_{e \in E} \tau_e\Big) + 1 + \tau_{\max} + \tau_{p_{\max}}$$

and

$$\Xi(f) \leq U^2 \cdot \Big(|E| + 2\sum_{e \in E} \tau_e + 4 + 4\tau_{\max} + 4\tau_{p_{\max}}\Big).$$

We also note that for small $\hat{\theta}$ the upper bound for the total travel time obtained by just multiplying the bound for the makespan by the total flow volume U (cf. Corollary 3.76) can be better than the above bound. However, for larger $\hat{\theta}$ the above bound is better and, more importantly, independent of $\hat{\theta}$.

6.2 Lower Bounds

In this section we will construct networks $\mathcal{N}$ and corresponding IDE with large makespan and total travel time compared to the network size $\tau(\mathcal{N})$ and the total flow volume $U(\mathcal{N})$. We will do this first for single-commodity networks and then for multi-commodity networks. To keep our notation simple, we will only consider networks with integer free flow travel times and capacities here.

6.2.1 Single-Commodity Networks

We start with a simple example showing that for acyclic networks our bounds from Theorem 6.5 are—in some sense—asymptotically tight:

Example 6.14. Consider a single-commodity network consisting of a single edge e of capacity $\nu_e = 1$ and free flow travel time $\tau_e \geq 1$ connecting a source node s with network inflow rate $u_s := U \cdot \mathbb{1}_{[\hat{\theta}-1,\hat{\theta}]}$ for some numbers $U \geq 1, \hat{\theta} \geq 1$ to the single sink node t (cf. Fig. 6.4 (top left)). Then it is easy to see that the unique Vickrey flow f gives us the following values for makespan and total travel times (cf. Fig. 6.4 (bottom)):

	value for f	upper bound from Theorem 6.5
$\Psi(f)$	$\hat{\theta} + U + \tau_e - 1$	$\hat{\theta} + U + \tau_e$
$\Xi(f)$	$\frac{1}{2} \cdot U^2 + (\tau_e - \frac{1}{2})U$	$U^2 + U\tau_e$

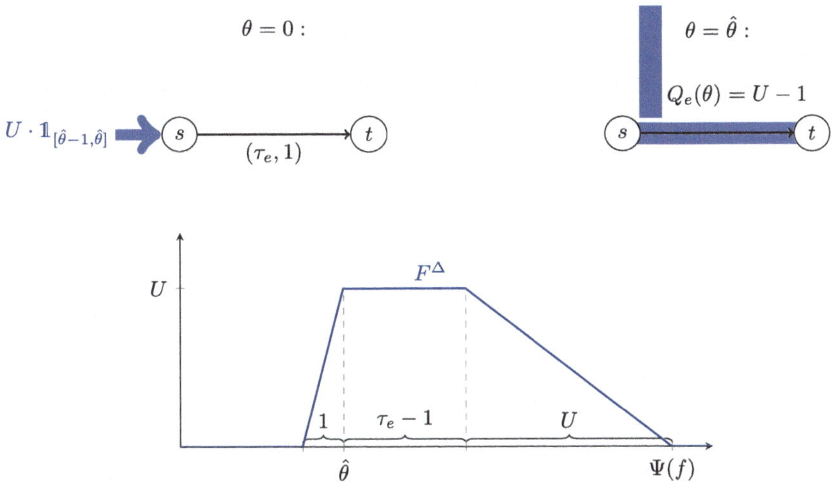

Fig. 6.4 A one edge network with large makespan and total travel time (top left). The picture in the top right shows the unique Vickrey flow in this network at time $\hat{\theta}$, i.e. at the time where the last particle enters the network. The graph at the bottom shows the network load for this flow over time. Thus, the area below this graph is exactly the total travel time incurred by this flow

Of course, this is a rather trivial example—while it almost matches our upper bounds (especially with respect to the makespan) it is also an instance where there is no difference between the optimal flow and an IDE as there is only a single source-sink path. We will, however, provide a different, more interesting lower bound example for acyclic graphs later on. First, however, we will turn to general single-commodity instances and construct a family of lower bound instances for this case.

Proof idea: The inspiration for our lower bound instances comes from our worst case analysis in the proof of the upper bound in the previous subsection: We saw there that if we have little inflow into the sink(s) during an extended time period (which is something we want in order to achieve a high makespan), then a sink-like subgraph starts to grow in the network and all flow inside this part of the subgraph leaves the network soon after. To prevent the whole flow from terminating, we must ensure that this growth stops before the whole graph becomes sink-like. Moreover, we want the main amount of flow to remain outside the subgraph which becomes sink-like.

Thus, we basically want to construct a network consisting of two parts: One (closer to the sink) which will (repeatedly) become a sink-like subgraph and one (further away from the sink) which will contain most of the flow. The second part then is essentially a large cycle wherein flow travels around without making any progress towards the sink while occasionally sending some small amount of flow into the first part of the network to prevent the sink-like subgraph from growing too large (cf. Fig. 6.5).

Now, if we were to take this idea literally and construct a simple four edge network like in Fig. 6.5, we would not achieve a better bound than with the one edge instance from Example 6.14. This is because in such a network every time the main part of the flow travels around the main cycle we have to loose flow with volume of the same order as the length of this cycle for diverting this flow away from the direct edges towards this sink. Thus, we have to construct the lower part of the network more carefully so that we can use the same flow multiple times for blocking the direct paths towards the sink. Hence, our actual lower bound instances will consist of two main parts: A cycling gadget and a blocking gadget.

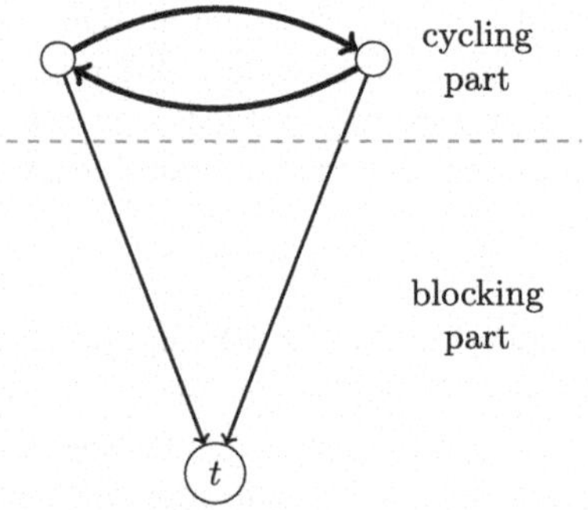

Fig. 6.5 The general structure of the network we use for the lower bound on makespan and total travel time in single-sink networks. The upper half consists of a large cycle where most of the flow should stay

Theorem 6.15. *There exists a family of networks $\mathcal{N}_{K,L}$ and associated IDE $f^{K,L}$ indexed by two natural numbers K and L such that*

- *The network size is asymptotically bounded by 3^{2K}, i.e. $\tau(\mathcal{N}_{K,L}) \in \mathcal{O}(3^{2K})$,*
- *the total flow volume is asymptotically bounded by $L3^K$, i.e. $U_{\mathcal{N}_{K,L}} \in \mathcal{O}(L3^K)$,*
- *we have $\Psi(f^{K,L}) \geq LK(3^K+1)$ and $\Xi(f^{K,L}) \geq KL(L-1) \cdot 3^{2K}$ and*
- *$\mathcal{N}_{K,L}$ has a single source node and a single sink node and a source, sink-path of free flow travel time $(K-1)(3^K+1) - 5 \cdot 3^{K-2} + 6$. Additionally, no flow enters the network after time $\theta = 1$.*

Here, we denote for any network $\mathcal{N}$ by $\tau(\mathcal{N}) := \sum_{e \in E} \tau_e$ the sum of all free flow travel times in $\mathcal{N}$ and by $U_{\mathcal{N}} := U(\hat{\theta})$ the total flow volume in $\mathcal{N}$.

Proof. Fix any two natural numbers $K, L \in \mathbb{N}^*$ with $K \geq 3$. Since we will now construct the network $\mathcal{N}_{K,L}$ as well as the IDE $f^{K,L}$ only for those fixed values K and L, we will drop the indices K and L from now on to keep the notation cleaner. As already indicated in Fig. 6.5 our network will be constructed from two parts: A cycling gadget C and a blocking gadget B. We will now build these two gadgets separately.

The blocking gadget: The blocking gadget B is constructed inductively out of smaller versions of itself. Its fundamental building block is the **delay gadget** D (cf. Fig. 6.6): It consists of three input nodes v_1, v_2 and v_3, one internal node y and one output node w. Each input node is connected to y via an edge $v_i y$ with free flow travel time 1 and capacity 1. Furthermore, there is an edge yz with free flow time 1 and capacity 1 connecting y to the output node z. We will later embedded D into a larger network such that the only edges entering D from the outside are edges e_1, e_2 and e_3 ending at the input nodes v_1, v_2 and v_3, respectively, and the only edge leaving D is edge e_0 starting at the output node z.

Claim 6.4. *Gadget D has the following two structural properties:*

(i) For any input node v_i of gadget D there exists a unique v_i, z-path which has a free flow travel time of 2.
(ii) The sum of all free flow travel times in D is 4.

Now assume that D is correctly embedded into a larger network $\mathcal{N}$. Let f be a Vickrey flow in this network, θ some time and $j \in \mathbb{N}_0$ such that

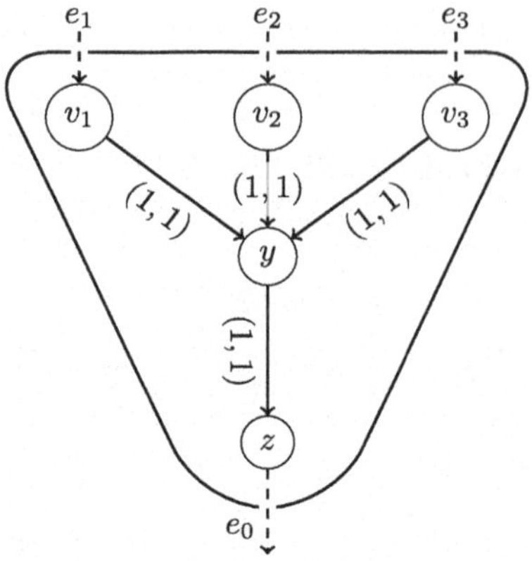

Fig. 6.6 The delay gadget D. Dashed edges are not part of the gadget and only indicate how this gadget will be embedded into a larger network

- *D carries no flow at time θ and*
- *the inflow into D via the edges e_i during $[\theta, \theta + 2 + 4.5 \cdot 3^{j+1}]$ satisfies*

$$\mathbb{1}_{[\theta+(i-1)3^j+0.5, \theta+(i-1)\cdot 3^j+3\cdot 3^j+0.5]} \leq f^-_{e_i} \leq \mathbb{1}_{[\theta+(i-1)\cdot 3^j, \theta+(i-1)\cdot 3^j+4.5\cdot 3^j]}$$

(cf. Fig. 6.7 (top)).

Then during $[\theta, \theta + 2 + 4.5 \cdot 3^{j+1}]$ this flow exhibits the following properties inside D (cf. Fig. 6.7 (middle and bottom)):

(iii) There are never any queues on the edges v_1y, v_2y and v_3y.
(iv) The queue on yz grows at a rate of at most 2 during $[\theta, \theta + 1 + 5.5 \cdot 3^j]$ and does not grow after that.
(v) The queue length on yz is upper bounded by $7 \cdot 3^j$ on the whole interval $[\theta, \theta + 4.5 \cdot 3^{j+1}]$.
(vi) The queue length on yz is upper bounded by $3^j - 0.5$ on $[\theta, \theta + 2 \cdot 3^j + 0.5]$.
(vii) The queue length on yz is lower bounded by $4 \cdot 3^j - 0.5$ on $[\theta + 4 \cdot 3^j + 1, \theta + 5 \cdot 3^j + 2]$.
(viii) The queue on yz is empty before time $\theta + 3^j + 1$ and after time $\theta + 4.5 \cdot 3^{j+1} + 1$.

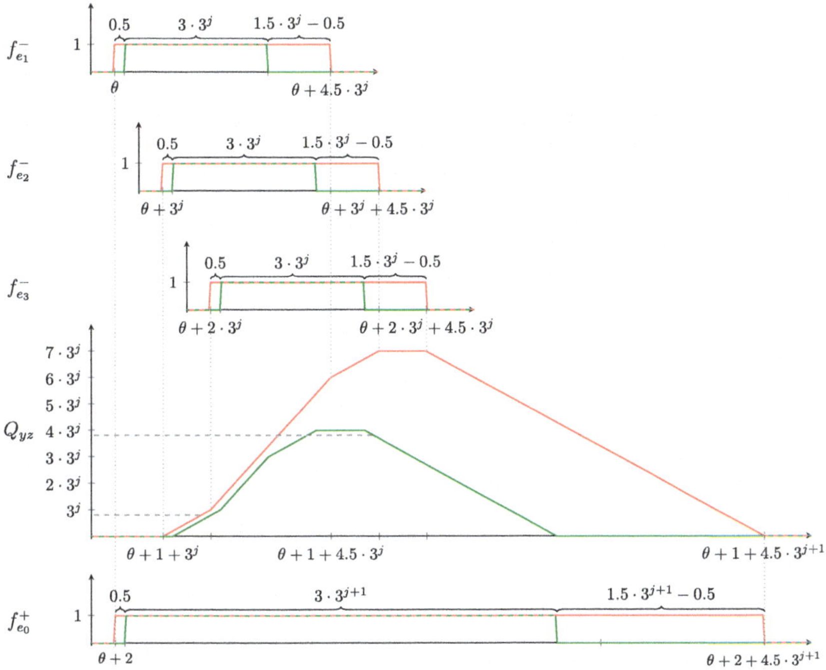

Fig. 6.7 A visual depiction of the flow bounds in gadget D stated in Claim 6.4: If the inflow into the gadget via edges e_1, e_2 and e_3 is bounded by the functions given in the three graphs at the top (red is the upper bound, green the lower bound), then the queue length on edge yz is bounded by the functions given in the middle graph and the outflow from the gadget via edge e_0 is bounded by the functions given in the graph at the bottom. Note, that the graphs a positioned horizontally in such a way that the time axis for the first three graphs align while the fourth one's is shifted by $-\tau_{v_i y} = -1$ and the fifth one's by $(\tau_{v_i y} + \tau_{yz}) = -2$)

(ix) *The outflow from the gadget over edge* e_0 *satisfies* $\mathbb{1}_{[\theta+2.5,\theta+2.5+3\cdot 3^{j+1}]} \leq f^+_{e_0} \leq \mathbb{1}_{[\theta+2,\theta+2+4.5\cdot 3^{j+1}]}$.

Proof. Properties (i) and (ii) are immediately clear from the construction.

If the inflow rates into the input nodes v_i never exceed the capacities of the following edges v_i (which is 1), no queues ever form on these edges. This shows (iii).

To show the remaining bounds on the flow (i.e. (iv) to (ix)) we first note that due to the monotonicity of the edge flow dynamics (Corollary 3.23) it suffices to show that if the inflow rates into D match the lower/upper bounds, then both the queue length function on edge yz and the outflow rate from D matches the respective lower/upper bounds. This can be deduced directly from Fig. 6.7 but we will also explain it in a bit more detail here.

First, assume that the inflow rates into v_i exactly match the upper bounds. Then flow starts to arrive at node y at a rate of 1 at time $\theta + 1$. This rate increases to 2 at time $\theta + 3^j + 1$ at which point a queue starts to grow on edge yz at a rate of 1. At time $\theta + 2 \cdot 3^j + 1$ the inflow rate into yz increases once more to 3 and the queue (currently at length 3^j) now growth at a rate of 2. At time $\theta + 4.5 \cdot 3^j + 1$ the inflow rate goes back to 2 and the queue (currently at length $3^j + 2 \cdot 2.5 \cdot 3^j = 6 \cdot 3^j$) now only grows at a rate of 1. At time $\theta + 5.5 \cdot 3^j + 1$ the inflow drops further to 1 and the queue now stays at a constant length of $7 \cdot 3^j$. Finally, at time $\theta + 6.5 \cdot 3^j + 1$ the inflow completely stops and the queue starts to deplete at a rate of 1. Thus, the queue is empty again by time $\theta + 6.5 \cdot 3^j + 1 + 7 \cdot 3^j = \theta + 13.5 \cdot 3^j + 1 = \theta + 4.5 \cdot 3^{j+1} + 1$. From this, it immediately follows that the outflow from the gadget exactly matches the given upper bound. Thus, we have shown (v), (vi), (viii) and the upper bound in (ix). Furthermore, this also shows that after time $\theta + 1 + 5.5 \cdot 3^j$ flow arrives at node y at a rate of at most 1. Thus, the queue on edge yz cannot grow any more after that. As flow can never arrive at y at a rate of more than 3, the queue also never grows at a rate of more than 2 before that time. This shows (iv).

Now, assume that the inflow rates into v_i exactly match the lower bounds. Then we can determine the flow evolution in exactly the same way as before: Flow starts to arrive at node y at time $\theta + 1.5$ at a rate of 1. This rate increases to 2 at time $\theta + 3^j + 1.5$ at which point a queue starts to grow on edge yz at a rate of 1. At time $\theta + 2 \cdot 3^j + 1.5$ the inflow rate increases once more to 2 and the queue (currently at length 3^j) now grows at a rate of 2. At time $\theta + 3 \cdot 3^j + 1.5$ the inflow rate goes back to 2 and the queue (currently at length $3^j + 3^j \cdot 2 = 3 \cdot 3^j$) now only growth at a rate of 1. At time $\theta + 4 \cdot 3^j + 1.5$, the inflow decreases further to 1 and the queue now remains at its current length of $3 \cdot 3^j + 3^j = 4 \cdot 3^j$. Finally, at time $\theta + 5 \cdot 3^j + 1.5$ the inflow completely stops and the queue starts to deplete (at a rate of 1). Thus, the queue is empty again at time $\theta + 5 \cdot 3^j + 1.5 + 4 \cdot 3^j = \theta + 3 \cdot 3^{j+1} + 1.5$. The outflow from edge yz then stops one time unit later. This shows (vii) and the lower bound in (ix). ∎

We now construct the **blocking gadget** B as follows: For any $k \in [K]$ we define B^k as follows: B^1 (Fig. 6.8) consists of three input nodes v_1, v_2 and v_3, one output node z and three edges v_1z, v_2z and v_3z with free flow travel time and capacity 1. B^k for $k \geq 2$ (Fig. 6.9) is then constructed by taking one copy of gadget B^{k-1} and 3^{k-1} copies of gadget D and connecting the output nodes of the latter with the input

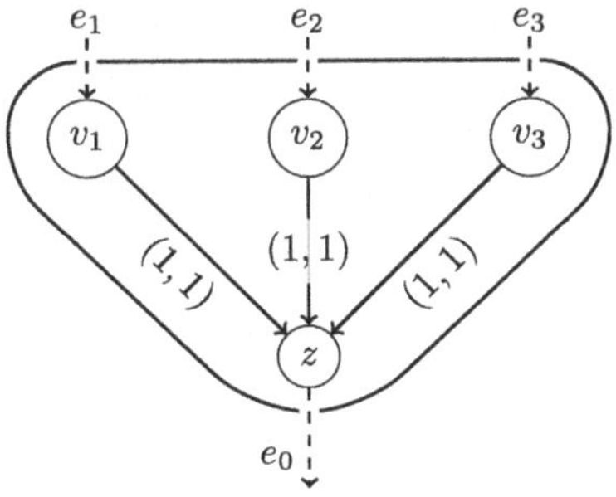

Fig. 6.8 Gadget B^1. Dashed edges are not part of the gadget and only indicate how to correctly embed this gadget into a larger network

nodes of the former via edges with free flow travel time $3^K - 10 \cdot 3^{K-k} - 1$ and capacity 1. Only for $k = 2$ we have to make an exception and set these free flow travel time to 1 (otherwise these edges would have a negative free flow travel time). The input nodes of the delay-gadgets D will then be the input nodes of B^k and we rename them to v_1 to v_{3^k}. The output node z of B^{k-1} will also be the output node of B^k. We will say that B^k is correctly embedded into a larger network if the only incoming edges $e_1, \ldots, e_{3^k}$ enter at its input nodes and the only outgoing edge e_0 leaves at its output node.

Claim 6.5. *Any gadget B^k satisfies the following structural properties:*

(i) For any input node v_i there exists a unique v_i, z-path p_i^k which has a free flow travel time of

$$\begin{cases} 1, & \text{if } k = 1 \\ (k-2)(3^K+1) - 5 \cdot (3^{K-2} - 3^{K-k}) + 4, & \text{if } k \geq 2 \end{cases}.$$

(ii) The sum of all free flow travel times of all edges in B^k is

$$\begin{cases} 3, & \text{if } k = 1 \\ \frac{1}{2}(3^K+3)(3^k-9) - 10 \cdot (k-2) \cdot 3^{K-1} + 18 & \text{if } k \geq 2 \end{cases}.$$

Now assume that B^k is correctly embedded into a larger network. Let f be a Vickrey flow in this larger network and $\theta \in \mathbb{R}_{\geq 0}$ some time such that

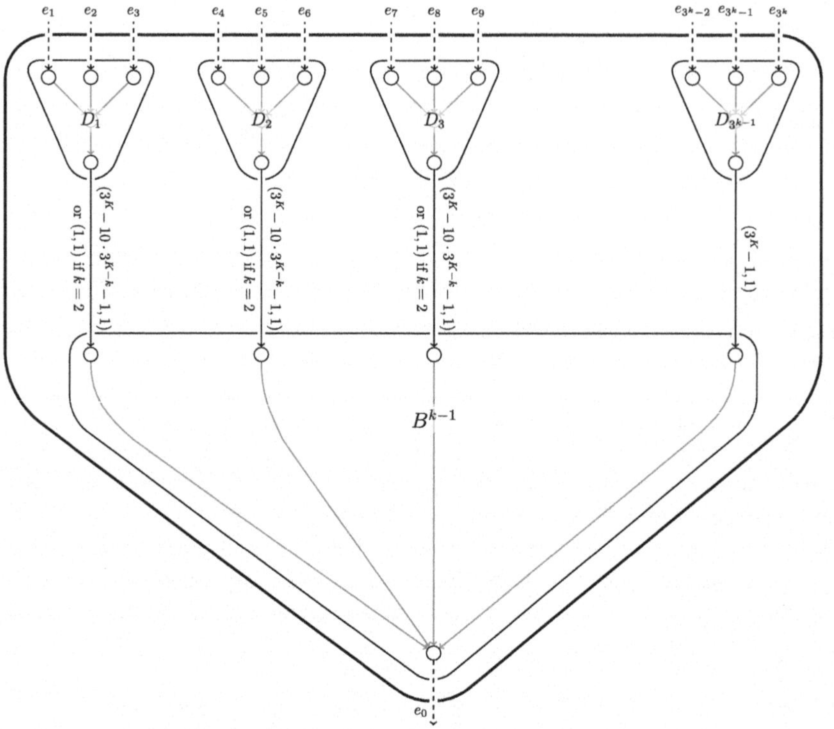

Fig. 6.9 Gadget B^k constructed from 3^{k-1} copies of the delay gadget D and one copy of gadget B^{k-1}. Dashed edges are not part of the gadget and only indicate how to correctly embed this gadget into a larger network

- *B^k carries no flow at time θ and*
- *the inflow into B^k via the edges e_i during $[\theta, (k-1) \cdot (3^K + 1)]$ satisfies*

$$f_{e_i}^- \geq \mathbb{1}_{[\theta+(i-1)3^{K-k}+0.5, \theta+(i-1)3^{K-k}+3 \cdot 3^{K-k}+0.5]} \text{ and}$$
$$f_{e_i}^- \leq \mathbb{1}_{[\theta+(i-1)3^{K-k}, \theta+(i-1)3^{K-k}+4.5 \cdot 3^{K-k}]}.$$

Then, for any path p_i^k the following properties hold:

(iii) At any time in $[\theta, (k-1) \cdot (3^K + 1)]$ there is at most one edge on p_i^k with a non-empty queue.

(iv) *The current travel time along p_i^k never grows at a rate of more than 2 during $[\theta, (k-1)\cdot(3^K+1)]$.*

(v) *There are no queues on path p_i^k before time $\theta + (i-2)\cdot 3^{K-k} + 1$ or after time $\theta + (i+4)\cdot 3^{K-k} + (k-2)(3^K+1) + 8.5\cdot 3^{K-2} + 1$. Moreover, all flow still on path p_i^k at the latter time cannot be part of any new queue on this path afterwards.*

Furthermore, for any $\ell \in [k-1]$, $m \in [3^{k-\ell}] - 1$ and $n \in [3^\ell]$ the waiting time along path $p^k_{m\cdot 3^\ell + n}$ satisfies the following properties:

(vi) *It is upper bounded by $7\cdot 3^{K-k+\ell-1}$ during $[\theta + (\ell-1)\cdot(3^K+1) + m\cdot 3^{K-k+\ell} + 5\cdot 3^{K-k} - 1, \theta + (\ell-1)\cdot(3^K+1) + m\cdot 3^{K-k+\ell} + 5\cdot 3^{K-k} - 0.5]$.*

(vii) *It is upper bounded by $3^{K-k+\ell-1} - 0.5$ during $[\theta + (\ell-1)\cdot(3^K+1) + (m-1)\cdot 3^{K-k+\ell} + 5\cdot 3^{K-k}, \theta + (\ell-1)\cdot(3^K+1) + (m-1)\cdot 3^{K-k+\ell} + 5\cdot 3^{K-k} + 0.5]$.*

(viii) *It is lower bounded by $4\cdot 3^{K-k+\ell-1} - 0.5$ during $[\theta + (\ell-1)\cdot(3^K+1) + m\cdot 3^{K-k+\ell} + 5\cdot 3^{K-k}, \theta + (\ell-1)\cdot(3^K+1) + m\cdot 3^{K-k+\ell} + 5\cdot 3^{K-k} + 0.5]$.*

Proof. We show this claim by induction on k:

Base Case ($k = 1$): In this case properties (i) and (ii) are directly clear from the construction while properties (vi) to (viii) are trivially true as $[k-1] = [0]$ is the empty set. Finally, properties (iii) to (v) hold because there are never any queues in B^1.

Induction Step ($k - 1 \to k$): Properties (i) and (ii) again follow from the construction of B^k:

(i) For any node v_i there exists the following unique v_i, z-path p_i^k: First through one of the copies of gadget D (free flow travel time 2 by Claim 6.4(i)), then over a connecting edge to one of the input nodes of B^{k-1} and from there the unique path $p^{k-1}_{\lceil i/3 \rceil}$ towards z which exists by induction.
If $k = 2$ this path p_i^k has a free flow travel time of $2 + 1 + 1 = 4$ while for $k \geq 3$ it has a free flow travel time of

$$\begin{aligned}
&2 + 3^K - 10\cdot 3^{K-k} - 1 + (k-3)(3^K+1) - 5\cdot(3^{K-2} - 3^{K-k+1}) + 4 \\
&= (k-2)(3^K+1) - 5\cdot(3^{K-2} - 3^{K-k}) + 4
\end{aligned}$$

where we used induction to get the free flow travel time for the part of p_i^k in B^{k-1}.

(ii) Similarly, using Claim 6.4(ii) and induction we get for the sum of all free flow travel times

$$\begin{aligned}
&3^{k-1} \cdot (4 + 3^K - 10 \cdot 3^{K-k} - 1) + \tfrac{1}{2}(3^K + 3)(3^{k-1} - 9) \\
&\quad - 10 \cdot (k-3) \cdot 3^{K-1} + 18 \\
&= 3^{k-1} \cdot (3^K + 3) - 10 \cdot 3^{K-1} + \tfrac{1}{2}(3^K + 3)(3^{k-1} - 9) \\
&\quad - 10 \cdot (k-3) \cdot 3^{K-1} + 18 \\
&= \tfrac{1}{2}(3^K + 3)(2 \cdot 3^{k-1} + 3^{k-1} - 9) - 10 \cdot (k-2) \cdot 3^{K-1} + 18 \\
&= \tfrac{1}{2}(3^K + 3)(3^k - 9) - 10 \cdot (k-2) \cdot 3^{K-1} + 18
\end{aligned}$$

if $k \geq 3$ and $3 + 3 + 3 \cdot 4 = 18$ if $k = 2$.

For the remaining properties, we first observe that the gadget B^{k-1} as well as all copies of D used to construct gadget B^k are correctly embedded in the larger network. Now, for any $m = 0, 1, \ldots, 3^{k-1} - 1$ the edges e_{3m+1}, e_{3m+2} and e_{3m+3} all enter the $(m+1)$-th copy of gadget D and carry flow matching exactly the properties of the inflow used in Claim 6.4 with $\theta = \theta + m \cdot 3^{K-k+1}$ and $j = K - k$. Thus, all the properties from Claim 6.4 hold for all copies of D. In particular, Claim 6.4(ix) ensures that the inflow rate into the connecting edge e from D_{m+1} to B^{k-1} satisfies

$$\begin{aligned}
f_e^+ &\geq \mathbb{1}_{[\theta+m\cdot 3^{K-k+1}+2.5,\theta+m\cdot 3^{K-k+1}+2.5+3\cdot 3^{K-k+1}]} \text{ and} \\
f_e^+ &\leq \mathbb{1}_{[\theta+m\cdot 3^{K-k+1}+2,\theta+m\cdot 3^{K-k+1}+2+4.5\cdot 3^{K-k+1}]}.
\end{aligned} \tag{6.5}$$

If $k \geq 3$, then the connecting edge from D_{m+1} to B^{k-1} has a free flow travel time of $3^K - 10 \cdot 3^{K-k} - 1 \geq 1$ and, thus, its outflow rate satisfies

$$\begin{aligned}
f_e^- &\geq \mathbb{1}_{[\theta'+m\cdot 3^{K-k+1}+0.5,\theta'+m\cdot 3^{K-k+1}+0.5+3\cdot 3^{K-k+1}]} \text{ and} \\
f_e^- &\leq \mathbb{1}_{[\theta'+m\cdot 3^{K-k+1},\theta'+m\cdot 3^{K-k+1}+4.5\cdot 3^{K-k+1}]}.
\end{aligned} \tag{6.6}$$

with $\theta' := \theta + 3^K - 10 \cdot 3^{K-k} + 1$. As this edge leads towards the $(m+1)$-th input node of B^{k-1}, this shows that the inflow into gadget B^{k-1} satisfies the bounds for the flow in Claim 6.5 (for $k-1$ instead of k and θ' instead of θ). Thus, all the properties from Claim 6.5 already hold in B^{k-1} by induction. Note, that this conclusion is also true in the case $k = 2$ since no queues ever form in $B^{k-1} = B^1$ and properties (vi) to (viii) are trivially true for B^{k-1} as we have $[k-1-1] = [0] = \emptyset$ then.

Now, for $k \geq 2$ let $p^k_{3^\ell m+n}$ with $\ell \in [k-1]$, $m \in [3^{k-\ell}] - 1$ and $n \in [3^\ell]$ be any path in B^K. Then we can split this path in three subpaths: A first part in D_{m+1}, a second part which is just the connecting edge from D_{m+1} to B^{k-1} and a third path in B^{k-1}. We note that this third part then also has the name $p^{k-1}_{3^{\ell-1}m+n'}$ where $n' = \lceil \frac{n}{3} \rceil \in [3^{\ell-1}]$. Moreover, we know from Claim 6.4(iii) and (viii) that there is no queue on the first part of path $p^k_{3^\ell m+n}$ after time $\theta + m \cdot 3^{K-k+1} + 4.5 \cdot 3^{K-k+1} + 1$ and from property (v) for B^{k-1} that there is no queue on the third part (i.e. $p^{k-1}_{3^{\ell-1}m+n'}$) before time $\theta' + (3^{\ell-1}m + n' - 2) \cdot 3^{K-k+1} + 1$. Finally, by construction there is never a queue on the middle part (i.e. the connecting edge).

With these observations we can now show the remaining properties for B^k:

(iii) For $k = 2$ this follows directly from Claim 6.4(iii) as there can never be any queues in B^1.
For $k \geq 2$ take any path p^k_{3m+n} with $m \in [3^{k-1}] - 1$, $n \in [3]$. By our previous observations it then suffices to show that the queue in D_{m+1} depletes before the first queue forms on p^{k-1}_{m+1}. This is in fact true since we have

$$\begin{aligned}
&\theta + m \cdot 3^{K-k+1} + 4.5 \cdot 3^{K-k+1} + 1 \\
&\quad = \theta + (m-1) \cdot 3^{K-k+1} + 5.5 \cdot 3^{K-k+1} + 1 \\
&\quad = \theta + (m-1) \cdot 3^{K-k+1} + 16.5 \cdot 3^{K-k} + 1 \\
&\quad \leq \theta + (m-1) \cdot 3^{K-k+1} + 27 \cdot 3^{K-k} - 10 \cdot 3^{K-k} + 1 \\
&\quad = \theta + 3^{K-k+3} - 10 \cdot 3^{K-k} + 1 + (m-1) \cdot 3^{K-k+1} \\
&\quad \overset{(*)}{\leq} \theta + 3^K - 10 \cdot 3^{K-k} + 1 + (m-1) \cdot 3^{K-k+1} + 1 \\
&\quad = \theta' + (m-1) \cdot 3^{K-k+1} + 1
\end{aligned}$$

where we use $k \geq 3$ at $(*)$. This shows that property (iii) holds for B^k.

(iv) We know from Claim 6.4(iv) that the waiting time inside any copy of D never grows at a higher rate than 2. Because of property (iv) for B^{k-1} the same is true inside B^{k-1}. Together with (iii) this shows that (iv) holds for B^k as well.

(v) As we have already shown that the first queue on any path p^k_{3m+n} forms inside D_{m+1} we can deduce from Claim 6.4(viii) that this does not happen before time $\theta+m\cdot 3^{K-k+1}+3^{K-k}+1=\theta+(3m+1)\cdot 3^{K-k}\geq\theta+(3m+n-2)\cdot 3^{K-k}$. If $k=2$, this is also the last queue and by Claim 6.4(viii) it is empty by time $\theta+m\cdot 3^{K-2+1}+4.5\cdot 3^{K-2+1}+1=\theta+(3m+5)\cdot 3^{K-2}+8.5\cdot 3^{K-2}\leq\theta+(3m+n+4)\cdot 3^{K-2}+8.5\cdot 3^{K-2}+1$. If, on the other hand, we have $k\geq 3$, then the last queue on p^k_{3m+n} is in B^{k-1} and, by induction, is empty by time

$$
\begin{aligned}
&\theta'+(m+1+4)\cdot 3^{K-k+1}+(k-3)(3^K+1)+8.5\cdot 3^{K-2}+1\\
&\quad=\theta+3^K-10\cdot 3^{K-k}+1+(3m+15)\cdot 3^{K-k}+(k-3)(3^K+1)\\
&\qquad+8.5\cdot 3^{K-2}+1\\
&\quad=\theta+(3m+1+4)\cdot 3^{K-k}+(k-2)(3^K+1)+8.5\cdot 3^{K-k}+1\\
&\quad\leq\theta+(3m+n+4)\cdot 3^{K-k}+(k-2)(3^K+1)+8.5\cdot 3^{K-k}+1.
\end{aligned}
$$

This shows property (v)

(vi) For $\ell=1$ the queue on path p^k_{3m+n} in D_{m+1} is upper bounded by $7\cdot 3^{K-k}$ during $[\theta+m\cdot 3^{K-k+1},\theta+m\cdot 3^{K-k+1}+4.5\cdot 3^{K-k+1}]$ according to Claim 6.4(v). Since we certainly have $4.5\cdot 3^{K-k+1}\geq 5\cdot 3^{K-k}-0.5$, this interval, in particular, includes $[\theta+m\cdot 3^{K-k+1}+5\cdot 3^{K-k}-1,\theta+m\cdot 3^{K-k+1}+5\cdot 3^{K-k}-0.5]$. Thus, the desired lower bound on the waiting time along p^k_{3m+n} holds, provided that there are no other queues on later parts of the path during this time. If $k=2$, this is immediately clear as there are never any queues in B^1. Otherwise this follows from our observation that there are no queues on p^{k-1}_{m+1} before $\theta'+(m-1)\cdot 3^{K-k+1}+1$ and we have

$$
\begin{aligned}
&\theta'+(m-1)\cdot 3^{K-k+1}+1\\
&\quad=\theta+3^K+1-10\cdot 3^{K-k}+(m-1)\cdot 3^{K-k+1}+1\\
&\quad=\theta+m\cdot 3^{K-k+1}+(3^k-10-3)\cdot 3^{K-k}+2\\
&\quad\overset{(*)}{\geq}\theta+m\cdot 3^{K-k+1}+5\cdot 3^{K-k}-0.5,
\end{aligned}
$$

where we use $k\geq 3$ at $(*)$.

For $\ell \geq 2$ we have

$$\begin{aligned}
&\theta + (\ell - 1)(3^K + 1) + m \cdot 3^{K-k+\ell} + 5 \cdot 3^{K-k} - 1 \\
&\quad \geq \theta + m \cdot 3^{K-k+\ell} + 3^K + 1 + 5 \cdot 3^{K-k} - 1 \\
&\quad = \theta + m \cdot 3^{K-k+1} + (3^k + 5) \cdot 3^{K-k} \\
&\quad \overset{(\#)}{\geq} \theta + m \cdot 3^{K-k+1} + 13.5 \cdot 3^{K-k} + 1 \\
&\quad \geq \theta + m \cdot 3^{K-k+1} + 4.5 \cdot 3^{K-k+1} + 1,
\end{aligned}$$

where we use $k > \ell \geq 2$ at (#). Hence, the only queues on path $p^k_{3^\ell m+n}$ during the relevant intervals are in B^{k-1}, i.e. on the subpath $p^{k-1}_{m3^{\ell-1}+n'}$. By induction the total waiting time on this path is upper bounded by $7 \cdot 3^{K-(k-1)+(\ell-1)-1}$ during $[\theta' + (\ell - 2) \cdot (3^K + 1) + m \cdot 3^{K-k+\ell} + 5 \cdot 3^{K-k+1} - 1, \theta' + (\ell - 2) \cdot (3^K + 1) + m \cdot 3^{K-k+\ell} + 5 \cdot 3^{K-k+1} - 0.5]$. Since

$$\begin{aligned}
&\theta' + (\ell - 2) \cdot (3^K + 1) + m \cdot 3^{K-k+\ell} + 5 \cdot 3^{K-k+1} \\
&\quad = \theta + 3^K + 1 - 10 \cdot 3^{K-k} + (\ell - 2) \cdot (3^K + 1) + m \cdot 3^{K-k+\ell} \\
&\qquad + 5 \cdot 3^{K-k+1} \\
&\quad = \theta + (\ell - 1)(3^K + 1) + m \cdot 3^{K-k+\ell} + 5 \cdot 3^{K-k},
\end{aligned}$$

this shows that property (vi) holds for B^k.

(vii) This can be proven in essentially the same way as the previous bound—only using Claim 6.4(vi) instead of Claim 6.4(v) for the upper bound in D. For $\ell = 1$ the queue on path p^k_{3m+n} in D_{m+1} is upper bounded by $3^{K-k} - 0.5$ during $[\theta + m \cdot 3^{K-k+1}, \theta + m \cdot 3^{K-k+1} + 2 \cdot 3^{K-k} + 0.5]$ according to Claim 6.4(vi). Since we have

$$\begin{aligned}
(m - 1) \cdot 3^{K-k+1} + 5 \cdot 3^{K-k} &= m \cdot 3^{K-k+1} - 3 \cdot 3^{K-k} + 5 \cdot 3^{K-k} \\
&= m \cdot 3^{K-k+1} + 2 \cdot 3^{K-k},
\end{aligned}$$

this interval, in particular, includes $[\theta + (m-1) \cdot 3^{K-k+1} + 5 \cdot 3^{K-k}, \theta + (m - 1) \cdot 3^{K-k+1} + 5 \cdot 3^{K-k} + 0.5]$. Thus, the desired lower bound on the waiting time along p^k_{3m+n} holds, provided that there are no other queues on later parts of the path during this time. If $k = 2$, this is immediately clear as there are never any queues in B^1. Otherwise this follows from our observation that there are no queues on p^{k-1}_{m+1} before $\theta' + (m - 1) \cdot 3^{K-k+1} + 1$ and we have

$$\begin{aligned}
&\theta' + (m-1)\cdot 3^{K-k+1} + 1\\
&\quad = \theta + 3^K + 1 - 10\cdot 3^{K-k} + (m-1)\cdot 3^{K-k+1} + 1\\
&\quad = \theta + (m-1)\cdot 3^{K-k+1} + (3^k - 10)\cdot 3^{K-k} + 2\\
&\quad \overset{(*)}{\geq} \theta + (m-1)\cdot 3^{K-k+1} + 5\cdot 3^{K-k} + 0.5,
\end{aligned}$$

where we use $k \geq 3$ at $(*)$.
For $\ell \geq 2$ we have

$$\begin{aligned}
&\theta + (\ell-1)(3^K+1) + (m-1)3^{K-k+\ell} + 5\cdot 3^{K-k}\\
&\quad \geq \theta + m\cdot 3^{K-k+\ell} + 3^K + 1 - 3^{K-k+\ell} + 5\cdot 3^{K-k}\\
&\quad \geq \theta + m\cdot 3^{K-k+1} + (3^k - 3^\ell + 5)\cdot 3^{K-k} + 1\\
&\quad \overset{(\#)}{\geq} \theta + m\cdot 3^{K-k+1} + 23\cdot 3^{K-k} + 1\\
&\quad \geq \theta + m\cdot 3^{K-k+1} + 4.5\cdot 3^{K-k+1} + 1,
\end{aligned}$$

where we use $k > \ell \geq 2$ at (#). Hence, the only queues on path $p^k_{3^\ell m+n}$ during the relevant intervals are in B^{k-1}, i.e. on the subpath $p^{k-1}_{m3^{\ell-1}+n'}$. By induction the total waiting time on this path is upper bounded by $3^{K-(k-1)+(\ell-1)-1} - 0.5$ during $[\theta' + (\ell-2)\cdot(3^K+1) + (m-1)\cdot 3^{K-k+\ell} + 5\cdot 3^{K-k+1}, \theta' + (\ell-2)\cdot(3^K+1) + (m-1)\cdot 3^{K-k+\ell} + 5\cdot 3^{K-k+1} + 0.5]$. Since

$$\begin{aligned}
&\theta' + (\ell-2)\cdot(3^K+1) + (m-1)\cdot 3^{K-k+\ell} + 5\cdot 3^{K-k+1}\\
&\quad = \theta + 3^K + 1 - 10\cdot 3^{K-k} + (\ell-2)\cdot(3^K+1) + (m-1)\cdot 3^{K-k+\ell}\\
&\qquad + 5\cdot 3^{K-k+1}\\
&\quad = \theta + (\ell-1)(3^K+1) + (m-1)\cdot 3^{K-k+\ell} + 5\cdot 3^{K-k},
\end{aligned}$$

this shows that property (vii) holds for B^k.

(viii) For $\ell = 1$ the queue on path p^k_{3m+n} in D_{m+1} is lower bounded by $4\cdot 3^{K-k} - 0.5$ during $[\theta + m\cdot 3^{K-k+1} + 5\cdot 3^{K-k}, \theta + m\cdot 3^{K-k+1} + 5\cdot 3^{K-k} + 0.5]$ according to Claim 6.4(vii).
For $\ell \geq 2$ we can use property (viii) for B^{k-1} to get a lower bound on the waiting time on $p^{k-1}_{3^{\ell-1}m+n'}$ of $4\cdot 3^{K-(k-1)+(\ell-1)+1} - 0.5$ during $[\theta' + (\ell-2)(3^K+1) + m\cdot 3^{K-k+\ell} + 5\cdot 3^{K-k+1}, \theta' + (\ell-2)(3^K+1) + m\cdot 3^{K-k+\ell} + 5\cdot 3^{K-k+1} + 0.5]$. Since

$$\begin{aligned}
&\theta' + (\ell - 2)(3^K + 1) + m \cdot 3^{K-k+\ell} + 5 \cdot 3^{K-k+1} \\
&\quad = \theta + 3^K + 1 - 10 \cdot 3^{K-k} + (\ell - 2)(3^K + 1) + m \cdot 3^{K-k+\ell} \\
&\qquad + 5 \cdot 3^{K-k+1} \\
&\quad = \theta + (\ell - 1)(3^K + 1) + m \cdot 3^{K-k+\ell} + 5 \cdot 3^{K-k}
\end{aligned}$$

this shows property (viii) for B^k. ■

The Cycling Gadget: Next, we construct the cycling gadget C. The basic building block of this gadget is the **injection gadget** I (Fig. 6.10) consisting of three node v, w and z and two edges: vw with free flow travel time 1 and capacity 3 and wz with free flow time 1 and capacity 1. This gadget will be embedded into a larger network in such a way that there are incoming and outgoing edges of equal capacity at node v and one more outgoing edge at node z leading towards the sink via some path p. Furthermore, one of the outgoing edges from v (which we will call e') has free flow travel time 1 and leads to some node v' from which a path p' leads towards the sink t. Finally, the free flow travel time along the paths p and p' is such that vw, wz, p is the shortest v, t-path and has the same (physical) length as p'. Every path starting with another edge leaving v has strictly longer free flow travel time.

Claim 6.6. *Gadget I satisfies the following property:*

(i) The sum of all free flow travel times of all edges in I is 2

Now assume that I is correctly embedded into a larger network. Let f be any IDE in the larger network, $\theta \in \mathbb{R}_{\geq 0}$ some time and $\bar{\beta} \geq \beta > 9$ and $0 \leq \alpha \leq \frac{1}{2} - \frac{3}{\beta-3}$ positive real numbers such that

- *gadget I and edge e' carry no flow at time θ,*
- *paths p and p' have no queues during $[\theta, \theta + 6.5]$ and*
- *the only inflow into node v during this time interval is over edge e and satisfies*

$$\beta \cdot \mathbb{1}_{[\theta+\alpha,\theta+0.5]} \leq f_e^- \leq \bar{\beta} \cdot \mathbb{1}_{[\theta,\theta+0.5]}.$$

Then the flow splits at v between e' and vw in such a way that during $[\theta, \theta + 6.5]$ the outflow from gadget I satisfies

(ii) $(\beta - 3) \cdot \mathbb{1}_{[\theta+\alpha+\frac{3}{\beta-3},\theta+0.5]} \leq f_{e'}^+ \leq \bar{\beta} \cdot \mathbb{1}_{[\theta,\theta+0.5]}$ *and*

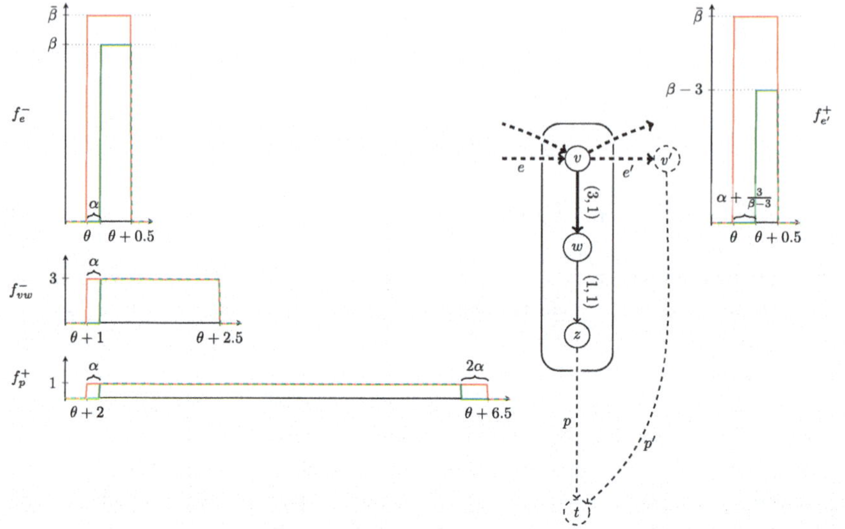

Fig. 6.10 The injection gadget I. Dashed edges are not part of the gadget and only indicate its correct embedding into a larger network. The graphs depict the upper and lower bounds on the flow from Claim 6.6: The graphs on the left show (from top to bottom) the bounds for the inflow into node v, the inflow into node w and the inflow into path p. Similarly to Fig. 6.7 the graphs are position horizontally in such a way that the time axis of the middle graph is shifted by -1 and the one of the bottom graph by -2. The graph on the right shows the bounds for the inflow into edge e'

(iii) $\mathbb{1}_{[\theta+2+\alpha,\theta+6.5-2\alpha]} \leq f_p^+ \leq \mathbb{1}_{[\theta+2,\theta+6.5]}$ *(where we use* f_p^+ *to denote the inflow into the first edge of path* p*).*

(iv) Moreover, the gadget is empty by time $\theta + 6.5$.

Proof. Property (i) is immediately clear from the construction as gadget I contains only two edges, both with free flow travel time 1.

To show the bounds on the outflow rates (i.e. (ii) to (iv)) we start with the case that the inflow into node v (i.e. f_e^-) exactly matches the lower bound. In this case the first particles arrive at node v at time $\theta+\alpha$. At this time, edge vw is the only active edge starting at v (as the path vw, wz, p is at least one shorter than any other v,t-path and we have no queues on p). Thus, at first all flow arriving at v enters the edge vw and a queue starts to grow at a rate of $\beta - 3$. At time $\theta + \alpha + \frac{3}{\beta-3} \leq \theta + 0.5$ the queue on edge vw reaches a length of 3 and, hence, edge e' becomes active as well (as the free flow travel time along e', p' is exactly one more than along vw, wz, p

and there is also no queue on p'). From here on the flow enters edge vw at a rate of 3 (to keep the queue length constant) and edge e' at a rate of $\beta - 3$. This already shows $f_{e'}^{+} = (\beta - 3) \cdot \mathbb{1}_{[\theta+\alpha+\frac{3}{\beta-3}, \theta+0.5]}$ on the relevant interval. Furthermore, the outflow of edge vw is described by $f_{vw}^{+} = 3 \cdot \mathbb{1}_{[\theta+1+\alpha, \theta+2.5]}$ (note that at time $\theta + 0.5$ the inflow into edge vw stops with a queue of length 3). From this, it immediately follows that we have $f_p^{+} = f_{wz}^{-} = \mathbb{1}_{[\theta+2+\alpha, \theta+6.5-2\alpha]}$.

Choosing $\alpha = 0$ and $\beta = \bar{\beta}$ now immediately implies that the outflow rates are upper bounded by the stated upper bounds if the inflow rate into node v matches the given upper bound. The bounds then also hold for all other inflow rates (obeying the stated bounds) due to the monotonicity of the edge flow dynamics (cf. Corollary 3.23). ■

Additionally, we need another gadget of very similar form which we will call **redirect gadget** R (Fig. 6.11). This gadget consists of two nodes v and z connected by an edge of free flow travel time and capacity 1. The gadget will be embedded into a larger network in such a way that there are incoming edges and one outgoing edge of equal capacity at node v and one more outgoing edge from node z leading towards the sink via a unique path p. Furthermore, the outgoing edge from v (which we will denote by e') has free flow time 1 and leads to some node v' from which there exists a path p' towards the sink. This path has the exact same free flow travel time as vw, p.

Claim 6.7. *Gadget R satisfies the following structural property:*

(i) The sum of all free flow travel times of all edges in R is 1

Now assume that R is correctly embedded into a larger network. Let f be some IDE in the larger network, $\theta \in \mathbb{R}_{\geq 0}$ some time and $\bar{\beta} \geq \beta \geq 3$, $\gamma \geq 0$ and $0 \leq \alpha \leq 0.5 - \frac{1+\gamma}{\beta-1}$ positive real numbers such that

- *gadget R and edge e' carry no flow at some time θ,*
- *the only inflow into the gadget during $[\theta, \theta+0.5]$ arrives over one edge $e \in \delta^{-}(v)$ and satisfies*

$$\beta \cdot \mathbb{1}_{[\theta+\alpha, \theta+0.5]} \leq f_e^{-} \leq \bar{\beta} \cdot \mathbb{1}_{[\theta, \theta+0.5]},$$

- *the waiting time along path p' is at most γ during all times in $[\theta, \theta + 0.5]$ and*
- *the waiting time along any v', t-path not containing v never grows at a rate of more than 2 during this interval.*

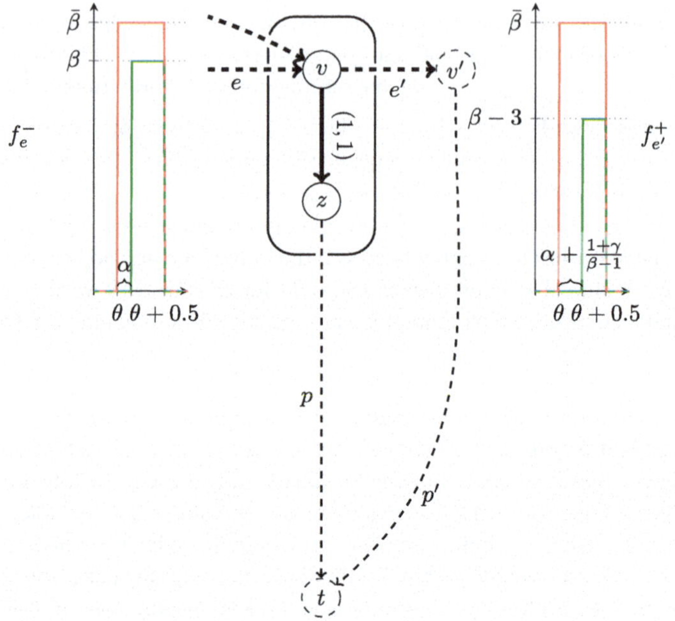

Fig. 6.11 The redirect gadget R. Dashed edges are not part of the gadget and only indicate how to embedded this gadget into a larger network. The graphs depict the upper and lower bounds on the flow from Claim 6.7 for the outflow rate from edge e (left) and the inflow rate into edge e' (right)

Then the flow splits at v in such a way between vz and e' that the inflow into edge e' satisfies

(ii) $(\beta - 3) \cdot \mathbb{1}_{[\theta+\alpha+\frac{1+\gamma}{\beta-1},\theta+0.5]} \leq f_{e'}^+ \leq \bar{\beta} \cdot \mathbb{1}_{[\theta,\theta+0.5]}$.

Proof. Property (i) follows directly from the construction. The upper bound in (ii) follows by flow conservation at node v. Due to the monotonicity of the edge flow dynamics it, therefore, suffices to show that the lower bound holds if the outflow from edge e exactly matches the given lower bound. So, assume that this is the case. Then flow starts to arrive at node v at a rate of β at time $\theta + \alpha$. At first all this flow enters edge vz and a queue starts to grow there at a rate of $\beta - 1$. This continues until edge e' becomes active. Since the free flow travel time along the path e', p' is exactly one more than along vz, p and the total waiting time along the former path is at most γ, this happens at the latest at time $\theta + \alpha + \frac{\gamma+1}{\beta-1} \leq \theta + 0.5$ (at which point the queue on edge vz would have reached a length of $\gamma + 1$). Since no queue ever forms on edge e' and the waiting time along any v', t-path grows at most at a rate of 2, the queue on edge vz can grow at most at this rate as well after this time. Thus, flow enters edge vz at a rate of at most 3 after time $\theta + \alpha + \frac{\gamma+1}{\beta-1}$ and, consequently, into edge e' at a rate of at least $\beta - 3$. Thus, the lower bound from (ii) holds as well. ■

From this we construct the **cycling gadget** C as follows (cf. Fig. 6.12): We take 3^K copies of the injection gadget I and one copy of the redirect gadget R. We rename the input nodes of the injection gadgets such that the input node of the i-th injection gadget I_i is called v_i and refer to the input node of the redirect gadget both by v_0 and v_{3^K+1}. We then connect them as follows: For $\ell \in [K] - 1$ and $m \in [3^{K-\ell}] - 1$ we add and edge of free flow travel time 3^ℓ and capacity $2U$ from $v_{3^\ell m+1}$ to $v_{3^\ell(m+1)+1}$. Additionally, there is one edge v_0v_1 of free flow travel time 1 and capacity $2U$. Here, we define $U := \frac{1}{2}(6L \cdot 3^{K+1} + 3K + L)$.

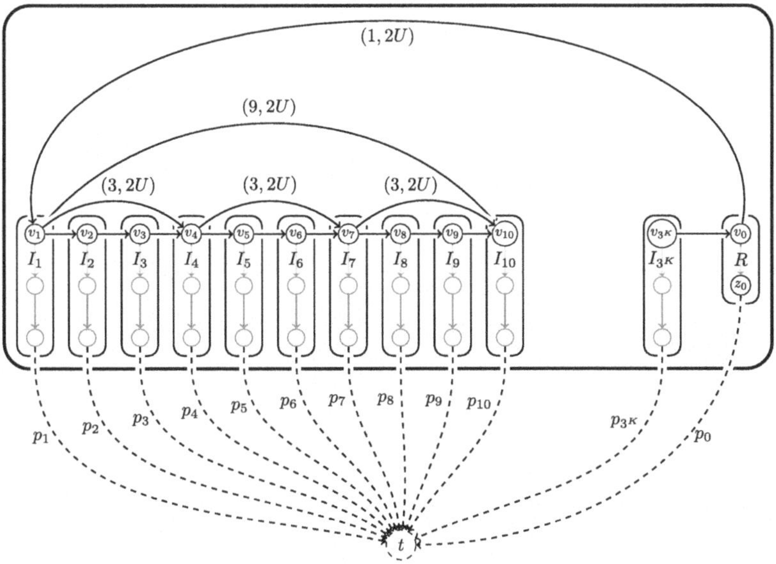

Fig. 6.12 The cycling gadget C

Finally, we get our complete network $\mathcal{N}_{K,L}$ by combining one cycling gadget C and one blocking gadget B^K in the following way (cf. Fig. 6.13): For every $i \in [3^K]$ we connect the output node of the i-th injection gadget in C with the i-th input node of gadget B^K via a direct edge e_i with free flow travel time $3^K - 6$ and capacity 1. We add an additional node t and add one edge of free flow travel time 1

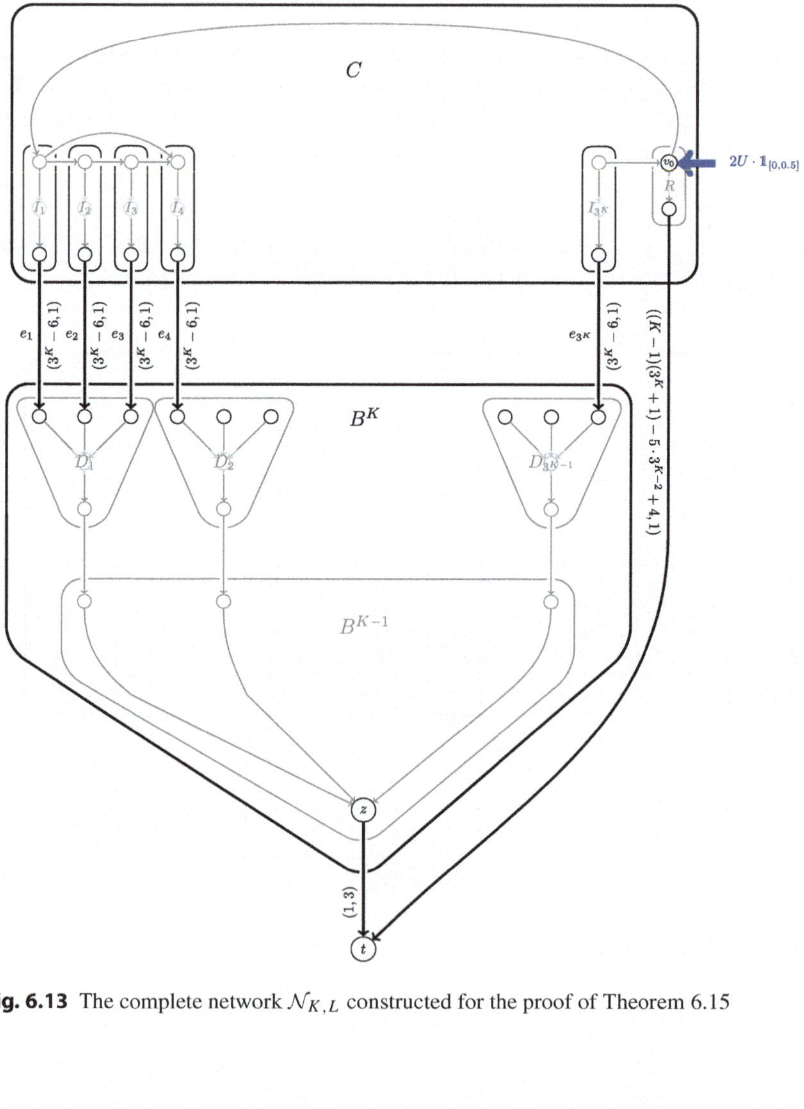

Fig. 6.13 The complete network $\mathcal{N}_{K,L}$ constructed for the proof of Theorem 6.15

and capacity 3 from the output node of B^K to t and one edge of free flow travel time $(K-1)(3^K+1)-5\cdot 3^{K-2}+4$ and capacity 1 from the output node of gadget R in C to t. Finally, we make t the only sink node in $\mathcal{N}_{K,L}$ and v_0 the only node with a non-zero network inflow rate which we define as $2U \cdot \mathbb{1}_{[0,0.5]}$.

Claim 6.8. *The network $\mathcal{N}_{K,L}$ then satisfies the following structural properties:*

(i) For any $i \in [3^K]$ there exists a unique path p_i connecting the output node of the i-th copy of gadget I to the sink node t. The free flow travel time along this path is $(K-1)(3^K+1)-5\cdot 3^{K-2}+3$.
(ii) Gadgets B^K, R and all copies of gadget I are embedded correctly in the network.
(iii) The sum of all free flow travel times in $\mathcal{N}_{K,L}$ is

$$\tau(\mathcal{N}_{K,L}) = \tfrac{3}{2}\cdot 3^{2K} + (2K-8)\cdot 3^K + (55-30K)\cdot 3^{K-2} + K + \tfrac{21}{2}.$$

Now, let (f,θ) be a partial IDE in $\mathcal{N}_{K,L}$ and $\alpha, \beta \in \mathbb{R}_{\geq 0}$ two constants such that

- *the constants satisfy $\beta \geq (5L+1)3^{K+1}+3K$ and $\alpha \leq 0.5 - \frac{1}{2L}$,*
- *at time θ the only edge in C carrying flow is one edge $e \in \delta^-(v_0)$,*
- *the inflow into node v_0 (via this edge e) during the interval $[\theta, \theta+1]$ satisfies*

$$\beta \cdot \mathbb{1}_{[\theta+\alpha,\theta+0.5]} \leq f_e^- \leq 2U \cdot \mathbb{1}_{[\theta,\theta+0.5]},$$

- *edge e is empty by time $\theta+1$ and*
- *in any extension of (f,θ) as a Vickrey flow any flow currently on the (unique) path from the i-th injection gadget to t for some $i \in [3^K]$ will not be part of any queues after time $\theta+(i-1)$.*

Then we can extend (f,θ) to a partial IDE up to time $\theta + K\cdot(3^K+1)$ satisfying the following properties:

(iv) At time $\theta+K\cdot(3^K+1)$ the only edge in C carrying flow is one edge $e \in \delta^-(v_0)$. Moreover, in any further extension of $(f,\theta+K\cdot(3^K+1))$ as a Vickrey flow any flow currently on the (unique) path from the i-th injection gadget to t for some $i \in [3^K]$ will not be part of any queues after time $\theta+K\cdot(3^K+1)+(i-1)$. Finally, edge e is empty by time $\theta+K\cdot(3^K+1)+1$.

(v) The inflow into node v_0 (via edge e) satisfies

$$f_e^- \geq (\beta - 3(3^K + K)) \cdot \mathbb{1}_{[\theta+K\cdot(3^K+1)+\alpha+\frac{1}{2L}-\varepsilon,\theta+K\cdot(3^K+1)+0.5]} \text{ and}$$
$$f_e^- \leq 2U \cdot \mathbb{1}_{[\theta+K\cdot(3^K+1),\theta+K\cdot(3^K+1)+0.5]}$$

during the interval $[\theta + K \cdot (3^K + 1), \theta + K \cdot (3^K + 1) + 1]$ *for some* $\varepsilon > 0$.

Proof. By construction the output node of any injection gadget I_i has exactly one outgoing edge e_i which connects it to the i-th input node of B^K. By Claim 6.5(i) there is then a unique path p_i^K to the output node of B^K from where there is a single edge leading directly to the sink. The free flow travel time of this path (which we will denote by p_i) is then

$$3^K - 6 + (K-2)(3^K+1) - 5\cdot(3^{K-2}-1) + 4 + 1 = (K-1)(3^K+1) - 5\cdot 3^{K-2} + 3.$$

This shows property (i).

With this it follows directly from the construction of $\mathcal{N}_{K,L}$ that all used gadget are embedded correctly, i.e. property (ii) is satisfied as well.

To get the sum of all free flow travel times in $\mathcal{N}_{K,L}$ we first calculate this sum in gadget C (using Claim 6.6(i) and Claim 6.7(i)):

$$3^K \cdot 2 + 1 + 1 + \sum_{\ell=0}^{K-1} 3^{K-\ell} \cdot 3^\ell = (2 + K) \cdot 3^K + 2.$$

Adding the sum of free flow travel times in B^K from Claim 6.5(ii) and the free flow travel times of the additional edges outside the two gadget C and B^K we get

$$\begin{aligned}\tau(\mathcal{N}_{K,L}) &= (2 + K) \cdot 3^K + 2 + \tfrac{1}{2}(3^K + 3)(3^K - 9) - 10 \cdot (K - 2) \cdot 3^{K-1} + 18 \\ &\quad + 3^K \cdot (3^K - 6) + (K - 1)(3^K + 1) - 5 \cdot 3^{K-2} + 4 + 1 \\ &= \tfrac{3}{2} \cdot 3^{2K} + (2K - 8) \cdot 3^K + (55 - 30K) \cdot 3^{K-2} + K + \tfrac{21}{2}.\end{aligned}$$

Thus, property (iii) holds as well.

Now, let (f, θ) be a partial IDE satisfying the assumptions in the claim. We construct the extension in K phases. First, there is a *charging phase* during which the main flow volume traverses once through the cycling gadget C sending a small amount of flow towards the blocking gadget B^K at every injection gadget I. This is then followed by $K - 1$ *blocking phases* during which the flow inside gadget B^K creates a sequence of increasing and decreasing waiting times on the paths p_i towards the sink which let the main amount of flow traverse the cycling gadget without loosing any additional flow (except for at the redirection gadget R). We now describe the extension in more detail and show that it is an IDE satisfying the required properties:

Charging phase: Here, we can use any extension of (f, θ) to a partial IDE up to time $\theta + 3^K + 1$. Any such extension then has the following form: Since the assumptions on (f, θ) guarantee, in particular, that there is no waiting time on p_1 during $[\theta, \theta + 1]$, we can apply Claim 6.7(ii)[1] (with $v' = v_1$) to deduce that the flow will arrive at node v_1 at a rate bounded by $(\beta - 3) \cdot \mathbb{1}_{[\theta+1+\alpha+\frac{1}{\beta-1}, \theta+1.5]} \leq f^+_{e'} \leq 2U \cdot \mathbb{1}_{[\theta,\theta+0.5]}$. Now, iteratively applying Claim 6.6 (together with our assumption guaranteeing that there is no waiting time on path p_i during $[\theta + (i-1), \theta + (i-1) + 6.5]$) shows that the flow will then travel from one injection gadget to the next, eventually arriving back at node v_0 at a rate bounded by $\beta' \cdot \mathbb{1}_{[\theta'+\alpha',\theta'+0.5]} \leq f^+_{e'} \leq 2U \cdot \mathbb{1}_{[\theta',\theta'+0.5]}$ where $\theta' = \theta + 3^K + 1$, $\alpha' := \alpha + \frac{1}{\beta-1} + \sum_{i=1}^{3^K} \frac{3}{\beta-3-3i}$ and $\beta' := \beta - 3(1+3^K)$. These bounds (with appropriate time shift) than also hold at every intermediate node v_i and, thus, Claim 6.6(iii) guarantees that the inflow into edge e_i connecting C and B^K is bounded by $\mathbb{1}_{[\theta+i+2+\alpha',\theta+i+6.5-2\alpha']} \leq f^+_{e_i} \leq \mathbb{1}_{[\theta+i+2,\theta+i+6.5]}$. This flow then arrives $3^K - 6$ time units later at the i-th input node of the blocking gadget B^K. Thus, the inflow into B^K satisfies the assumptions of Claim 6.5 with $\theta = \theta + 3 + 3^K - 6$ and $k = K$. Thus, during $[\theta + 3^K - 3, \theta + K \cdot (3^K + 1)]$ the flow inside B^K satisfies the properties (iv) to (viii) from Claim 6.5 (with $k = K$ and a time shift of $3^K - 3$), provided that there is no additional inflow into B^K. We will construct such an extension in the following blocking phases.

Blocking phases: We inductively construct $K - 1$ blocking phases of length $3^K + 1$ each: At the beginning of the first blocking phase at time $\theta' = \theta + 3^K + 1$ we know from the previous charging phase that flow arrives at node v_0 via edge $v_{3^K} v_0$ at a rate bounded by $\beta' \cdot \mathbb{1}_{[\theta'+\alpha',\theta'+0.5]} \leq f^+_{e'} \leq 2U \cdot \mathbb{1}_{[\theta',\theta'+0.5]}$. As

[1] Note that we have not checked the requirements on α and β yet. We will do that for all application of both Claims 6.6 and 6.7 all at once at the end of the proof.

the waiting time on path p_1 is bounded by $7 \cdot 3^0 = 7$ according to Claim 6.5(vi) and grows at a rate of at most 2 according to Claim 6.5(iv), we can apply Claim 6.7(ii) to ensure that in any IDE extension flow splits in gadget R in such a way as to arrive at node v_1 one time unit later at a rate bounded by $(\beta' - 3) \cdot \mathbb{1}_{[\theta'+1+\alpha'+\frac{8}{\beta'-1}, \theta'+1.5]} \leq f^+_{v_0 v_1} \leq 2U \cdot \mathbb{1}_{[\theta'+1, \theta'+1.5]}$. Since during

$$[\theta'+1, \theta'+1.5] = [\theta+3^K+2, \theta+3^K+2.5] = [\theta+3^K-3+5, \theta+3^K-3+5+0.5]$$

the waiting time on path p_1 is at least 3.5 (according to Claim 6.5(viii)) while the waiting time is at most 0.5 on path p_4 (according to Claim 6.5(vii)), sending all this flow along edge $v_1 v_4$ satisfies the IDE property. This pattern then continues, i.e. we can send all the flow along the path $v_1, v_4, v_7, \ldots, v_{3K+1} = v_0$ until it arrives back at v_0 at a rate bounded by

$$(\beta' - 3) \cdot \mathbb{1}_{[\theta'+3^K+1+\alpha'+\frac{8}{\beta'-1}, \theta'+3^K+1.5]} \leq f^+_{v_{3K-2} v_0} \leq 2U \cdot \mathbb{1}_{[\theta'+3^K+1, \theta'+3^K+1+0.5]}.$$

Now, for $\ell = 2, 3, \ldots, K-1$ we can assume (by induction) that flow arrives at node v_0 via some edge $e \in \delta^-(v_0)$ at a rate bounded by $\beta_\ell \cdot \mathbb{1}_{[\theta_\ell+\alpha_\ell, \theta_\ell+0.5]} \leq f^+_e \leq 2U \cdot \mathbb{1}_{[\theta_\ell, \theta_\ell+0.5]}$ where $\theta_\ell := \theta' + (\ell-1)(3^K+1)$, $\alpha_\ell := \alpha' + \sum_{j=2}^{\ell} \frac{1+7\cdot 3^{j-2}}{\beta'-3(j-2)-1}$ and $\beta_\ell := \beta' - 3(\ell-1)$ while the rest of gadget C as well as all connecting edges e_i are empty at time θ_ℓ and no additional flow has entered the blocking gadget after the flow send towards it during the blocking phase. Note that we have just shown this to be true for $\ell = 2$. Then, according to Claim 6.5(iv) and (vi) the waiting time on p_1 is at most $7 \cdot 3^{\ell-1}$ and grows at a rate of at most 2 during

$$\begin{aligned}
[\theta_\ell, \theta_\ell + 0.5] &= [\theta' + (\ell-1)(3^K+1), \theta' + (\ell-1)(3^K+1) + 0.5] \\
&= [\theta + 3^K + 1 + (\ell-1)(3^K+1), \theta + 3^K + 1 + (\ell-1)(3^K+1) + 0.5] \\
&= [\theta + 3^K - 3 + (\ell-1)(3^K+1) + 4, \theta + 3^K - 3 + (\ell-1)(3^K+1) + 4.5].
\end{aligned}$$

Thus, we can apply Claim 6.7(ii) to ensure that in any IDE extension flow splits in gadget R in such a way as to arrive at node v_1 one time unit later at a rate bounded by $\beta_{\ell+1} \cdot \mathbb{1}_{[\theta_\ell+1+\alpha_{\ell+1}, \theta_\ell+1.5]} \leq f^+_e \leq 2U \cdot \mathbb{1}_{[\theta_\ell+1, \theta_\ell+1.5]}$. Here, we use $\alpha_\ell + \frac{1+7\cdot 3^{\ell-1}}{\beta_\ell - 1} = \alpha_{\ell+1}$. Now, by the time this flow arrives at node v_1 the waiting time on path p_1 is at least $4 \cdot 3^{\ell-1} - 0.5$ (according to Claim 6.5(viii)) while the waiting time is at most $3^{\ell-1} - 0.5$ on path $p_{3\ell+1}$ (according to Claim 6.5(vii)).

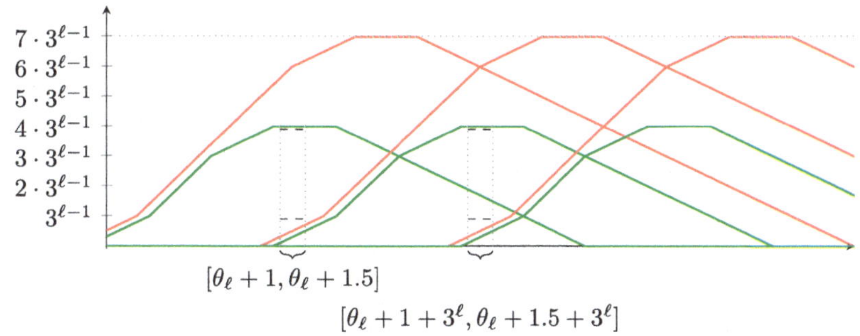

Fig. 6.14 The upper and lower bounds on the waiting times along consecutive paths p_1, $p_{3^\ell+1}$ and so on during the ℓ-th blocking phase

Thus, sending all this flow along edge $v_1 v_{3^\ell+1}$ satisfies the IDE property. This pattern then continues (see Fig. 6.14), i.e. we can send all the flow along the path $v_1, v_{3^\ell+1}, v_{2\cdot 3^\ell+1}, \dots, v_{3^K+1} = v_0$ until it arrives back at v_0 at a rate bounded by $\beta_{\ell+1} \cdot \mathbb{1}_{[\theta_\ell+3^K+1+\alpha_{\ell+1}, \theta_\ell+3^K+1.5]} \leq f_e^+ \leq 2U \cdot \mathbb{1}_{[\theta_\ell+3^K+1, \theta_\ell+3^K+1.5]}$ and, therefore, satisfying our assumption for the next blocking phase $\ell + 1$.

After the final blocking phase (i.e. at time $\theta_K = \theta + K\cdot(3^K+1)$) flow then arrives at node v_0 at a rate bounded by $\beta_K \cdot \mathbb{1}_{[\theta_K+\alpha_K, \theta_K+0.5]} \leq f_e^+ \leq 2U \cdot \mathbb{1}_{[\theta_K, \theta_K+0.5]}$ for some edge $e \in \delta^-(v)$. Moreover, all other edges in C as well as all connecting edges e_i carry no flow at that time. Finally, according to Claim 6.5(v) the flow currently on any path p_i will not be part of any queues after time

$$\begin{aligned}
&\theta + 3^K - 3 + (i+4) + (K-2)(3^K+1) + 8.5 \cdot 3^{K-2} + 1 \\
&\quad = \theta + (3^K + 1) + (K-2)(3^K+1) + (9 \cdot 3^{K-2} + 1) + i - 0.5 \cdot 3^{K-2} \\
&\quad \leq \theta + K \cdot (3^K + 1) + (i - 1)
\end{aligned}$$

where we use $K \geq 3$ for the last inequality. This shows that our extension satisfies property (iv). To show that property (v) holds as well, we observe that

$$\beta_K = \beta' - 3(K-1) = \beta - 3(3^K+1) - 3(K-1) = \beta - 3(3^K + K)$$

and

$$\begin{aligned}
\alpha_K &= \alpha' + \sum_{j=2}^{K} \frac{1 + 7 \cdot 3^{j-2}}{\beta' - 3(j-2) - 1} \\
&= \alpha + \frac{1}{\beta - 1} + \sum_{i=1}^{3^K} \frac{3}{\beta - 3 - 3i} + \sum_{j=2}^{K} \frac{1 + 7 \cdot 3^{j-2}}{\beta - 3(3^K + 1) - 3(j-2) - 1} \\
&\leq \alpha + \frac{1}{\beta - 1} + \sum_{i=1}^{3^K} \frac{3}{\beta - 3 - 3^{K+1}} + \sum_{j=2}^{K} \frac{8 \cdot 3^{j-2}}{\beta - 3^{K+1} - 3K} \\
&= \alpha + \frac{1}{\beta - 1} + \frac{3^{K+1}}{\beta - 3 - 3^{K+1}} + \frac{4 \cdot (3^{K-1} - 1)}{\beta - 3^{K+1} - 3K} \\
&\leq \alpha + \frac{1 + 3^{K+1} + 4 \cdot 3^{K-1} - 4}{5L \cdot 3^{K+1}} = \alpha + \frac{2.5 \cdot 3^{K+1} - 9.5 \cdot 3^{K-1} - 3}{5L \cdot 3^{K+1}} \\
&\leq \alpha + \frac{1}{2L} - \frac{9.5}{45L} < \alpha + \frac{1}{2L}.
\end{aligned}$$

Finally, this then also allows us to show that at every time in the above proof where we applied Claim 6.6 or Claim 6.7, the required bounds on α and β for the flow arriving at the respective input node v_i were satisfied: Since those values are always lower bounded by α_K and β_K it suffices to show the requirements for those: Indeed, we have

$$\beta_K = \beta - 3(3^K + K) \geq (5L + 1)3^{K+1} + 3K - 3(3^K + K) = 5L \cdot 3^{K+1} \geq 9$$

as well as

$$\begin{aligned}
\alpha_K + \frac{3}{\beta_K - 3} &\leq \alpha + \frac{1}{2L} - \frac{9.5}{45L} + \frac{3}{5L \cdot 3^{K+1} - 3} \\
&\leq \alpha + \frac{1}{2L} + \frac{27 - 9.5(3^{K+1} - 1)}{45L(3^{K+1} - 1)} \leq \alpha + \frac{1}{2L} \leq 0.5
\end{aligned}$$

and, finally,

$$\begin{aligned}
\alpha_K + \frac{1 + \gamma}{\beta_K - 1} &\leq \alpha + \frac{1}{2L} - \frac{9.5}{45L} + \frac{1 + 7 \cdot 3^{K-1}}{5L \cdot 3^{K+1} - 1} \\
&\leq \alpha + \frac{1}{2L} + \frac{9 + 7 \cdot 3^{K+1} - 9.5 \cdot (3^{K+1} - 1)}{45L \cdot (3^{K+1} - 1)} \leq \alpha + \frac{1}{2L} \leq 0.5,
\end{aligned}$$

which concludes our proof. ■

We now construct an IDE with the desired makespan and total travel time as follows: We start with the zero-flow up to time 0 (0, 0) which clearly satisfies the assumptions on the given partial IDE in Claim 6.8 (except that the inflow into node v_0 does not arrive via some edge $e \in \delta^-(v_0)$ but instead as network inflow—however, this clearly does not change anything). We can, therefore, apply Claim 6.8 to get a partial IDE until time $K \cdot (3^K + 1)$ which again satisfies the assumptions of this claim on the given partial IDE. Thus, we can iteratively apply this claim until, after L extensions, we have a partial IDE $(f, LK \cdot (3^K + 1))$ which still has not terminated (as flow still arrives at least at a rate of $2U - 3L(3^K + K)) = 6L \cdot 3^{K+1} + 3K + L - 3L(3^K + K) = 5L \cdot 3^{K+1} + L > 0$ during at least the interval $[LK \cdot (3^K + 1) + \frac{L}{2L} - \varepsilon, LK \cdot (3^K + 1) + 0.5]$ for some $\varepsilon > 0$). Extending this flow to an IDE for all times (which we can always do by Theorem 4.15) gives us the desired IDE f. We have $\Psi(f) \geq LK(3^K + 1)$ and

$$\Xi(f) \overset{\text{Prop. 3.75}}{=} \int_0^{\Psi(f)} F^{\Delta}(\zeta)\mathrm{d}\zeta \geq \sum_{j=1}^{L} K(3^K + 1) \cdot (\tfrac{1}{2} - \tfrac{j}{2L}) \cdot (2U - 3j(3^K + K))$$

$$\geq K \cdot 3^K \sum_{j=1}^{L} \tfrac{L-j}{2L} \cdot 5L \cdot 3^{K+1} \geq K \cdot 3^{2K} \cdot L(L-1).$$

Finally, we get $\tau(\mathcal{N}_{K,L}) \in \mathcal{O}(3^{2K})$ from Claim 6.8(iii) and $U_{\mathcal{N}_{K,L}} \in \mathcal{O}(L3^K)$ from our definition of U and our subsequent choice of the network inflow rate. Moreover, v_0 is the only source node, the last network inflow happens before time 1 and, by Claim 6.8(i) there exists a v_0, t-path of length $(K - 1)(3^K + 1) - 5 \cdot 3^{K-2} + 6$. □

Unfolding the above network into an acyclic network now also provides us with a more interesting lower bound for acyclic networks (compared to the one from Example 6.14)

Corollary 6.16. *There exists a family of acyclic networks $\tilde{\mathcal{N}}_{K,L}$ and associated IDE $f^{K,L}$ indexed by two natural numbers K and L such that*

- *The network size is asymptomatically bounded by $3^{2K} + LK3^K$, i.e. $\tau(\tilde{\mathcal{N}}_{K,L}) \in \mathcal{O}(3^K(3^K + KL))$,*
- *the total flow volume is asymptomatically bounded by $L3^K$, i.e. $U_{\tilde{\mathcal{N}}_{K,L}} \in \mathcal{O}(L3^K)$,*
- *we have $\Psi(f^{K,L}) \geq LK(3^K + 1)$ and $\Xi(f^{K,L}) \geq KL(L-1) \cdot 3^{2K}$ and*
- *it holds that $\tau_{p_{\min}}(\tilde{\mathcal{N}}_{K,L}) \in \mathcal{O}(K3^K)$ as well as $\tau_{p_{\max}}(\tilde{\mathcal{N}}_{K,L}) \in \Omega(LK3^K)$.*

Here, we denote by $\tau_{p_{\min}}(\tilde{\mathcal{N}}_{K,L})$ and $\tau_{p_{\max}}(\tilde{\mathcal{N}}_{K,L})$ the free flow times along the physically shortest and longest source-sink-path in $\tilde{\mathcal{N}}_{K,L}$, respectively.

Proof. We construct $\tilde{\mathcal{N}}_{K,L}$ in a similar way as $\mathcal{N}_{K,L}$ in the proof of Theorem 6.15 (cf. Fig. 6.15). Only, this time, we use LK copies of gadget C and a single copy of gadget B^K. Furthermore, we add one sink node t and 3^K+1 nodes $w_0, w_1, \dots, w_{3^K}$. Then, instead of connecting the gadgets C directly to gadget B^K or the sink (as before) we connect the i-th output node of every copy of gadget C with an edge of free flow travel time and capacity 1 to node w_i. From there we add one edge to the i-th input node of B^K or directly to the sink (for $i = 0$). These edges all have a capacity of 1 and a free flow travel time of $3^K - 7$ and $(K-1)(3^K+1) - 5 \cdot 3^{K-2} + 3$, respectively. Finally, we remove the backwards edges v_0v_1 from every copy of gadget C and instead connect node v_0 of any gadget C with node v_1 of the next gadget C. Using the same inflow rate as in network $\mathcal{N}_{K,L}$, but only in the first copy of gadget C, the

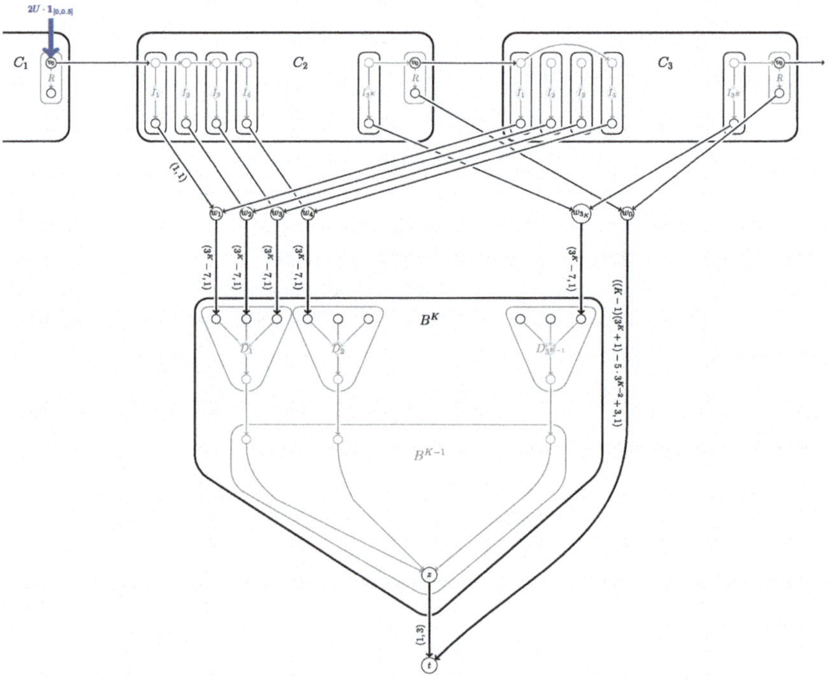

Fig. 6.15 Part of the network $\tilde{\mathcal{N}}_{K,L}$ constructed for the proof of Corollary 6.16

"same" flow as in network $\mathcal{N}_{K,L}$ is also an IDE in this new network and satisfies the same lower bounds on the three quality measures.

Note, that in any single copy of gadget C this flow then only uses one of the K parallel v_1, v_0-paths through this gadget. Thus, we can remove $K-1$ of those paths from every gadget without making the flow infeasible. After this change, the sum of free flow travel times in any copy of gadget C is in $\mathcal{O}(3^K)$. Thus, together with the bound from Theorem 6.15 we get $\tau(\tilde{\mathcal{N}}_{K,L}) \in \mathcal{O}(3^{2K} + LK3^K)$. □

Remark 6.17. Note that, after the deletion of the unused parallel path in the construction $\tilde{\mathcal{N}}_{K,L}$ the flow $f^{K,L}$ actually becomes the *unique* IDE in this network. Thus, the lower bounds from Corollary 6.16 hold not just for one but for all IDE in this network. A similar strengthening of Theorem 6.15 can be achieved (without increasing the asymptotic bound on the size of the network) by using K copies of the cycling gadget instead of just 1 and then connecting them in a similar way to the construction of $\tilde{\mathcal{N}}_{K,L}$.

6.2.2 Multi-Commodity Networks

In the last subsection we saw that in a single-commodity IDE a small amount of flow can keep a much larger amount of flow from making any progress towards the sink for quite a long time. However, eventually, such an IDE still must terminate as the part of the flow used for blocking the rest of the flow is necessarily closer to the sink and, thus, cannot be blocked itself (this is exactly the main insight that allowed us to prove termination of single-commodity flows in Subsection 6.1.2).

For multi-commodity IDE this need not be true anymore: If the blocking flow belongs to a different commodity than the cycling flow, it could itself be separated from its sink by that cycling flow. Thus, the cycling flow could, simultaneously, act as blocking flow for its blocking flow. Hence, conceptually, a two commodity network with two cycles inbetween two sink nodes could be able to trap an IDE flow forever (cf. Fig. 6.16). In fact, the instance used by Ismaili to show that GPS-agents in a packet routing game with current information setting may cycle forever ([Ism17, Theorem 8]) has almost exactly this form. However, despite the apparent similarity between this discrete and our continuous model, this result can not be directly transferred to our setting. The main difference here is that in Ismaili's model packets increase the current travel time on an edge (as perceived by other agents) even if they themselves do not actually experience any increased travel time (because the capacity of the edge they are currently traversing is not exceeded). This allows Ismaili to construct an instance wherein individual agents observe a constantly fluctuating current travel time (which allows them to travel around a cycle forever) even though in reality there never are any queues anywhere in the network.

In contrast, for our model we require actual queues to form (and disperse) in some repeated pattern in order to affect the behaviour of our flow particles. This, in turn, requires for the flow particles waiting in such a queue to be able to catch up again to the particles leaving the queue earlier in order to be able to build up the same queue again later. However, the general idea of constructing a network consisting of two parts—each of them containing flow of one commodity and separating the other flow from its sink—is exactly what we will use in the proof of the following theorem:

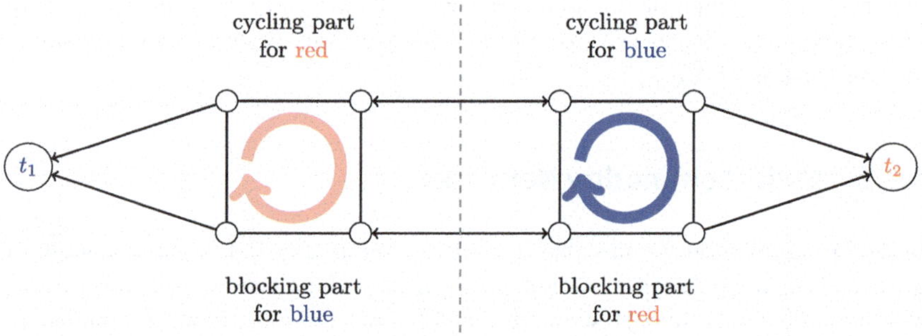

Fig. 6.16 The general structure of the network we use to construct a 2-commodity network in which IDE never terminate

Theorem 6.18. *There exists a two-commodity network with finitely lasting network inflow rates and all free flow times and capacities equal to* 1 *wherein no IDE ever terminates.*

Proof. We, once again, construct the network out of several smaller gadgets. Our main building block will be a **cycling gadget**[2] C_i^k (Fig. 6.17) consisting of two cycles vw, wy_1, y_1y_2, y_2v and $vw, wx_1, x_1x_2, x_2x_3, x_3v$ of length 4 and 5, respectively, which share one common edge vw. Additionally, there are three nodes z_1, z_2 and z_3 and edges wz_1m y_2z_2 and x_3y_3. The two parameters are the commodity $i \in I =$

[2] Note that, while this gadget serves a similar function to and bears the same name as the cycling gadget C from the proof of Theorem 6.15, its structure is quite different.

$\{1, 2\}$ and a time shift $k \in \mathbb{N}_0$. The only node with positive network inflow is node v with a network inflow rate of 2 for commodity i during the interval $[k, k+1]$.

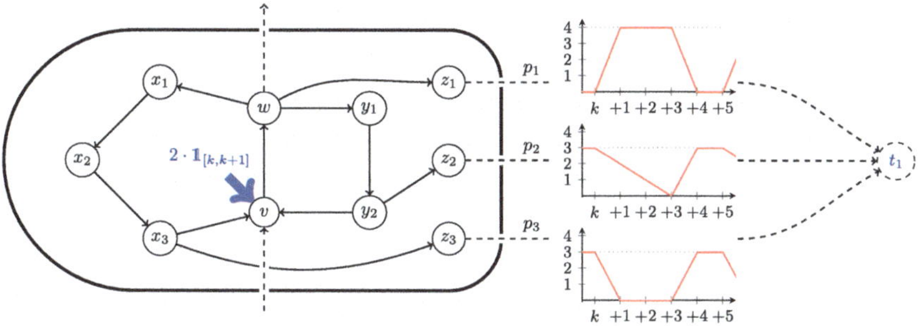

Fig. 6.17 The cycling gadget $C^k_{flowOneTextCol1}$. Dashed edges and nodes are not part of the gadget and only indicate its correct embedding into a larger network. The graphs on the paths p_1, p_2 and p_3 indicate the waiting times assumed to be on those paths in Claim 6.9

We will embed this gadget into a larger network in such a way that the only edge entering this gadget ends at node v and the only outgoing edges start at nodes w, z_1, z_2 and z_3. Furthermore, we have a physically shortest z_j, t_i-path p_j (where t_i will be the unique sink node of commodity i) for each $j \in [3]$ which share the same free flow travel time. Finally, any other path towards the sink t_i starting at either z_1, z_2 or z_3 has a free flow travel time of at least 4 more than the paths p_j and any other w, t_i-path has a free flow travel time of at least 5 more.

Claim 6.9. *Assume that gadget C^k_i is correctly embedded into a larger network $\mathcal{N}$ and f is an IDE in the larger network. Additionally, assume that f satisfies the following properties:*

- *No flow enters gadget C^k_i from outside the gadget.*
- *The waiting time on path p_1 is at least 4 during each interval $[5n+k+1, 5n+k+3]$ and 0 during each interval $[5n+k+4, 5n+k+5]$ for $n \in \mathbb{N}_0$ (e.g. as in the graph on path p_1 in Fig. 6.17).*
- *The waiting time on path p_2 is more than 1 during each interval $[5n+k+1, 5n+k+2)$, less than 1 during $(5n+k+2, 5n+k+3]$ and at least 3 during each interval $[5n+k+4, 5n+k+5]$ for $n \in \mathbb{N}_0$.*
- *The waiting time on path p_3 is 0 during each interval $[5n+k+1, 5n+k+3]$ and at least 3 during each interval $[5n+k+4, 5n+k+5]$ for $n \in \mathbb{N}_0$.*

Then the flow inside the gadget exhibits the pattern displayed in Fig. 6.18. In particular,

(i) No flow ever leaves the gadget.
(ii) The waiting time on edge vw increases linearly from 0 to 1 during each interval $[5n+k, 5n+k+1]$ and decreases linearly from 1 to 0 during $[5n+k+1, 5n+k+2]$ for $n \in \mathbb{N}_0$. Otherwise the waiting time on this edge is 0.
(iii) Moreover, if the assumed waiting time pattern holds only until some time $\xi \in \mathbb{R}_{\geq 0}$, then the waiting time pattern described in (ii) holds until time $\xi + 1$.

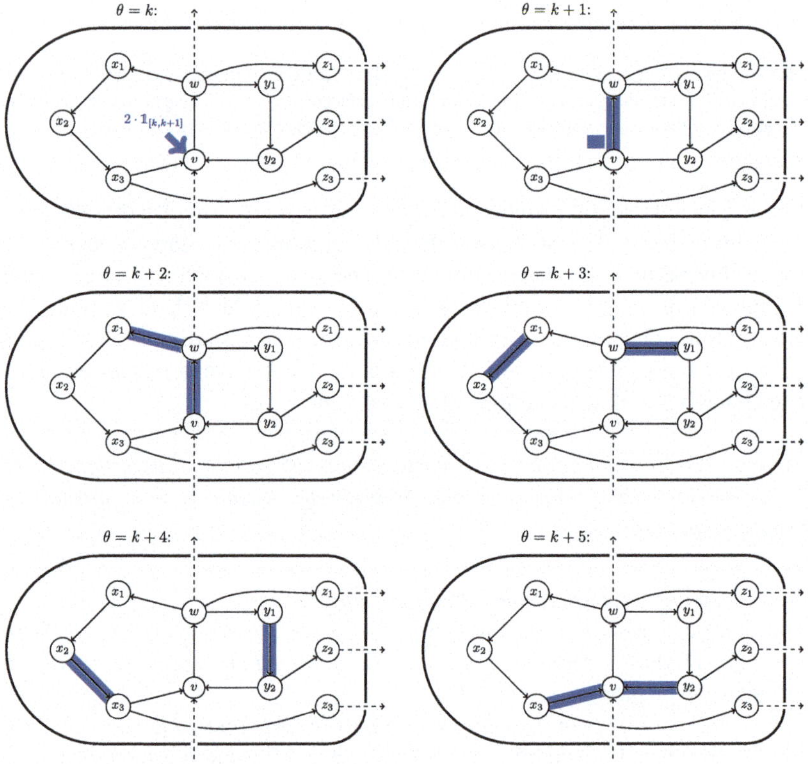

Fig. 6.18 The flow pattern inside gadget C_1^k as stated in Claim 6.9. Note that after time $\theta = k + 5$ this pattern repeats with a periodicity of 5

Proof. We show this claim by describing the flow evolution inside this gadget: Since no flow enters the gadget from outside, there is no flow inside the gadget before time $\theta = k$. After that, flow of commodity i enters the network at node v at a rate of 2 for a unit time interval. As vw is the only edge leaving node v all flow enters this edge, forming a queue of size 1 by time $k + 1$. After that the network inflow stops and, thus, the queue depletes and is empty again at time $k + 2$. Thus, there is an outflow from edge vw at a rate of 1 during $[k + 1, k + 3]$. By our assumption the waiting time along path p_1 is at least 4 and exactly 0 along path p_3 during this interval. Thus, edge wz_1 is never active during this interval. Moreover, during $[k + 1, k + 2)$ we have a waiting time of more than 1 along path p_3. Hence, edge wx_1 is the only active edge leaving w during this time and all flow arriving at node w enters this edge. During $(k + 2, k + 3]$ the waiting time on path p_2 is less than 1, edge wy_1 becomes the unique active edge leaving w and, therefore, all flow enters this edge. During $[k + 4, k + 5]$ the waiting times along both paths p_2 and p_3 are at least 3 while the waiting time on path p_1 is 0. Thus, both edges x_3z_3 and y_2z_2 are inactive and flow enters both edges x_3v and y_2v at a rate of 1 during this interval. Finally, this flow arrives back at node v at a combined rate of 2 during $[k + 5, k + 6]$ at which point the described pattern repeats. This shows properties (i) and (ii).

For property (iii) we observe that the queue on edge vw until some time $\xi + 1$ is completely determined by the edge inflow rate on edges x_3v and y_2w until time ξ. Therefore, if the waiting time pattern on the paths p_j is as assumed until time ξ, the inflow into those two edges follows the pattern described before at least until time ξ as well and, thus, the queue on edge vw follows the desired pattern at least until time $\xi + 1$. ■

By connecting multiple copies of gadget C_i^k in series (with different values of k) we can now essentially create paths with essentially any 5-periodic waiting time pattern. In particular, we can build paths exhibiting the waiting time patterns required on the paths p_1, p_2 and p_3. We will, thus, create three types of **blocking gadgets**[3] $B_i^{1,k}$, $B_i^{2,k}$ and $B_i^{3,k}$—parametrised again by the commodity $i \in \{1, 2\}$ in the cycling gadgets used to construct the blocking gadget and a time shift $k \in \mathbb{N}_0$.

Gadget $B_i^{3,k}$ is constructed as follows (Fig. 6.19): We take three copies of gadget C_i^{k+3} and three copies of gadget C_i^{k+4}. We then connect these gadgets in series with 3 edges between each of them, i.e. we add a path consisting of three edges starting at node w of the first copy of gadget C_i^{k+3} to the node v of the second copy and so on. Then, we add two new nodes v and w and add an edge from the new node v to

[3] Again, this gadget should not be confused with its namesake from the proof of Theorem 6.15.

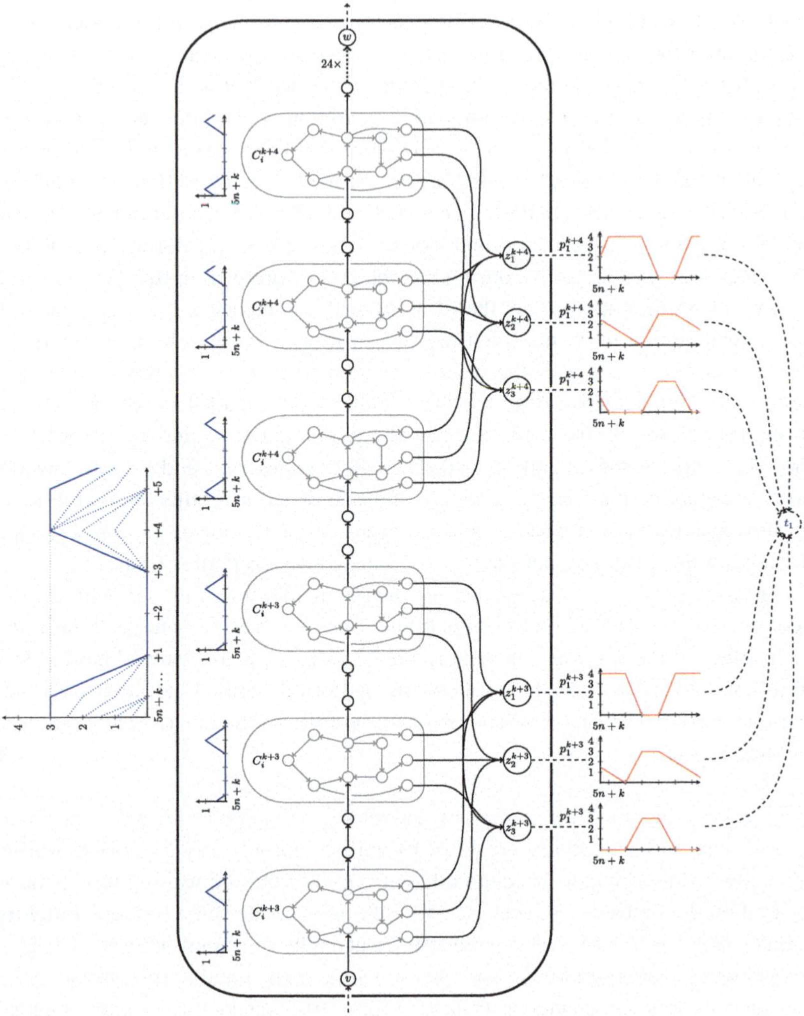

Fig. 6.19 The gadget $B_1^{3,k}$. The graphs next to the gadgets $C_1^{k'}$ indicate the waiting time pattern on the central edge vw of the respective gadget. The graph to the left of the whole gadget $B_1^{3,k}$ indicates the resulting waiting time pattern on the whole central v, w-path through this gadget. The red graphs on the paths $p_j^{k'}$ indicate the required waiting time pattern on those in order to keep the flow inside $B_1^{3,k}$ stable (i.e. the assumption in Claim 6.10)

the node v of the first copy of C_i^{k+3} and a path consisting of 25 edges from node w of the third copy of C_i^{k+4} to the new node w. Finally, we add six more new nodes z_j^{k+m} for $j \in [3]$ and $m \in \{3, 4\}$. We then connect each node z_j of each copy of gadget C_i^{k+m} with a direct edge to node z_j^{k+m}.

Gadgets $B_i^{1,k}$ and $B_i^{2,k}$ are constructed analogously: For gadget $B_i^{1,k}$ we use four copies of each gadget C_i^{k+0}, C_i^{k+1} and C_i^{k+2} and connect them in series with three edges inbetween each consecutive pair. This time, however, we connect node w of the last copy of C_i^{k+3} to the new node w of gadget $B_i^{1,k}$ with only a single edge. Additionally, we have nodes z_j^{k+m} for $j \in [3]$ and $m \in \{0, 1, 2\}$ which have incoming edges from the nodes z_j of the gadgets C_i^{k+m} as before. For gadget $B_i^{2,k}$ we use two copies of C_i^{k+0} one copy of C_i^{k+1} and three copies each of C_i^{k+3} and C_i^{k+4}. We connect those again in series (always with three edges inbetween) and connect the last node w to the new node w with a path of length 13. Finally, we have nodes z_j^{k+m} for $j \in [3]$ and $m \in \{0, 1, 3, 4\}$ which have incoming edges from the nodes z_j of the gadgets C_i^{k+m}.

We will later embed these gadgets into a larger network in such a way that the only incoming edge to gadget $B_i^{\ell,k}$ enters at node v and the only outgoing edges start at node w and at the nodes z_j^{k+m}. Furthermore, we will ensure that for every z_j^{k+m} there is a unique physically shortest z_j^{k+m}, t_i-path p_j^{k+m} with equal length while every other path towards t_i starting at at some node z_j^{k+m} has a free flow travel time of at least 4 more than the paths p_j^{k+m}, Moreover, there exists no w, t_i-path.

Claim 6.10. *Any gadget $B_i^{\ell,k}$ has the following property:*

(i) There exists a unique v, w-path and it has a free flow travel time of 47.

Assume that a gadget $B_i^{\ell,k}$ is correctly embedded into a larger network $\mathcal{N}$ and f is an IDE in the larger network. Additionally, assume that f satisfies the following properties:

- *No flow enters gadget $B_i^{\ell,k}$ from outside the gadget.*
- *The waiting time on path any p_1^{k+m} is at least* 4 *during each interval* $[5n + k + m + 1, 5n + k + m + 3]$ *and* 0 *during each interval* $[5n + k + m + 4, 5n + k + m + 5]$ *for $n \in \mathbb{N}_0$.*

- *The waiting time on any path p_2^{k+m} is more than* 1 *during each interval* $[5n + k + m + 1, 5n + k + m + 2)$*, less than* 1 *during* $(5n + k + m + 2, 5n + k + m + 3]$ *and at least* 3 *during each interval* $[5n + k + m + 4, 5n + k + m + 5]$ *for* $n \in \mathbb{N}_0$.
- *The waiting time on any path p_3^{k+m} is* 0 *during each interval* $[5n+k+m+1, 5n+k+m+3]$ *and at least* 3 *during each interval* $[5n + k + m + 4, 5n + k + m + 5]$ *for* $n \in \mathbb{N}_0$.

Then the flow inside the gadget exhibits the pattern displayed in Fig. 6.19. In particular,

(ii) *No flow ever leaves the gadget.*
(iii) *For $\ell = 1$ the waiting time along the unique v, w-path through gadget $B_i^{1,k}$ is* 4 *during each interval* $[5n + k + 1, 5n + k + 3]$ *and* 0 *during each interval* $[5n + k + 4, 5n + k + 5]$ *for* $n \in \mathbb{N}_0$.
iv *For $\ell = 2$ the waiting time along the unique v, w-path through gadget $B_i^{2,k}$ is more than* 1 *during each interval* $[5n + k + 1, 5n + k + 2)$*, less than* 1 *during each interval* $(5n + k + 2, 5n + k + 3]$ *and* 4 *during each interval* $[5n + k + 4, 5n + k + 5]$ *for* $n \in \mathbb{N}_0$.
(v) *For $\ell = 3$ the waiting time along the unique v, w-path through gadget $B_i^{3,k}$ is* 0 *during each interval* $[5n + k + 1, 5n + k + 3]$ *and* 3 *during each interval* $[5n + k + 4, 5n + k + 5]$ *for* $n \in \mathbb{N}_0$.
(vi) *Moreover, if the assumed waiting time pattern holds only until some time $\xi \in \mathbb{R}_{\geq 0}$, then the waiting time pattern described in (iii), (iv) or (v), respectively, holds until time $\xi + 1$.*

Proof. First, it is clear from the construction that there is a unique v, w-path in any gadget $B_i^{n,k}$ via the edges vw of every gadget C_i^k used in the construction. The number of edges in this path is

- $1 + 12 \cdot 1 + 11 \cdot 3 + 1 = 47$ for gadget $B_i^{1,k}$,
- $1 + 9 \cdot 1 + 8 \cdot 3 + 13 = 47$ for gadget $B_i^{2,k}$ and
- $1 + 6 \cdot 1 + 5 \cdot 3 + 25 = 47$ for gadget $B_i^{3,k}$.

This shows property (i).

For the remaining properties we start by observing that all gadget C_i^k used in the construction are embedded correctly in the larger network: For any node z_j in any copy of gadget C_i^{k+m} there exists a z_j, t_i-path $p_j := z_j z_j^{k+m}, p_j^{k+m}$ of equal length. Any other z_j, t_i-path also has to go over node z_j^{k+m} first and, thus, has a free flow travel time of at least 4 more by our assumption that $B_i^{n,k}$ is correctly embedded. Any w, t_i-path p' which is not $wz_1, z_1z_1^{k+m}, p_1^{k+m}$ either also leaves gadget C_i^{k+m} via z_1 (free flow travel time 1 before even arriving at z_1) or it leaves directly at node w and traverses another gadget $C_i^{k+m'}$ (free flow travel time of at least 5 before arriving at some node $z_{j'}$ in that next gadget). In both cases our assumption that $B_i^{n,k}$ is correctly embedded into the larger network guarantees that the free flow travel time along p' is at least 5 more than along a shortest z_j, t_i-path.

Next, we note that the claim's assumptions on the waiting times on the path p_j^{k+m} exactly match the assumptions on the waiting times along path p_j in Claim 6.9. Thus, within each gadget C_i^{k+m} the flow exhibits the properties stated in this claim. In particular, no flow leaves any of those gadgets and, hence no flow leaves gadget $B_i^{n,k}$ (i.e. property (i) holds). Moreover, the waiting time along the central v, w-path through gadget $B_i^{n,k}$ is just the sum of the waiting times specified in Claim 6.9(ii). It is now easy to see that these waiting times satisfy properties (iii), (iv) or (v), respectively.

Moreover, Claim 6.9(iii) implies that this waiting time pattern holds until at least $\xi + 1$ if the waiting time pattern on the paths p_j^{k+m} satisfies our assumptions until at least time ξ. Thus, property (iv) holds as well. ■

Now, to build the full network $\mathcal{N}$ (cf. Fig. 6.20) we take one copy of each gadget $B_i^{n,k}$ for $n \in [3]$, $k \in [5] - 1$ and $i \in [2]$ and one sink node t_1 and t_2 for each commodity 1 and 2. We then connect these as follows: For any gadget $B_i^{n,k}$ we connect its node w with a direct edge to commodity $(i + 1)$'s sink node t_{i+1} and any node z_j^{k+m} with a direct edge to the the node v of gadget $B_{i+1}^{j,k+m}$ (where both the index $i + 1$ in t_{i+1} and $B_{i+1}^{j,k+m}$ and the superscript $k + m$ in $B_i^{j,k+m}$ are to be read modulo 5).

We first show that every gadget $B_i^{n,k}$ is correctly embedded in $\mathcal{N}$. For any node z_j^{k+m} in gadget $B_i^{n,k}$ there exists the following z_j^{k+m}, t_i-path p_j^{k+m} of length 49: First the direct edge towards node v of gadget $B_{i+1}^{j,k+m}$, then the direct v, w-path through this gadget (with length 47 according to Claim 6.10(i)) and then the direct edge from node w to the sink t_i. Any other z_j^{k+m}, t_i-path must start the same way but then leave gadget $B_{i+1}^{j,k+m}$ towards some gadget $B_i^{n',k'}$. By the time the path enters this gadget the alternative path has already accumulated a length of at least 6 and still has to traverse one complete gadget $B_{i+1}^{n'',k''}$ in order to reach the sink t_i, giving us a total free flow travel time of at least $47 + 6 = 49 + 4$. Additionally, it is immediately clear from the construction of network $\mathcal{N}$ that there exists no w, t_i-path from node w in gadget gadget $B_i^{n,k}$. Thus, gadget $B_i^{n,k}$ is correctly embedded in network $\mathcal{N}$.

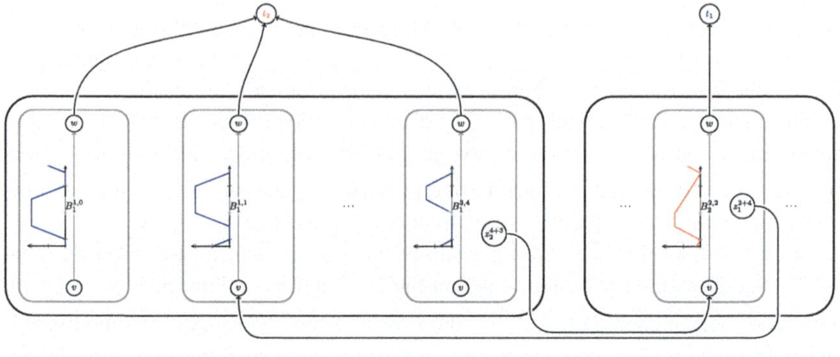

Fig. 6.20 A sketch of the whole network constructed in the proof of Theorem 6.18. Commodity 1 is blue and commodity 2 is red. Not all copies of gadgets $B_i^{n,k}$ and not all connecting edges are drawn

Now let f be any IDE in $\mathcal{N}$. Let gadget $B_i^{n,k}$ be any blocking gadget used in the construction of $\mathcal{N}$. We then show via induction on $\xi \in \mathbb{N}_0$ that f exhibits the pattern described in Claim 6.10 until at least ξ.

Base Case ($\xi = 0$): This is trivially satisfied.

Induction Step ($\xi \to \xi + 1$): By construction and induction the waiting time pattern on all paths p_j^{k+m} leaving gadget $B_{i+1}^{n,k}$ satisfy the assumption on the waiting pattern in Claim 6.10 until at least time ξ. Then we know from Claim 6.10(vi) that the desired waiting time pattern inside gadget $B_i^{n,k}$ until at least time $\xi + 1$.

This now shows, in particular, that f never terminates (as all waiting times would be forever zero after the flow terminates). □

Remark 6.19. An interesting use for the instance constructed in the proof above has been proposed by Hagenmaier in [Hag23]: Since this instance has a unique IDE which one can describe for all times, it can be used as a (quite complex) benchmark instance for numerical algorithms that compute (approximate) IDE when testing how close the computed flows stay to an exact IDE over time.

6.3 The Price of Anarchy

A good way of summarizing the results from this chapter is by turning them into asymptotic bounds for the Price of Anarchy of IDE. In general, the Price of Anarchy (PoA) of any game, as introduced by Koutsoupias and Papadimitriou in [KP99, Pap01], is the worst case ratio of the objective in an equilibrium to the optimal objective. For a single network we define the following PoA:

Definition 6.20. defindexl Let $\mathcal{N}$ be a feasible network with finitely lasting network inflow rates. Then we define the *makespan Price of Anarchy for* $\mathcal{N}$ by

$$\Psi\text{−PoA}_{\mathcal{N}} := \sup\left\{ \frac{\Psi(f)}{\Psi(g)} \;\middle|\; f \text{ an IDE in } \mathcal{N} \text{ and } g \text{ any feasible flow in } \mathcal{N} \right\}.$$

Analogously, we define the *total travel time Price of Anarchy for* $\mathcal{N}$ by

$$\Xi\text{−PoA}_{\mathcal{N}} := \sup\left\{ \frac{\Xi(f)}{\Xi(g)} \;\middle|\; f \text{ an IDE in } \mathcal{N} \text{ and } g \text{ any feasible flow in } \mathcal{N} \right\}.$$

Now, instead of calculating the PoA for individual networks, we will consider whole classes of networks and derive asymptotic bounds for the PoA parametrised in the total flow volume U and the size of the networks τ.

Definition 6.21. Let $\mathfrak{N}$ be some class of feasible networks with finitely lasting inflow rates. Then we define for any two natural numbers $U, \tau \in \mathbb{N}_0$ the *makespan Price of Anarchy for* $\mathfrak{N}$ by

$$\Psi\text{−PoA}_{\mathfrak{N}}(U,\tau) := \sup\left\{ \Psi\text{−PoA}_{\mathcal{N}} \;\middle|\; \begin{array}{l} \mathcal{N} = ((V,E),\tau,\nu,I,u,T) \in \mathfrak{N} \text{ with} \\ \sum_{e\in E} \tau_e \leq \tau \text{ and } \sum_{i\in I} U_i(\hat{\theta}_i) \leq U \end{array} \right\}.$$

Analogously, we define the *total travel time Price of Anarchy for* $\mathfrak{N}$ by

$$\Xi\text{−PoA}_{\mathfrak{N}}(U,\tau) := \sup\left\{ \Xi\text{−PoA}_{\mathcal{N}} \;\middle|\; \begin{array}{l} \mathcal{N} = ((V,E),\tau,\nu,I,u,T) \in \mathfrak{N} \text{ with} \\ \sum_{e\in E} \tau_e \leq \tau \text{ and } \sum_{i\in I} U_i(\hat{\theta}_i) \leq U \end{array} \right\}.$$

In order to get good lower bounds on these PoA we need one final construction which allows us to transform a network into one with a very simple optimal flow (and low makespan/total travel time) but still essentially the same IDE.

Lemma 6.22. *Let $\mathcal{N}$ be a feasible network with finitely lasting bounded network inflow rates, size $\tau(\mathcal{N})$ and total flow volume $U_{\mathcal{N}}$. Let $M \geq 0$ be an essential bound to all network inflow rates $S := \{ (s, i) \in V \times I \mid u_{s,i} \neq 0 \}$ the set of source nodes and for any $(s, i) \in S$ the path $p_{s,i}$ a (physically) shortest s, T_i-path with free flow travel time $\tau_{p_{s,i}}$. Then, there exists a network $\tilde{\mathcal{N}}$ which includes $\mathcal{N}$ as a subnetwork and satisfies the following properties:*

(i) The size of $\tilde{\mathcal{N}}$ is $\tau(\mathcal{N}) + |S| \cdot (5 + 3\hat{\theta})$ and the total flow volume in $\tilde{\mathcal{N}}$ is $U_{\mathcal{N}} + \sum_{(s,i) \in S}(2\hat{\theta} + \tau_{p_{s,i}})$

(ii) There exists a Vickrey flow g in $\tilde{\mathcal{N}}$ with $\Psi(g) = 4 + 2\hat{\theta}$ and $\Xi(g) = (3 + \hat{\theta})\big(U_{\mathcal{N}} + \sum_{(s,i) \in S}(\tau_{p_{s,i}} + 2\hat{\theta}\big)$.

(iii) There exists a one to one correspondence between IDE f in $\mathcal{N}$ and IDE $\tilde{f}$ in $\tilde{\mathcal{N}}$ such that we have $\tilde{f}^+_{e,i}(\theta + 2 + 2\hat{\theta}) = f^+_{e,i}(\theta)$ for all $\theta \in \mathbb{R}_{\geq 0}$, $i \in I$ and all edges e from the original network.

Proof. We transform $\mathcal{N}$ into $\tilde{\mathcal{N}}$ as follows: For any $(s, i) \in S$ let $t_{s,i} \in T_i$ be the the end node of $p_{s,i}$. We then add the following nodes and edges to $\mathcal{N}$ (cf. Fig. 6.21): Two new nodes $\tilde{s}$ and $\tilde{w}$ and four edges:

- an edge $\tilde{s}t_{s,i}$ with free flow travel time $3 + \hat{\theta}$ and a capacity of $\tilde{M} := \max\{ M + 1, \tau_{p_{s,i}} + \hat{\theta} \}$,
- an edge $\tilde{w}t_{s,i}$ with free flow travel time 1 and capacity 1,
- an edge $\tilde{s}\tilde{w}$ with free flow travel time $1 + \hat{\theta}$ and capacity $\tilde{M}$ and
- an edge $\tilde{w}s$ with free flow travel time $\hat{\theta}$ and capacity M.

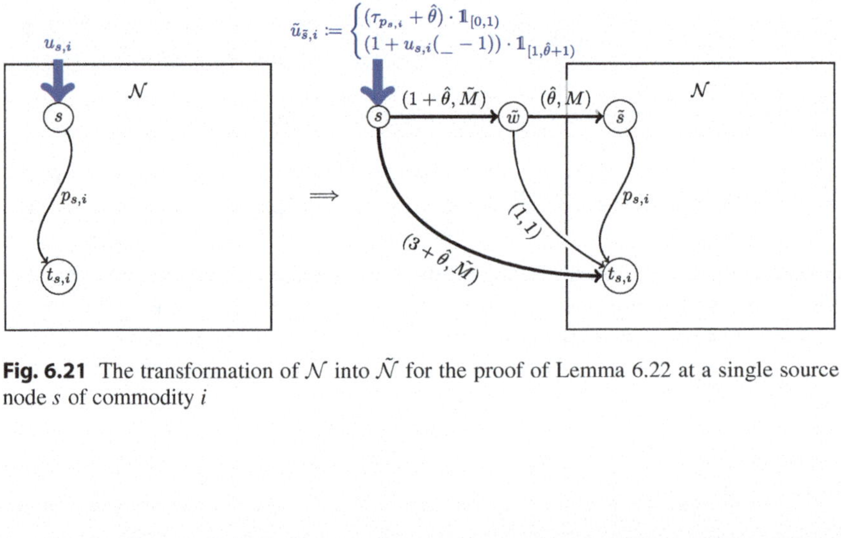

Fig. 6.21 The transformation of $\mathcal{N}$ into $\tilde{\mathcal{N}}$ for the proof of Lemma 6.22 at a single source node s of commodity i

Finally, we set the network inflow of commodity i at node s to zero and instead define the following network inflow rate at node $\tilde{s}$:

$$\tilde{u}_{\tilde{s},i}(\theta) := \begin{cases} \tau_{p_{s,i}} + \hat{\theta}, & \text{if } \theta \in [0, 1) \\ 1 + u_{s,i}(\theta - 1), & \text{if } \theta \in [1, \hat{\theta} + 1] \\ 0, & \text{else.} \end{cases}$$

The size of the new network is then

$$\tau(\tilde{\mathcal{N}}) = \tau(\mathcal{N}) + \sum_{(s,i) \in S} (5 + 3\hat{\theta})$$

while the total flow volume is

$$U_{\tilde{\mathcal{N}}} = U_{\mathcal{N}} + \sum_{(s,i) \in S} (2\hat{\theta} + \tau_{p_{s,i}}).$$

This shows property (i).

Now let $\tilde{f}$ be any IDE in the newly constructed network. First, we observe that there will be no flow inside the part from the original network $\mathcal{N}$ before time $1 + 2\hat{\theta}$. Thus, $\tilde{f}$ starts in the following way: During the interval $[0, 1 + \hat{\theta})$ the only active source-sink-paths are the paths $\tilde{s}\tilde{w}$, $\tilde{w}t_{s,i}$. Thus, all flow entering the network enters theses paths. During $[1 + \hat{\theta}, 2 + \hat{\theta}]$ the flow arriving at node $\tilde{w}$ enters the direct edge towards $t_{s,i}$ building up a queue there. At time $2 + \hat{\theta}$ this queue reaches a length of $\tau_{p_{s,i}} + \hat{\theta} - 1$ and, hence, the edge $\tilde{w}s$ becomes active as well. From here on, the flow arriving at node $\tilde{w}$ enters the edge towards $t_{s,i}$ at a rate of 1 (to keep the queue length constant) while the rest of the flow enters the edge towards s. Thus, flow arrives at node s exactly at a rate of $u_{s,i}(\_ + 2 + 2\hat{\theta})$ during $[2 + 2\hat{\theta}, 2 + 3\hat{\theta}]$. From here on $\tilde{f}$ can behave in exactly the same way as any IDE in the original network $\mathcal{N}$ just with a time shift of $2 + 2\hat{\theta}$. This shows property (iii).

For property (ii) we consider the flow g in $\tilde{\mathcal{N}}$ where we send all network inflow along the direct edges $\tilde{s}t_{s,i}$. As the capacity on these edges are large enough such that no queues ever form there, this flow then terminates by time $1 + \hat{\theta} + 3 + \hat{\theta}$ and, thus, satisfies

$$\Psi(f) = 4 + 2\hat{\theta} \quad \text{and} \quad \Xi(g) = (3 + \hat{\theta})\Big(U_{\mathcal{N}} + \sum_{(s,i) \in S} (\tau_{p_{s,i}} + 2\hat{\theta}\Big).$$

□

With this Lemma we can now derive the following asymptotic bounds for the PoA of IDE:

Theorem 6.23. *Let $\mathfrak{N}^{\text{SC}}$ be the class of feasible single-commodity networks with finitely lasting, essentially bounded network inflow rates, strictly positive, integer free flow travel times and capacities and no dead-end nodes. Then we have*

$$\Psi\text{−PoA}_{\mathfrak{N}^{\text{SC}}}(U, \tau),\ \Xi\text{−PoA}_{\mathfrak{N}^{\text{SC}}}(U, \tau) \in \mathcal{O}(U\tau).$$

Furthermore, for large enough τ and $U \geq \sqrt{\tau} \log \tau$ we have instances $\mathcal{N} \in \mathfrak{N}^{\text{SC}}$ with

$$\Psi\text{−PoA}_{\mathcal{N}},\ \Xi\text{−PoA}_{\mathcal{N}} \in \Omega(U \log \tau).$$

Let $\mathfrak{N}^{\text{ac}}$ be the class of feasible acyclic networks with finitely lasting, essentially bounded network inflow rates, strictly positive, integer free flow travel times and capacities and no dead-end nodes. Then we have

$$\Psi\text{−PoA}_{\mathfrak{N}^{\text{ac}}}(U, \tau),\ \Xi\text{−PoA}_{\mathfrak{N}^{\text{ac}}}(U, \tau) \in \mathcal{O}(U + \tau)$$

Furthermore, there exist instances $\mathcal{N} \in \mathfrak{N}^{\text{ac}}$ with

$$\Psi\text{−PoA}_{\mathcal{N}},\ \Xi\text{−PoA}_{\mathcal{N}} \notin \mathcal{O}(U + \tau_{p_{\min}}).$$

Let $\mathfrak{N}^{\text{MC}}$ be the class of feasible (multi-commodity) networks with finitely lasting inflow rates and non-zero integer capacities and free flow travel times. Then we have

$$\Psi\text{−PoA}_{\mathfrak{N}^{\text{MC}}}(U, \tau) = \Xi\text{−PoA}_{\mathfrak{N}^{\text{MC}}}(U, \tau) = \infty$$

for large enough U and τ.

Proof. We start with $\mathfrak{N}^{\text{SC}}$: Let $\mathcal{N} \in \mathfrak{N}^{\text{SC}}$ be any such network with flow volume $U_{\mathcal{N}}$ and size $\tau(\mathcal{N})$. Since we do not allow any edges of zero free flow time, edges leaving a sink node will always be inactive (cf. Proposition 2.67k)) and, thus, can be removed from the network without changing any IDE. Then, according to Theorem 6.12/Remark 6.13, we have $\Psi(f) \leq \hat{\theta} + (U_{\mathcal{N}} + 1) \cdot 3\tau(\mathcal{N})$ and $\Xi(f) \leq (U_{\mathcal{N}})^2 \cdot (11\tau(\mathcal{N}) + 4)$ for any IDE f in $\mathcal{N}$. On the other hand any Vickrey flow g in $\mathcal{N}$ certainly satisfies $\Psi(g) \geq \hat{\theta} + 1$ and $\Xi(g) \geq U_{\mathcal{N}}$. This gives us the desired asymptotic upper bounds.

For the lower bound we apply the transformation from Lemma 6.22 to the networks $\mathcal{N}_{K,L}$ from Theorem 6.15. This way we obtain networks $\tilde{\mathcal{N}}_{K,L}$ with the following properties:

- $\tau(\tilde{\mathcal{N}}_{K,L}) \in \mathcal{O}(3^{2K})$,
- $U_{\tilde{\mathcal{N}}_{K,L}} \in \mathcal{O}(L3^K + K3^K)$,
- there is an IDE f in $\tilde{\mathcal{N}}_{K,L}$ with $\Psi(f) \in \Omega(LK3^K)$ and $\Xi(f) \in \Omega(KL^2 3^{2K})$ and
- there is a Vickrey flow g in $\tilde{\mathcal{N}}_{K,L}$ with $\Psi(g) \in \mathcal{O}(1)$ and $\Xi(g) \in \mathcal{O}(L3^K + K3^K)$.

Now take any τ, U with $\tau \geq 3^6$ and $U \geq \sqrt{\tau} \cdot \log \tau$. Then we can choose $K \in \mathbb{N}_0$ with $K \geq 3$ such that $\tau \approx 3^{2K}$ and $L \geq K$ such that $U \approx L3^K$. The network $\tilde{\mathcal{N}}_{K,L}$ then has an IDE f with $\Psi(f) \in \Omega(LK3^K) = \Omega(U \log \tau)$ and $\Xi(f) \in \Omega(KL^2 3^{2K}) = \Omega(U^2 \log \tau)$ and a Vickrey flow g with $\Psi(g) \in \mathcal{O}(1)$ and $\Xi(g) \in \mathcal{O}(L3^K) = \mathcal{O}(U)$. This shows the lower bound.

Next, we consider $\mathfrak{N}^{\mathrm{ac}}$. The upper bound then follows directly from Theorem 6.5. For the lower bound we apply the transformation from Lemma 6.22 to the networks $\mathcal{N}_{K,L}$ from Corollary 6.16 to obtain networks $\tilde{\mathcal{N}}_{K,L}$ with the following properties:

- $\tau_{p_{\min}}(\tilde{\mathcal{N}}_{K,L}) \in \mathcal{O}(1)$,
- $U_{\tilde{\mathcal{N}}_{K,L}} \in \mathcal{O}(L3^K + K3^K)$,
- there is an IDE f in $\tilde{\mathcal{N}}_{K,L}$ with $\Psi(f) \in \Omega(LK3^K)$ and $\Xi(f) \in \Omega(KL^2 3^{2K})$ and
- there is a Vickrey flow g in $\tilde{\mathcal{N}}_{K,L}$ with $\Psi(g) \in \mathcal{O}(1)$ and $\Xi(g) \in \mathcal{O}(L3^K + K3^K)$.

This proves the lower bound.

Finally, the statement for $\mathfrak{N}^{\mathrm{MC}}$ follows directly from Theorem 6.18 since this Theorem provides a network in $\mathfrak{N}$ in which makespan and total travel time of any IDE are infinite. At the same time there clearly exists a Vickrey flow where both these quality measures are finite. Thus, for all U and τ larger than the total flow volume and size in this network, we get an unbounded PoA. □

Remark 6.24. The above bounds on the Price of Anarchy are equally true for the *Price of Stability* ([Ans+08]) which is defined in the same way as the Price of Anarchy with the only difference being that we always take the best instead of the worst equilibrium flow. This is because we can choose the lower bound instances such that the IDE in them are unique (cf. Remark 6.17).

6.4 Bibliographic Notes and Open Questions

The results presented in this chapter are mostly based on joint work with Tobias Harks published in [GH22]. The non-termination result for multi-commodity networks (Theorem 6.18) is from our paper [GH19].[4] There we also already showed that IDE in single-commodity networks are guaranteed to terminate, but could not give any explicit bounds on the termination time. The required strengthening of that proof (presented here in Theorem 6.12) was inspired by a discussion of the original termination result with Kathrin Gimmi.

Natural goals for future research arising from this chapter are to close the gap between the upper and lower bounds for makespan/total travel time for single-commodity networks and finding a characterisation of multi-commodity networks that guarantee termination in terms of the topological structure of the underlying graphs. For the former question it seems quite likely that the upper bounds in particular are not tight as several of the estimations used in the proof of Theorem 6.15 are rather rough. For the second question we already know from Theorem 6.5 that the class of graphs guaranteeing termination includes at least all acyclic graphs but it seems quite likely that this class is much larger. For example, graphs with only a single cycle should certainly still be contained in this class. A bolder, but still plausible seeming, conjecture would be that planar graphs also guarantee termination. Our non-termination instance is certainly not planar (note that already in a single gadget $B_i^{3,k}$ one can easily find a $K_{3,3}$ as a graph minor—see Fig. 6.19) and it seems unlikely that the complex interlinking structure needed for non-termination would be possible in a planar graph. Studying (non-)termination for planar graph would, of course, also be interesting from an application point of few as real world instances of road networks typically are planar.

Finally, it would also be interesting to know which of the results from this chapter can be extended to other physical flow models. This seems particularly likely for the upper bound results as their proofs do not rely to heavily on the exact details of our physical flow model. More precisely, the only two main places where we use specific properties of the deterministic queuing model in Sect. 6.1 are in the induction step of the proof of Claim 6.2 (when proving the upper bound for acyclic networks) and, in form of the no-idling property (Corollary 3.24), in the proofs of Corollary 6.9 and Proposition 6.10.

---

[4] On a slightly personal note: This was in fact my first ever mathematical result and, thus, makes for a very fitting theorem to conclude this thesis.

Conclusion 7

We conclude this thesis with a short summary of its results, a comparison to some corresponding results in the full information setting and some ideas for potential future research.

7.1 Summary and Comparison to Full Information Equilibria

In this thesis we introduced the concept of instantaneous dynamic equilibria as a model for dynamic traffic flows with adaptive (selfish) route choice based on current information. We then studied several key properties of such equilibrium flows: We showed existence in a very general setting and provided alternative existence proofs for more specialized cases like right-constant flow rates and single-commodity networks. For these special cases we then investigated the computational complexity of finding such flows: Here we found that individual extensions can be computed in finite or even polynomial time, respectively. Moreover, we showed that for the case of single-commodity networks (with right-constant network inflow) a finite number of such extensions suffices to reach any given finite time horizon. On the other hand we gave an example for a multi-commodity network with zero free flow travel times where a certain finite time horizon cannot be reached by any finite number of (constant) extensions. Additionally, we provided NP-hardness results for several decision problems involving IDE. Finally, we turned to the quality of IDE. Here, we derived upper and lower bounds for makespan and total travel time and, in particular, showed that IDE are guaranteed to terminate in single-commodity networks whereas in multi-commodity networks it is possible for particles in an IDE to be trapped in a cycle forever.

L. Graf, *Dynamic Network Flows with Adaptive Route Choice based on Current Information*, Mathematische Optimierung und Wirtschaftsmathematik | Mathematical Optimization and Economathematics,
https://doi.org/10.1007/978-3-658-44948-3_7

In the following table we summarize our main results for IDE (left) and contrast them to some results for the analogous equilibria in the full information setting (right). We use the abbreviations sc = single-commodity, mc = multi-commodity and rc = right-constant.

	Current information	Full information
existence		
for sc	Yes: for rc inflow using thin flows and convex optimization (Thm. 4.36)	Yes: for rc inflow using thin flows and a fixed point theorem ([CCL15, Thm. 5])
for mc	Yes: for rc inflow using thin flows and a fixed point theorem (Thm. 4.31), for p-integrable inflows using a fixed point theorem (Thm. 4.15) or a variational inequality ([GHS20, Thm. 5.6])	Yes: for rc inflow using thin flows and a variational inequality ([Ser20, Thm. 5.10]), for p-integrable inflows using a variational inequality ([CCL15, Thm. 8])
uniqueness (of $L_{v,i}$)		
for sc	No: Ex. 3.65[1]	Yes: [CCL15, Thm. 6], [OSK22, Thm. 3.1]
for mc	No: Ex. 3.65	No: [Iry11, Sec. 4]
computation (rc only)		
of single ext.	in polynomial time for sc (Cor. 5.7), in finite time for mc with $\tau_e > 0$ (Cor. 5.2)	in polynomial time for series-parallel networks ([Kai22a, Cor. 32]), in finite time for sc ([Ser20, Sec. 3.6.1])
of entire flow	in finite time for sc (Cor. 5.16), may require $\Omega(T)$ many extensions to reach time T for sc (Theorem 5.21b)) and infinitely many extensions for mc with $\tau_e = 0$ (Ex. 5.17)	may require exponentially many extensions wrt network size ([CCO22a, Sec. 5.2]); otherwise unknown
complexity	decision problems involving IDE can be NP-hard (Thm. 5.25)	computing thin flows is in PPAD ([CCL15, Sec. 3])
steady state (sc only)	No: Thm. 5.21c)	Yes: [CCO22a, Thm. 3], [OSV22, Thm. 3.3]
termination		
for sc	Yes: Thm. 6.12	Yes: trivial
for mc	No: Thm. 6.18 (Yes in acyclic networks: Thm. 6.5)	Yes: trivial
quality (sc only)	makespan and total travel time PoA in $\mathcal{O}(U\tau) \cap \Omega(U \log \tau)$ (Thm. 6.23)	makespan PoA $= \frac{e}{e-1}$ ([CCO22b, Lem. 6], assuming monotonicity conjecture), makespan and total travel time PoA $=$ 1 for shortest-path networks ([KS11, Thm. 4])

We can draw two main observation from this comparison: On the one hand, IDE tend to be less well behaved when it comes to quality (i.e. termination and PoA) and

[1] Note that turning the network from Ex. 3.65 into a single-commodity instance does not change the possible flow patterns in an IDE.

stability (i.e. uniqueness and long term behaviour). This is, to some extent, to be expected due to the fact that agents make their decisions based on incomplete information and could, thus, be seen as a "price of short-sightedness". The contribution of this thesis to that aspect is then a study of this price of short-sightedness. On the other hand, IDE appear to be more open to a certain type of local analysis both when it comes to positive results (e.g. computation) and when it comes to negative results (i.e. the gadget-based construction and analysis of quite complicated networks). This allowed us to answer questions for IDE that are still open for the full information setting (e.g. the number of required extension steps to construct equilibrium flows). Despite these differences, there are of course also many similarities between these two types of equilibria—especially when it comes to the model itself. Because of this, we were able to reuse and adapt several ideas from the full information setting for IDE (e.g. the description based on node labels or the concept of thin flows).

7.2 Potential Directions for Future Research

As we already discussed several open questions directly related to the results of this thesis in the "Open Questions"-sections of the respective chapters, we will only discuss three more general topics for future research in the context of IDE here:

A unified framework for current and full information equilibria: Considering how much of the model introduced by [KS11] and [CCL15] for the full information setting we were able to adapt and reuse for the current information setting here, this raises the question whether there could be a more general model that encompasses both these settings. Such a general model would ideally allow a generalization of results that hold for both models (e.g. existence results) and also facilitate a better understanding of which results can be transferred from one to the other and which ones cannot (and why). Moreover, one could potentially use such a general model to study a parametrized price of short-sightedness, i.e. how the quality of equilibrium flows is affected by giving the agents some but not complete knowledge about the future (which might be more realistic than either of the two extremes of no and full knowledge).

One such general framework has already been proposed by Graf, Harks, Kollias and Markl in the form of *dynamic prediction equilibria* [GHKM23], though no theoretical results about the fully general model have been shown yet. In particular, there is no general existence result yet that would apply to both the current and the full information setting. One of the main obstacle in unifying the known existence results for these two models seems to be the difference in what is the natural definition of a partial equilibrium flow in the two settings: A flow where all edge flow rates are fixed until some common time ξ versus one where all edge flow rates are fixed until the earliest entrance time for particles starting from the source at time ξ. This same difference can also be seen in the exact definition of thin flows for both settings.

Extensions of the IDE-model: Extending the model presented in this thesis could be interesting for at least two reasons: Firstly, depending on the application one has in mind, different or more powerful physical models could be more realistic (e.g. using linear edge delays instead of the Vickrey point queue model or including spillback effects). Several such extensions have already been studied in the full information setting (see e.g. [ZM00, SV18]) that could also be studied in the current information setting. Moreover, as already pointed out throughout this thesis, several of our results do not rely to heavily on the particular details of the physical model used here and should, therefore, also hold for other models.

Secondly, one might see the rather large price of anarchy of IDE as a motivation for extensions of the behavioural model that try to improve the quality of the resulting equilibrium. This could be done via tolls, by changing the network parameters (which turns the game into a Stackelberg model—see [BFA15] for such an extension of the full information model) or by letting the agents make their decisions using more complex predictions based on current (and past) information. The latter setting has already been studied in [GHKM23] and numerical experiments indicate that this can indeed lead to an improvement of the overall travel times.

Computational studies: Our positive results with respect to the computation of IDE suggest that they could be a good target for computational studies. In particular, it should be interesting to contrast the theoretical worst case quality bounds shown in this thesis with the actual quality of IDE in realistic instances. Two implementation of dynamic flows that allow for the computation of IDE are already available: One by Michael Markl (available at https://github.com/ArbeitsgruppeTobiasHarks/dynamic-prediction-equilibria) and one by Johannes Hagenmaier (available at https://github.com/johanneshage/ide-repository). The latter also includes a discrete variant of IDE, i.e. current information equilibria in the competitive packet routing model. Thus, it could also be used to investigate whether for small enough packet sizes the discrete and continuous version of IDE are close approximations of each other (similarly to the experimental study [Zie+21] conducted by Ziemke, Sering, Vargas Koch, Zimmer, Nagel and Skutella for the full information setting).

References

[AB06] Charalambos D. Aliprantis and Kim C. Border. *Infinite Dimensional Analysis. A Hitchhiker's Guide*. 3rd ed. Springer Berlin, Heidelberg, 2006. https://doi.org/10.1007/3-540-29587-9.

[Ans+08] Elliot Anshelevich, Anirban Dasgupta, Jon Kleinberg, Éva Tardos, TomWexler and Tim Roughgarden. "The Price of Stability for Network Design with Fair Cost Allocation". In: *SIAM Journal on Computing* 38.4 (2008), pp. 1602–1623. https://doi.org/10.1137/070680096. URL: https://doi.org/10.1137/070680096.

[BFA15] Umang Bhaskar, Lisa Fleischer and Elliot Anshelevich. "A Stackelberg strategy for routing flow over time". In: *Games and Economic Behavior* 92 (2015), pp. 232–247. ISSN: 0899-8256. URL: https://www.sciencedirect.com/science/article/pii/S0899825613001334.

[Bom+06] Enrico Bombieri, Stephen Cook, Pierre Deligne, Charles L. Fefferman, Jeremy Gray, Arthur Jaffe, John Milnor, Andrew Wiles and Edward Witten. *The Millennium Prize Problems*. Ed. by James Carlson, Arthur Jaffe and Andrew Wiles. Clay Mathematics Institute, 2006. Also available at https://www.claymath.org/publications/onlinebooks.

[BRL93] David E. Boyce, Bin Ran and Larry J. Leblanc. "A New Class of Instantaneous Dynamic User-Optimal Traffic Assignment Models". In: *Operations Research* 41.1 (1993), pp. 192–202. ISSN: 0030364X, 15265463. https://doi.org/10.1287/opre.41.1.192. URL: http://www.jstor.org/stable/171953.

[BRL95] David E. Boyce, Bin Ran and Larry J. Leblanc. "Solving an Instantaneous Dynamic User-Optimal Route Choice Model". In: *Transportation Science* 29.2 (1995), pp. 128–142. ISSN: 00411655, 15265447. https://doi.org/10.1287/trsc.29.2.128. URL: http://www.jstor.org/stable/25768680.

[BS09] Nadine Baumann and Martin Skutella. "Earliest Arrival Flows with Multiple Sources". In: *Mathematics of Operations Research* 34.2 (2009), pp. 499–512. ISSN: 0364765X, 15265471. https://doi.org/10.1287/moor.1090.0382. URL: http://www.jstor.org/stable/40538395.

[BS20] Vladimir I. Bogachev and Oleg G. Smolyanov. *Real and Functional Analysis*. Springer Cham, 25th Feb. 2020. https://doi.org/10.1007/978-3-030-38219-3.

[BV04] Stephen Boyd and Lieven Vandenberghe. *Convex Optimization*. 28th priting. Cambridge University Press, 2004. https://doi.org/10.1017/CBO9780511804441. URL: https://web.stanford.edu/~boyd/cvxbook/.

L. Graf, *Dynamic Network Flows with Adaptive Route Choice based on Current Information*, Mathematische Optimierung und Wirtschaftsmathematik | Mathematical Optimization and Economathematics,
https://doi.org/10.1007/978-3-658-44948-3

[CCCW22] Zhigang Cao, Bo Chen, Xujin Chen and Changjun Wang. "Bounding Residence Times for Atomic Dynamic Routings". In: *Mathematics of Operations Research* 47.4 (2022), pp. 3261–3281. https://doi.org/10.1287/moor.2021.1242. eprint: https://doi.org/10.1287/moor.2021.1242. URL: https://doi.org/10.1287/moor.2021.1242. Also available at http://wrap.warwick.ac.uk/160244/.

[CCL15] Roberto Cominetti, José Correa and Omar Larré. "Dynamic Equilibria in Fluid Queueing Networks". In: *Operations Research* 63.1 (2015), pp. 21–34. https://doi.org/10.1287/opre.2015.1348. eprint: https://doi.org/10.1287/opre.2015.1348. URL: https://doi.org/10.1287/opre.2015.1348.

[CCO22a] Roberto Cominetti, José Correa and Neil Olver. "Long-Term Behavior of Dynamic Equilibria in Fluid Queuing Networks". In: *Operations Research* 70.1 (2022), pp. 516–526. https://doi.org/10.1287/opre.2020.2081. eprint: https://doi.org/10.1287/opre.2020.2081. URL: https://doi.org/10.1287/opre.2020.2081. Also available on arXiv: https://arxiv.org/abs/2112.10412v1.

[CCO22b] José Correa, Andrés Cristi and Tim Oosterwijk. "On the Price of Anarchy for Flows over Time". In: *Mathematics of Operations Research* 47.2 (2022), pp. 1394–1411. https://doi.org/10.1287/moor.2021.1173. URL: https://doi.org/10.1287/moor.2021.1173.

[CLRS22] Thomas H. Cormen, Charles E. Leiserson, Ronald L. Rivest and Clifford Stein. *Introduction to Algorithms*. 4th ed. MIT Press, 2022. ISBN: 9780262367509.

[CM02] Malachy Carey and Mark McCartney. "Behaviour of a whole-link travel time model used in dynamic traffic assignment". In: *Transportation Research Part B: Methodological* 36.1 (2002), pp. 83–95. ISSN: 0191-2615. https://doi.org/10.1016/S0191-2615(00)00039-4. URL: https://www.sciencedirect.com/science/article/pii/S0191261500000394.

[Con19] John B. Conway. *A Course in Functional Analysis*. Springer New York, NY, 19th Mar. 2019. https://doi.org/10.1007/978-1-4757-4383-8.

[CS03] Vincent Conitzer and Tüomas Sandholm. "Complexity Results about Nash Equilibria". In: IJCAI'03. Acapulco, Mexico: Morgan Kaufmann Publishers Inc., 2003, pp. 765–771. Also available on arXiv: https://arxiv.org/abs/cs/0205074.

[DGP10] Constantinos Daskalakis, Paul W. Goldberg and Christos H. Papadimitriou. "The Complexity of Computing a Nash Equilibrium". In: *SIAM Journal on Computing* 39.1 (2010), pp. 195–259. https://doi.org/10.1137/070699652.

[FF56] Lester Randolph Ford and Delbert Ray Fulkerson. "Maximal Flow Through a Network". In: *Canadian Journal of Mathematics* 8 (1956), pp. 399–404. https://doi.org/10.4153/CJM-1956-045-5.

[FF58] Lester Randolph Ford and Delbert Ray Fulkerson. "Constructing Maximal Dynamic Flows from Static Flows". In: *Operations Research* 6.3 (1958), pp. 419–433. ISSN: 0030364X, 15265463. URL: http://www.jstor.org/stable/167028.

[FF62] Lester Randolph Ford and Delbert Ray Fulkerson. *Flows in Networks*. Princeton: Princeton University Press, 1962. ISBN: 9781400875184. https://doi.org/10.1515/9781400875184. URL: https://doi.org/10.1515/9781400875184.

[FH19] Terry L. Friesz and Ke Han. "The mathematical foundations of dynamic user equilibrium". In: *Transportation Research Part B: Methodological* 126 (2019), pp. 309–328. ISSN: 0191-2615. https://doi.org/10.1016/j.trb.2018.08.015. URL: https://www.sciencedirect.com/science/article/pii/

S0191261517301960. Also available at https://par.nsf.gov/biblio/10122277-mathematical-foundationsdynamic-user-equilibrium.

[FLTW89] Terry L. Friesz, Javier Luque, Roger L. Tobin and Byung-Wook Wie. "Dynamic Network Traffic Assignment Considered as a Continuous Time Optimal Control Problem". In: *Operations Research* 37.6 (1989), pp. 893–901. ISSN: 0030364X, 15265463. https://doi.org/10.1287/opre.37.6.893. URL: http://www.jstor.org/stable/171471.

[Fri+93] Terry L. Friesz, David Bernstein, Tony E. Smith, Roger L. Tobin and Byung-Wook Wie. "A Variational Inequality Formulation of the Dynamic Network User Equilibrium Problem". In: *Operations Research* 41.1 (1993), pp. 179–191. ISSN: 0030364X, 15265463. https://doi.org/10.1287/opre.41.1.179. URL: http://www.jstor.org/stable/171952.

[FT98] Lisa Fleischer and Éva Tardos. "Efficient continuous-time dynamic network flow algorithms". In: *Operations Research Letters* 23.3 (1998), pp. 71–80. ISSN: 0167-6377. https://doi.org/10.1016/S0167-6377(98)00037-6. URL: https://www.sciencedirect.com/science/article/pii/S0167637798000376.

[Gai+22] Martin Gairing, Carolina Osorio, Britta Peis, David Watling and Katharina Eickhoff. "Dynamic Traffic Models in Transportation Science (Dagstuhl Seminar 22192)". In: *Dagstuhl Reports* 12.5 (2022). Ed. by Martin Gairing, Carolina Osorio, Britta Peis, David Watling and Katharina Eickhoff, pp. 92–111. ISSN: 2192-5283. https://doi.org/10.4230/DagRep.12.5.92. URL: https://drops.dagstuhl.de/opus/volltexte/2022/17444.

[GH19] Lukas Graf and Tobias Harks. "Dynamic Flows with Adaptive Route Choice". In: *Integer Programming and Combinatorial Optimization*. Ed. by Andrea Lodi and Viswanath Nagarajan. Cham: Springer International Publishing, 2019, pp. 219–232. ISBN: 978-3-030-17953-3. https://doi.org/10.1007/978-3-030-17953-3_17.

[GH20] Lukas Graf and Tobias Harks. "The Price of Anarchy for Instantaneous Dynamic Equilibria". In: *Web and Internet Economics*. Ed. by Xujin Chen, Nikolai Gravin, Martin Hoefer and Ruta Mehta. Cham: Springer International Publishing, 2020, pp. 237–251. ISBN: 978-3-030-64946-3. https://doi.org/10.1007/978-3-030-64946-3_17.

[GH21] Lukas Graf and Tobias Harks. "A Finite Time Combinatorial Algorithm for Instantaneous Dynamic Equilibrium Flows". In: *Integer Programming and Combinatorial Optimization*. Ed. by Mohit Singh and David P. Williamson. Cham: Springer International Publishing, 2021, pp. 104–118. ISBN: 978-3-030-73879-2. https://doi.org/10.1007/978-3-030-73879-2_8.

[GH22] Lukas Graf and Tobias Harks. "The Price of Anarchy for Instantaneous Dynamic Equilibria". In: *Mathematics of Operations Research* (2022). https://doi.org/10.1287/moor.2022.1336. Also available on arXiv: https://arxiv.org/abs/2007.07794v1.

[GH23] Lukas Graf and Tobias Harks. "A finite time combinatorial algorithm for instantaneous dynamic equilibrium flows". In: *Mathematical Programming* 197 (1st Feb. 2023), pp. 761–792. https://doi.org/10.1007/s10107-022-01772-0. Also available on arXiv: https://arxiv.org/abs/2007.07808v3.

[GHKM23] Lukas Graf, Tobias Harks, Kostas Kollias and Michael Markl. "Prediction Equilibrium for Dynamic Network Flows". In: *Journal of Machine Learning Research* 24.310 (2023), pp. 1–33. URL: http://jmlr.org/papers/v24/22-1446.html.

[GHP22] Lukas Graf, Tobias Harks and Prashant Palkar. *Dynamic Traffic Assignment for Electric Vehicles*. 2022. https://doi.org/10.48550/ARXIV.2207.04454. URL: https://arxiv.org/abs/2207.04454.

[GHS20] Lukas Graf, Tobias Harks and Leon Sering. "Dynamic flows with adaptive route choice". In: *Mathematical Programming* 183 (1st Sept. 2020), pp. 309–335. https://doi.org/10.1007/s10107-020-01504-2. Also available on arXiv https://arxiv.org/abs/1811.07381v4.

[GJ79] Michael R. Gare and David S. Johnson. *Computers and Intractability*. W. H. Freeman and Company, 1979.

[Hag23] Johannes Hagenmaier. "Approximation dynamischer Flüsse mit adaptiver Routenwahl in Netzwerken mit mehreren Senken". German. Master's Thesis. University of Augsburg, 2023.

[HFY13] Ke Han, Terry L. Friesz and Tao Yao. "Existence of simultaneous route and departure choice dynamic user equilibrium". In: Transportation Research Part B: Methodological 53 (2013), pp. 17–30. https://doi.org/10.1016/j.trb.2013.01.009. URL: https://doi.org/10.1016%2Fj.trb.2013.01.009.

[Hir76] Morris W. Hirsch. *Differential Topology*. Springer-Verlag, 1976. https://doi.org/10.1007/978-1-4684-9449-5.

[How08] Rodney R. Howell. *On Asymptotic Notation with Multiple Variables*. Technical Report. Kansas State University, 18th Jan. 2008. URL: https://people.cis.ksu.edu/~rhowell/asymptotic.pdf.

[Hun13] Dirk Hundertmark. *Lecture Notes Functional Analysis*. lecture notes. 13th Feb. 2013. URL: https://www.math.kit.edu/iana1/lehre/funcana2012w/media/fa-lecturenotes.pdf.

[Iry11] Takamasa Iryo. "Multiple equilibria in a dynamic traffic network". In: *Transportation Research Part B: Methodological* 45.6 (2011), pp. 867–879. ISSN: 0191-2615. https://doi.org/10.1016/j.trb.2011.02.010. URL: https://www.sciencedirect.com/science/article/pii/S0191261511000324.

[Ism17] Anisse Ismaili. "Routing Games over Time with FIFO Policy". In: *Web and Internet Economics*. Ed. by Nikhil R. Devanur and Pinyan Lu. Cham: Springer International Publishing, 2017, pp. 266–280. https://doi.org/10.1007/978-3-319-71924-5_19. Also available on arXiv https://arxiv.org/abs/1709.09484.

[Jän84] Klaus Jänich. *Topology*. Springer-Verlag New York, 1984.

[Jar+22] P. Jaramillo, S. Kahn Ribeiro, P. Newman, S. Dhar, O.E. Diemuodeke, T. Kajino, D.S. Lee, S.B. Nugroho, X. Ou, A. Hammer Strømman and J. Whitehead. "Transport". In: *Climate Change 2022: Mitigation of Climate Change. Contribution of Working Group III to the Sixth Assessment Report of the Intergovernmental Panel on Climate Change*. Ed. by P. R. Shukla, J. Skea, R. Slade, A. Al Khourdajie, R. van Diemen, D. McCollum, M. Pathak, S. Some, P. Vyas, R. Fradera, M. Belkacemi, A. Hasija, G. Lisboa, S. Luz and J. Malley. Cambridge, United Kingdom and New York, NY, USA: Cambridge University Press, 2022, pp. 1049–1160. https://doi.org/10.1017/9781009157926.012. Also available at https://www.ipcc.ch/report/ar6/wg3/.

[Jun12] Dieter Jungnickel. *Graphs, Networks and Algorithms*. 4th ed. Heidelberg: Springer Berlin, 8th Nov. 2012. https://doi.org/10.1007/978-3-642-32278-5.

[Jun14] Dieter Jungnickel. *Optimierungsmethoden. Eine Einführung*. deutsch. 3rd ed. Heidelberg: Springer Spektrum Berlin, 27th Nov. 2014. https://doi.org/10.1007/978-3-642-54821-5.

[Kai22a] Marcus Kaiser. "Computation of Dynamic Equilibria in Series-Parallel Networks". In: *Mathematics of Operations Research* 47.1 (2022), pp. 50–71. https://doi.org/10.1287/moor.2020.1108. URL: https://doi.org/10.1287/moor.2020.1108. Also available on arXiv https://arxiv.org/abs/2002.11428.

[Kai22b] Marcus Kaiser. *Disjoint Paths, Dynamic Equilibria, and the Design of Networks*. en. 2022. URL: https://nbn-resolving.de/urn/resolver.pl?urn:nbn:de:bvb:91-diss-20220401-1632255-1-1.

[Koc12] Ronald Koch. "Routing Games over Time". Doctoral Thesis. Berlin: Technische Universität Berlin, Fakultät II—Mathematik und Naturwissenschaften, 2012. https://doi.org/10.14279/depositonce-3347. URL: http://dx.doi.org/10.14279/depositonce-3347.

[KP99] Elias Koutsoupias and Christos Papadimitriou. "Worst-Case Equilibria". In: *STACS 99*. Ed. by Christoph Meinel and Sophie Tison. Berlin, Heidelberg: Springer Berlin Heidelberg, 1999, pp. 404–413. ISBN: 978-3-540-49116-3. https://doi.org/10.1007/3-540-49116-3_38.

[Kra20] Georg Kraus. "Calculation of IDE-Flows in SP-Networks". Master's Thesis. University of Augsburg, 13th Jan. 2020.

[KS11] Roland Koch and Martin Skutella. "Nash Equilibria and the Price of Anarchy for Flows over Time". In: *Theory of Computing Systems* 49 (1st July 2011), pp. 71–97. https://doi.org/10.1007/s00224-010-9299-y. URL: https://doi.org/10.1007/s00224-010-9299-y.

[LHY20] Zhi-Chun Li, Hai-Jun Huang and Hai Yang. "Fifty years of the bottleneck model: A bibliometric review and future research directions". In: *Transportation Research Part B: Methodological* (2020). https://doi.org/10.1016/j.trb.2020.06.009. Also available at https://www.ncbi.nlm.nih.gov/pmc/articles/PMC7333998/.

[Mar20] Michael Markl. "Berechnung von Nash-Gleichgewichten in dynamischen Flüssen". German. Bachelor's Thesis. University of Augsburg, 8th Jan. 2020. URL: https://github.com/michael-markl/dynamicequilibrium-flows/tree/master/bachelorarbeit.

[Mar21] Michael Markl. "Computation of Dynamic Prediction Equilibria". software project. University of Augsburg, 2021. URL: https://github.com/Schedulaar/dynamic-prediction-equilibria/tree/main/university-report.

[Mar22] Michael Markl. "Predicting Equilibria in Dynamic Traffic Assignment". Master's Thesis. University of Augsburg, 2022. URL: https://michael-markl.de/master-thesis-prediction-equilibriain-dynamic-traffic-assignment.pdf.

[Nas51] John Nash. "Non-Cooperative Games". In: *Annals of Mathematics* 54.2 (1951), pp. 286–295. ISSN: 0003486X. URL: http://www.jstor.org/stable/1969529.

[nLa23a] nLab authors. *total order*. Version 16. Revision 16. 13th Feb. 2023. URL: https://ncatlab.org/nlab/show/total+order.

[nLa23b] nLab authors. *Zorn's lemma*. Version 29. Revision 29. 13th Feb. 2023. URL: https://ncatlab.org/nlab/show/Zorn%27s+lemma.

[NRTV07] Noam Nisan, Tim Roughgarden, Éva Tardos and Vijay V. Vazirani, eds. *Algorithmic Game Theory*. Cambridge University Press, 2007. ISBN: 9780521872829.

[OSK22] Neil Olver, Leon Sering and Laura Vargas Koch. "Continuity, Uniqueness and Long-Term Behavior of Nash Flows Over Time". In: *2021 IEEE 62nd Annual Symposium on Foundations of Computer Science (FOCS)*. 2022, pp. 851–860. https://doi.org/10.1109/FOCS52979.2021.00087. full version available on arXiv: https://arxiv.org/abs/2111.06877.

[OSK23] Neil Olver, Leon Sering and Laura Vargas Koch. "Convergence of Approximate and Packet Routing Equilibria to Nash Flows Over Time". In: *2023 IEEE 64th Annual Symposium on Foundations of Computer Science (FOCS)*. 2023, pp. 123–133. https://doi.org/10.1109/FOCS57990.2023.00016. Full version on arXiv https://arxiv.org/abs/2402.04935.

[OSS22] Tim Oosterwijk, Daniel Schmand and Marc Schröder. "Bicriteria Nash Flows over Time". In: Web and Internet Economics—*18th International Conference, WINE 2022, Troy, NY, USA, December 12–15, 2022, Proceedings*. Ed. by Kristoffer Arnsfelt Hansen, Tracy Xiao Liu and Azarakhsh Malekian. Vol. 13778. Lecture Notes in Computer Science. Springer, 2022, p. 368. URL: https://link.springer.com/content/pdf/bbm:978-3-031-22832-2/1?pdf=chapter%5C%20toc. full version available on arXiv: https://arxiv.org/abs/2111.08589.

[OSV22] Neil Olver, Leon Sering and Laura Vargas Koch. "Continuity, Uniqueness and Long-Term Behavior of Nash Flows Over Time". In: *2021 IEEE 62nd Annual Symposium on Foundations of Computer Science (FOCS)*. 2022, pp. 851–860. https://doi.org/10.1109/FOCS52979.2021.00087.

[Pap01] Christos Papadimitriou. "Algorithms, Games, and the Internet". In: *Proceedings of the Thirty-Third Annual ACM Symposium on Theory of Computing*. STOC '01. Hersonissos, Greece: Association for Computing Machinery, 2001, pp. 749–753. ISBN: 1581133499. https://doi.org/10.1145/380752.380883. URL: https://doi.org/10.1145/380752.380883.

[Pap94] Christos H. Papadimitriou. "On the complexity of the parity argument and other inefficient proofs of existence". In: *Journal of Computer and System Sciences* 48.3 (1994), pp. 498–532. ISSN: 0022-0000. https://doi.org/10.1016/S0022-0000(05)80063-7. URL: https://www.sciencedirect.com/science/article/pii/S0022000005800637.

[Pap95] Christos H. Papadimitriou. *Computational Complexity*. Addison-Wesley Publishing Company, 1995.

[PS20] Hoang Minh Pham and Leon Sering. "Dynamic Equilibria in Time-Varying Networks". In: *Algorithmic Game Theory*. Ed. by Tobias Harks and Max Klimm. Cham: Springer International Publishing, 2020, pp. 130–145. ISBN: 978-3-030-57980-7. https://doi.org/10.1007/978-3-030-57980-7_9. full version available on arXiv: https://arxiv.org/abs/2007.01525.

[PVZ23] Christos H. Papadimitriou, Emmanouil-Vasileios Vlatakis-Gkaragkounis and Manolis Zampetakis. *The Computational Complexity of Multiplayer Concave Games and Kakutani Fixed Points*. 2023. arXiv: 2207.07557 [cs.CC].

[PZ01] Srinivas Peeta and Athanasios K. Ziliaskopoulos. "Foundations of Dynamic Traffic Assignment: The Past, the Present and the Future". In: *Networks and Spatial Economics* 1 (1st Sept. 2001), pp. 233–265. https://doi.org/10.1023/A:1012827724856.

[RB96] Bin Ran and David Boyce. *Modeling Dynamic Transportation Networks*. Heidelberg: Springer Berlin, 1996. https://doi.org/10.1007/978-3-642-80230-0.

[RF10] Halsey L. Royden and Patrick Fitzpatrick. *Real Analysis*. 4th ed. Pearson Education, 2010. ISBN: 9780135113554.

[Sch21] Daniel Schmand. "Recent Developments in Mathematical Traffic Models". In: *Dynamics in Logistics: Twenty-Five Years of Interdisciplinary Logistics Research in Bremen, Germany*. Ed. by Michael Freitag, Herbert Kotzab and Nicole Megow. Cham: Springer International Publishing, 2021, pp. 71–87. ISBN: 978-3-030-88662-2. https://doi.org/10.1007/978-3-030-88662-2_4. URL: https://doi.org/10.1007/978-3-030-88662-2_4.

[Ser20] Leon Sering. "Nash flows over time". Doctoral Thesis. Berlin: Technische Universität Berlin, 2020. https://doi.org/10.14279/depositonce-10640. URL: http://dx.doi.org/10.14279/depositonce-10640.

[Sku09] Martin Skutella. "An Introduction to Network Flows over Time". In: Research Trends in Combinatorial Optimization: Bonn 2008. Ed. by William Cook, László Lovász and Jens Vygen. Berlin, Heidelberg: Springer Berlin Heidelberg, 2009, pp. 451–482. https://doi.org/10.1007/978-3-540-76796-1_21. URL: https://doi.org/10.1007/978-3-540-76796-1_21.

[SS18] Leon Sering and Martin Skutella. "Multi-Source Multi-Sink Nash Flows over Time". In: *18th Workshop on Algorithmic Approaches for Transportation Modelling, Optimization, and Systems (ATMOS 2018)*. Ed. by Ralf Borndörfer and Sabine Storandt. Vol. 65. OpenAccess Series in Informatics (OASIcs). Dagstuhl, Germany: Schloss Dagstuhl-Leibniz-Zentrum fuer Informatik, 2018, 12:1–12:20. ISBN: 978-3-95977-096-5https://doi.org/10.4230/OASIcs.ATMOS.2018.12. URL: http://drops.dagstuhl.de/opus/volltexte/2018/9717.

[SV18] Leon Sering and Laura Vargas Koch. "Nash Flows over Time with Spillback". In: *CoRR* abs/1807.05862 (2018). arXiv: 1807.05862. URL: http://arxiv.org/abs/1807.05862.

[SV69] James Serrin and Dale E. Varberg. "A General Chain Rule for Derivatives and the Change of Variables Formula for the Lebesgue Integral". In: *The American Mathematical Monthly* 76.5 (1969), pp. 514–520. https://doi.org/10.1080/00029890.1969.12000249. eprint: https://doi.org/10.1080/00029890.1969.12000249. URL: https://doi.org/10.1080/00029890.1969.12000249.

[SVZ22] Leon Sering, Laura Vargas Koch and Theresa Ziemke. "Convergence of a Packet Routing Model to Flows over Time". In: *Mathematics of Operations Research* (2022). https://doi.org/10.1287/moor.2022.1318. URL: https://doi.org/10.1287/moor.2022.1318. Also available on arXiv: https://arxiv.org/abs/2105.13202.

[Var20] Laura Vargas Koch. "Competitive variants of discrete and continuous flows over time". Veröffentlicht auf dem Publikationsserver der RWTH Aachen University 2021; Dissertation, RWTH Aachen University, 2020. Dissertation. Aachen: RWTH Aachen University, 2020. https://doi.org/10.18154/RWTH-2020-11648. URL: https://publications.rwth-aachen.de/record/807846.

[Var65] Dale E. Varberg. "On Absolutely Continuous Functions". In: *The American Mathematical Monthly* 72.8 (1965), pp. 831–841. ISSN: 00029890, 19300972. URL: http://www.jstor.org/stable/2315025.

[Vic69] William S. Vickrey. "Congestion Theory and Transport Investment". In: The *American Economic Review* 59.2 (1969), pp. 251–260. ISSN: 00028282. URL: http://www.jstor.org/stable/1823678.

[Yag71] Samuel Yagar. "Dynamic traffic assignment by individual path minimization and queuing". In: Transportation Research 5.3 (1971), pp. 179–196. ISSN: 0041-1647. https://doi.org/10.1016/0041-1647(71)90020-7. URL: https://www.sciencedirect.com/science/article/pii/0041164771900207.

[Zie+21] Theresa Ziemke, Leon Sering, Laura Vargas Koch, Max Zimmer, Kai Nagel and Martin Skutella. "Flows Over Time as Continuous Limits of Packet-Based Network Simulations". In: *Transportation Research Procedia* 52 (2021). 23rd EURO Working Group on Transportation Meeting, EWGT 2020, 16–18 September 2020, Paphos, Cyprus, pp. 123–130. ISSN: 2352-1465. https://doi.org/10.1016/j.trpro.2021.01.014. URL: https://www.sciencedirect.com/science/article/pii/S2352146521000284.

[ZM00] Daoli Zhu and Patrice Marcotte. "On the Existence of Solutions to the Dynamic User Equilibrium Problem". In: *Transportation Science* 34.4 (2000), pp. 402–414. ISSN: 00411655, 15265447. URL: http://www.jstor.org/stable/25768930.

The manufacturer's authorised representative in the EU is Springer Nature Customer Service Centre GmbH, Europaplatz 3, 69115 Heidelberg, Germany. If you have any concerns regarding our products, please contact ProductSafety@springernature.com

Printed and bound by CPI Group (UK) Ltd, Croydon, CR0 4YY
01/10/2026
02225893-0001